# Physical Properties of Amorphous Materials

# Institute for Amorphous Studies Series

Series editors

David Adler
*Massachusetts Institute of Technology*
*Cambridge, Massachusetts*

and

Brian B. Schwartz
*Institute for Amorphous Studies*
*Bloomfield Hills, Michigan*
*and Brooklyn College of the City University of New York*
*Brooklyn, New York*

PHYSICAL PROPERTIES OF AMORPHOUS MATERIALS
Edited by David Adler, Brian B. Schwartz, and Martin C. Steele

A Continuation Order Plan is available for this series. A continuation order will bring delivery of each new volume immediately upon publication. Volumes are billed only upon actual shipment. For further information please contact the publisher.

# Physical Properties of Amorphous Materials

Edited by
## David Adler
*Massachusetts Institute of Technology*
*Cambridge, Massachusetts*

## Brian B. Schwartz
*Institute for Amorphous Studies*
*Bloomfield Hills, Michigan*
*and Brooklyn College of the City University of New York*
*Brooklyn, New York*

and
## Martin C. Steele
*Institute for Amorphous Studies*
*Bloomfield Hills, Michigan*

Plenum Press · New York and London

Library of Congress Cataloging in Publication Data

Physical properties of amorphous materials.
  (Institute for amorphous studies series)

  Includes bibliographical references and index.
  1. Amorphous substances. I. Adler, David. II. Schwartz, Brian B., 1938–    . III.
Steele, Martin C. (Martin Carl), 1919–    . IV. Series.
QC176.8.A44P49   1985                    530.4'1                    84-26370
ISBN 0-306-41907-6

Proceedings of a lecture series on Fundamentals of Amorphous Materials
and Devices, held during the period 1982–1983 at the Institute for Amorphous
Studies, Bloomfield Hills, Michigan

©1985 Plenum Press, New York
A Division of Plenum Publishing Corporation
233 Spring Street, New York, N.Y. 10013

Printed in the United States of America

PREFACE

      The Institute for Amorphous Studies was founded in 1982 as the international center for the investigation of amorphous materials. It has since played an important role in promoting the understanding of disordered matter in general. An Institute lecture series on "Fundamentals of Amorphous Materials and Devices" was held during 1982-83 with distinguished speakers from universities and industry. These events were free and open to the public, and were attended by many representatives of the scientific community. The lectures themselves were highly successful inasmuch as they provided not only formal instruction but also an opportunity for vigorous and stimulating debate. That last element could not be captured within the pages of a book, but the lectures concentrated on the latest advances in the field, which is why their essential contents are here reproduced in collective form. Together they constitute an interdisciplinary status report of the field. The speakers brought many different viewpoints and a variety of background experiences to bear on the problems involved, but though language and conventions vary, the essential unity of the concerns is very clear, as indeed are the ultimate benefits of the many-sided approach.

      The impact of amorphous materials on the worlds of science and technology has been enormous, covering such diverse applications as solar energy, image processing, energy storage, computer and telecommunication technology, thermoelectric energy conversion, and new materials synthesis. If the subject is to reach its full fruition, its role in educational programs will have to be commensurate. To an astonishing extent, the teaching of solid state physics is still restrictively tied to crystalline concepts, and the time has come for a change and a balance. Both will undoubtedly come but reform and reorientation take time. It is one of the Institute's purposes to accelerate the process.

We reconfirm our belief and confidence that amorphous materials will serve humankind in a variety of important ways, and we recognize that the scientists who are now in the forefront of this field are in effect social as well as technological pioneers. My special thanks are due to the contributors to this volume, to whom I wish continued success in the realms of new scientific ideas and technological achievement.

S.R. Ovshinsky
Troy, Michigan

# CONTENTS

Fundamentals of Amorphous Materials
  S.R. Ovshinsky

PART TWO:  STRUCTURE

The Constraint of Discord
  D. Weaire

Dislocation Mediated Pseudo-Melting at
    Silicon-Metal Interfaces
B.K. Chakraverty

PART THREE:  PHONONS

Vibrational Properties of Amorphous Solids
    G. Lucovsky

Light-Induced Effects in Hydrogenated Amorphous
     Silicon Alloys
  S. Guha

# INTRODUCTION

It is no surprise to note that the technological importance of amorphous solids has mushroomed over the past few years. Of course, this is not to say that they have been insignificant in the past. In fact, the use of glass in decorative and packaging applications has a history of more than 10,000 years. However, it was not until the advent of electrophotographic copying only about 25 years ago that the electronic properties of amorphous materials began to be exploited. In the very brief time since this breakthrough, we have seen the development of commercial computer memories, television pick-up tubes, solar cells, x-ray mirrors, thermoelectric devices, and imaging films based on amorphous materials, and we are on the verge of adding batteries, catalysts, video disks, displays, transient suppressors, and an array of other applications to this list. Yet most solid-state scientists appear to be unaware of even the existence of these materials, much less of their importance. Few textbooks deal with anything but crystalline solids and there is little evidence that even graduate-level courses at all but a handful of universities devote much time to their properties.

There has never been a dearth of technical review papers dealing with specialized properties of amorphous solids, but the same cannot be said of coherent books suitable for student texts. It is true that Mott and Davis's work, <u>Electronic Properties of Non-Crystalline Materials</u>, has gone through two editions and Zallen's recent <u>The Physics of Amorphous Solids</u> has just appeared, but neither of these is exhaustive in scope.

With the creation of the Institute for Amorphous Studies, devoted to both the training of scientists and the dissemination of information relating to amorphous materials and devices, we now have the opportunity to fill all of the gaps in the written as well as

1

oral avenues of pedagogic instruction. This volume represents
the first step in the direction of making the state of the art accessible to those who were not able to attend either the lecture series
or the various mini-courses held at the Institute. This book can
serve either as an initial introduction to the field or as the basis
of a comprehensive in-depth course. Much of the cited literature
is quite recent, reaching to the late 1984 publication of the Proceedings of the International Conference on Transport and Defects
in Amorphous Semiconductors (J. Noncryst. Solids, Vol. 66) held
at the Institute.

The 14 papers in this volume are arranged in five parts.
Part One, dealing with all aspects of the materials, can serve as a
general introduction to the whole field. The Adler paper is an attempt at outlining a new approach to the study of solid-state physics, in which periodicity is not considered to be fundamental.
Both the conventional and alternative viewpoints are discussed, and
the effects of disorder are analyzed in detail. Brief reviews of the
properties of amorphous silicon-based and chalcogenide alloys are
included. This paper is followed by one by Ovshinsky, primarily
dealing with his broad view of the chemical basis for controlling
the properties of amorphous materials and the tailoring of these
materials to optimize their use in specific applications. This
paper provides the rare opportunity to understand some aspects of
the original thinking which revolutionized the field.

Any problem in solid-state physics can be broken down into three sub-problems. The first is that of the equilibrium positions of the atoms constituting the solid, i.e., its structure. The
structure not only determines the mechanical properties of the material but also controls the other two sub-problems. One of these is
the nature of the vibrations of the atoms around their equilibrium
positions, i.e., the phonon modes of the solid. These determine
the thermal properties of the material. The other problem is that
of the states available to the electrons as they propagate through
the solid, i.e., the electronic structure. This controls the electrical and optical properties of the material. Part Two of this
volume is concerned with the structure of amorphous solids.
Weaire's paper discusses the conceptual basis of disordered structures in general, while those of Bienenstock, Stern, and Boolchand
deal with detailed experimental results on real systems.
Bienenstock provides a general introduction to studies of both simple and complex systems, pointing out the contrasts between the
analyses of crystalline and amorphous materials. Stern concen-
trates on the use of extended x-ray absorption fine structure (EXAFS)

to determine the structure of disordered solids, while Boolchand discusses the results of Mössbauer experiments in detail. Finally, Chakraverty presents a novel approach to the structure of silicon-metal interfaces, an important subject in many device applications of silicon-based materials.

Part Three, dealing with the analysis of phonon modes in amorphous materials, consists of the comprehensive paper by Lucovsky. This paper analyzes the vibrational properties of both network glasses and amorphous silicon-based alloys in detail.

The important problem of electronic structure is the basis of both Parts Four and Five. Part Four is primarily concerned with the density of states available to electrons as they move through the disordered system, $g(E)$, as a function of their energy, $E$. Once $g(E)$ is known, the near-equilibrium electrical and optical properties of the material as functions of the temperature can be derived from the known probability of occupation of these states, the Fermi-Dirac distribution function, $f(E)$. Fritzsche's paper presents an elementary introduction to the various experimental methods that have been developed for the determination of $g(E)$. In contrast, the paper of Cohen is primarily concerned with the theoretical basis for understanding the behavior of $g(E)$ and how the observed transport and optical properties can be explained.

One of the major advances of the past decade has been a detailed understanding of the approach to re-equilibration after the perturbation of a semiconductor away from its equilibrium distribution function, $f(E)$. This perturbation can be accomplished, e.g., by the application of an intense light pulse that creates excess free electrons and holes or an applied electric field that injects free carriers from one or both contacts. The resulting time dependence of the response, e.g., $i(t)$, can be quite complex, but recent theoretical models have been developed which can sort it out and, in fact, use this response as a sensitive probe of the density of localized electronic states, $g(E)$, in the material under investigation. Part Five is devoted to several aspects of these nonequilibrium phenomena. Henisch discusses the processes in general, paying particular attention to the complex issue of carrier injection and extraction via the contacts. Kastner's paper provides an elementary introduction to the subject of excess-carrier trapping and recombination, and shows how a simple model can explain the complex transient photoconductivity data. Silver is primarily concerned with the additional information that can be learned from sorting out the primary excess-carrier recombination mechanisms and the effects of high-level injection, in particular the use of such experiments to obtain a handle on the mobility of carriers in extended states. Finally, Guha discusses the important problem of light-induced variations in properties of amorphous semiconductors, paying special attention to hydrogenated amorphous silicon.

All of the lectures reproduced here stimulated a great deal of lively discussion, both during and after their presentation. We hope this volume does the same.

> David Adler
> Cambridge, Massachusetts
>
> Brian B. Schwartz
> Bloomfield Hills, Michigan
>
> Martin C. Steele
> Bloomfield Hills, Michigan

# CHEMISTRY AND PHYSICS OF COVALENT AMORPHOUS SEMICONDUCTORS

David Adler

Department of Electrical Engineering and Computer
Science, Massachusetts Institute of Technology
Cambridge, Massachusetts 02139

## I.    INTRODUCTION

Although, as Ovshinsky [1] has often noted, the history of
civilization is intimately connected with advances in materials
research (e.g., the Stone, Bronze, Iron, and Steel Ages), a quantitative theory of materials did not evolve until after the development of quantum mechanics in 1925. When Bloch [2] developed
the quantum theory of solids in 1930, physicists were unduly influenced by the discovery less than 20 years previous that many
solids were periodic arrays of atoms. By making use of the
mathematical simplifications resulting from periodicity, Bloch was
able to derive some general properties of the electronic states in
crystals, from which Wilson [3] developed the band theory of electronic transport. There is no question that this theory has been
very successful in providing a detailed understanding of the properties of crystalline solids, including the electrical and optical
properties of a wide array of commercially important semiconductors and metals. However, it has long been known that many
solids are not crystalline but are amorphous, and do not exhibit
any long-range periodicity. Furthermore, these amorphous solids
have been shown to exhibit the same range of electrical and
optical properties as do crystalline materials, strongly suggesting
that the band theory of solids is much more general than its original derivation indicates. Nevertheless, even now, over 15
years after the publication of Ovshinsky's landmark paper [4],

almost all textbooks and courses on "solid-state physics" restrict quantitative discussion to the special case of periodic crystals.

It is significant that Ovshinsky, as the prime mover in the development of both new amorphous solids and novel commercial applications of these materials, has also created the Institute for Amorphous Studies, one of whose major functions is to teach the fundamentals of noncrystalline solids to those who want a more accurate view of solid-state physics. Both the lecture series, upon which this monograph is based, and the monograph itself represent initial steps in this direction. There is no question that our present ideas will undergo a great deal of modification over the next few decades. But there is absolutely no doubt that the general viewpoint presented herein represents the wave of the future, and that solid-state theory must eventually adapt to the realities of the world. In this paper, I will try to outline the proper perspective for a course in the physics of solids, in which periodicity is not fundamental and the theory of crystalline materials represents a small, albeit important, subset of the general development.

## II.    SOLID-STATE THEORY

### A.    General Approach

After the development of quantum mechanics, it became possible to understand the properties of atoms, molecules, and solids in a much more fundamental manner than was previously imagined. In principle, the physics could be determined from the solution of a partial differential equation, which could be symbolically written:

$$H\Psi \left(\underline{r}_1, \ldots \underline{r}_n\right) = E\Psi \left(\underline{r}_1, \ldots \underline{r}_n\right). \tag{1}$$

In Eq. (1), E represents a number which gives the energy of the state characterized by the wave function, $\Psi \left(\underline{r}_1, \ldots \underline{r}_n\right)$, a complex function of the positions of the N particles in the system. The physical significance of the wave function is that its magnitude squared, $\left|\Psi \left(\underline{r}_1, \ldots \underline{r}_n\right)\right|^2$ yields the probability density for finding the particles near $\underline{r}_1, \underline{r}_2, \ldots, \underline{r}_n$, respectively. The symbol H in Eq. (1), called the Hamiltonian of the system, is a differential operator which can always be determined provided the forces between the particles are known. H can be written:

$$H = \sum_{i=1}^{N} \left( -\frac{\hbar^2}{2m_i} \nabla_i^2 \right) + V(\underline{r}_1, \ldots \underline{r}_n), \tag{2}$$

where $\nabla_i^2 \equiv \dfrac{\partial^2}{\partial x_i^2} + \dfrac{\partial^2}{\partial y_i^2} + \dfrac{\partial^2}{\partial z_i^2}$ is the Laplacian operator corresponding to particle $i$, $m_i$ is the mass of particle $i$, and $V(\underline{r}_1, \ldots, \underline{r}_n)$ is the potential energy corresponding to the forces between the particles; $\hbar$ is Planck's constant divided by $2\pi$, and is equal to $1.05 \times 10^{-34}$ joule-seconds in the MKS system of units.

The great simplification which enables the solution, in principle, of all nonrelativistic problems involving atoms, molecules, and solids is the fact that the only important forces are the electrostatic interactions between the positively-charged nuclei and the negatively-charged electrons. For example, if there are N nuclei of charge $+Ze$ and ZN electrons of charge $-e$ present, then:

$$V(\underline{R}_1, \ldots, \underline{R}_N; \underline{r}_1, \ldots, \underline{r}_{ZN}) =$$

$$\frac{1}{2} \sum_{\alpha, \beta}{}' \frac{Z^2 e^2}{|\underline{R}_\alpha - \underline{R}_\beta|} + \frac{1}{2} \sum_{i,j}{}' \frac{e^2}{|\underline{r}_i - \underline{r}_j|} \tag{3}$$

$$- \sum_{\alpha=1}^{N} \sum_{i=1}^{ZN}{}' \frac{Ze^2}{|\underline{r}_i - \underline{R}_\alpha|}$$

where the $\underline{R}_\alpha$ represents the positions of the nuclei, the $r_i$ represent the positions of the electrons, and the primes on the summation symbols mean that the terms $\alpha = \beta$ and $i = j$ are omitted from the double sums (since particles do not interact with themselves). The first set of terms of Eq. (3) represent the mutual repulsions between the nuclei, the second set the mutual repulsions between the electrons, and the third set the attractions between the nuclei and the electrons.

The quantitative mathematical problem represented by Eqs. (1)-(3) can be solved to a high degree of accuracy for atoms and simple molecules, but is much more difficult for solids, which typically contain $\sim 10^{23}$ nuclei and $\sim 10^{24}$ electrons. Ideally, we should like to calculate the lowest possible energy levels for any system, together with their corresponding wave functions, since these would yield the electrical, optical, and thermal properties of the system.

We know that, by definition, solids retain their shape. Physically, this means that the nuclei possess equilibrium positions in which the forces on them vanish. Since all materials (except helium, a very anomalous case because of its light mass and very weak interatomic forces) become solids at sufficiently low temperature, the lowest-energy state of any collection of atoms must possess equilibrium positions for the nuclei. This set of equilibrium positions for the lowest-energy state is called the structure of the solid. It would be a great accomplishment if we could calculate the structure of solids from the nature of the constituent atoms, but thus far the problem is much too difficult to solve. Instead, an empirical approach is used, in which the structure is determined or inferred from experimental observations.

B.    Structure

X-rays, whose wavelengths are of the order of interatomic separations, were discovered in 1895, and soon afterwards were used to study the structure of solids. In 1912, von Laue suggested that periodic crystals would act as a diffraction grating for x-rays, and the next year Bragg was the first to use this effect for determining crystal structures, viz. NaCl and ZnS. More recently, the diffraction of electrons and neutrons with similar wavelengths have also been used to study the structure of crystals, whose long-range periodicity makes the interpretation of such experiments relatively straightforward. On the other hand, the structure of amorphous solids is much difficult to determine. Diffraction experiments can be used to discover the interatomic separations of the neighboring atoms from a central atom, from which a structure can often be inferred. This procedure can be ambiguous in complex materials. However, in an elemental solid, the results are relatively straightforward. An important example is Si, for which the electron-diffraction results of Moss and Graczyk [5] are shown in Fig. 1. The plot is that of $4\pi r^2$ times the probability, $\rho(r)$, of finding an atom a distance r from a central Si atom, as a function of r for both amorphous and crystalline films. The positions of the peaks in $\rho(r)$ represent the interatomic separations and the areas under each peak yield the number of atoms at each such separation. The well-known structure of crystalline Si (c-Si) can be described as a central Si atom surrounded by four nearest-neighboring Si atoms 2.35Å away forming a regular tetrahedron. The regular tetrahedron restricts the <u>bond angle</u> (the angle between two Si-Si bonds emanating from the same atom) to a fixed value, 109.5°. Since each of the neighboring Si atoms must itself be surrounded by such a tetrahedron, there are 12 second-neighboring Si atoms 3.84Å away. In order to form a periodic crystal which fills space, the orientation of two neighboring tetrahedra must be fixed, thus yielding a single value for the <u>dihedral angle</u> (the angle between the planes containing two adjoining bond angles). Two types of crystal are possible, depending on whether all the di-hedral angles are 60° (diamond structure) or 25% of them are 0° (wurzite structure); the former represents the lowest-energy state for c-Si, and leads to a third-neighbor distance of 4.50Å. Such a configuration forms a cube 5.43Å on a side, which is the fourth-

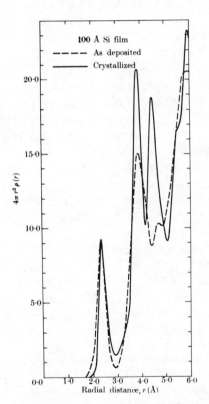

Fig. 1    Radial distribution function of an evaporated amorphous silicon film (---) 100Å thick compared with that of crystallized silicon (———). (From ref. 5.)

neighbor separation. The solid line in Fig. 1 shows the first four peaks of the c-Si structure. It is clear that the first peak of the dashed line is essentially identical to that of the solid line, indicating that the nearest-neighbor environment of a Si atom in a-Si is also a tetrahedron of atoms 2.35Å away. There is also a similarity between the two second peaks, although the greater width of the dashed line suggests that a-Si has a disorder-induced spread of second-neighbor distances. The extent of the broadening indicates that the bond angles in a-Si are not all 109.5° as in c-Si, but rather are distributed in the range 100°-120°. However, there is only the slightest indication of a third-neighbor peak in a-Si, suggesting that a great deal of dihedral-angle disorder exists.

Clearly, the origin of the similarities between the structures of c-Si and a-Si are not accidental, but rather reflect the chemistry of the Si atom. Si is a tathogen element, from Column IV of the Periodic Table. It possesses four electrons in its outermost (M) shell. In the isolated atom, two of these fill the 3s subshell, while the other two are in the 3p subshell, about 6eV higher in energy. However, to form covalent bonds, Si ordinarily promotes one of its 3s electrons to the 3p subshell; in this configuration, the atom can possess four singly-occupied $3sp^3$ hybrid orbitals which are preferentially directed along the axes of a regular tetrahedron. Using these orbitals, Si can form four covalent bonds with 109.5° bond angles, the maximum number of bonds possible with only s and p orbitals. Since the optimum bond length for Si-Si $sp^3$ bond is 2.35Å, both c-Si and a-Si are structures which yield low total energies. The c-Si structure optimizes the number of bonds, $z$, the bond length, $a$, and the bond angles, $\theta$. It is, thus, clearly the lowest-energy state for a collection of $\sim 10^{23}$ Si atoms. The a-Si structure also optimizes both $z$ and $a$, although the values of $\theta$ are not quite optimal in general. It represents an excited metastable state of the system and forms only when the solid is deposited directly from the gas phase onto a relatively cold substrate, a process which retards collective atomic motion. If a-Si is heated above about 1000K, it spontaneously crystallizes.

When the three parameters, $z$, $a$, and $\theta$ have well-determined values in a narrow range, the material is said to exhibit short-range order. Given the short-range order, it is not difficult to construct a model for the structure which does not have any long-range crystalline periodicity. Such models are called random networks.

The hypothesis that an amorphous solid could be modelled by a random network of atoms with near-perfect short-range order was first proposed by Zachariasen [6] over 50 years ago, with particular reference to oxide glasses. Many years later, Polk [7] constructed such a model for tetrahedral amorphous semiconductors. However, this approach to the structure of amorphous solids has remained controversial, and alternative models based on microcrystalline arrays have continually been suggested [8].

A recent neutron diffraction study [9] of one of the tetrahedrally bonded elemental amorphous solids, a-Ge, has elucidated the structure considerably. The slightly larger atomic radius of Ge relative to that of Si results in an optimal Ge-Ge bond separation of 2.45Å, the nearest-neighbor separation in c-Ge. The nearest-neighbor separation in a-Ge is 2.46Å, a deviation of less than 1%. The second-neighbor separation in a-Ge is 4.00Å, also within 1% of the crystalline value of 4.02Å. The average bond angle in a-Ge is 108.5°, again within 1% of the ideal tetrahedral value, but disorder-induced broadening exists with a bond-angle distortion of ± 10°. The number of nearest-neighbors in a-Ge is about 3.7, significantly lower than the ideal value of 4.0, suggesting the existence of a great deal of internal voids in the films investigated. The third-neighbor peak in c-Ge occurs at 4.7Å, reflecting the 60° dihedral-angle order of the diamond structure. This peak is completely absent in a-Ge. The minimum possible dihedral angle (0°) leads to a third-neighbor separation of 4.02Å, which is very close to the second-neighbor distance. The fact that some of these types of configurations are present in a-Ge is confirmed by the anomalously large experimental value for the number of second-neighbors, 12.1, which is greater than the 12.0 expected in the diamond structure. No crystalline model, even those invoking the high-pressure polymorphs of either Si or Ge, can account for the observed results on a-Ge. In contrast, many of the random network models fit the observations very well, at least up to the 5.7Å fourth-neighbor peak [10]. These models are based on a nearest-neighbor environment of 4 atoms at the optimal bond length, with second-neighbors in configurations allowing for a ± 10° spread in bond angles around the optimal tetrahedral value. At this point, the weight of evidence is strongly in favor of the random network model for the structure of tetrahedrally bonded amorphous solids.

In general, the main result discussed previously, viz. the similarity of the parameters z, a, and θ in corresponding crystalline and amorphous solids, carries over to nontetrahedral materials as well. Of course, these three parameters simply reflect the nature of the chemical forces which characterize the constituent atoms. These chemical forces are themselves just parameterizations of the complex interactions given in Eqs. (2) and (3), and arise from the balance between the kinetic energies of the electrons and the nuclei and the electrostatic couplings between them. The strongest chemical forces are ionic, covalent, or a resonant admixture of both of them. In the presence of a large number of atoms, there can be a contribution from a slightly weaker force called metallic, which basically has the same origin as that of an unsaturated covalent bond. Other more exotic chemical forces exist and will be discussed subsequently, but ultimately their origins must be the same as those of the common forces, since they reflect the same electrostatic interactions between the electrons and the nuclei.

Ionic bonding takes advantage of the attractions between oppositely-charged atoms and tends to predominate whenever at least one type of atom with a relatively low ionization potential (i.e., the minimum energy necessary to remove an electron) and another type of atom with a relatively high electron affinity (i.e., the maximum energy reduction from forming the negatively-charged ion) are both present. The lowest energy ionic solid maximizes the electrostatic attractions between the positively and negatively charged ions while minimizing the mutual repulsions among similarly charged ions. For two similarly sized ions, the optimal value of z is 8, and an optimal value for the bond length, a, also exists. As the ratio of the ionic radii begins to deviate from unity, lower values of z (six or four) become optimal for geometric reasons. In any case, the only constraints on the bond angles are that they maximize the distance between similarly charged second-neighboring ions. Consequently, it is not too difficult to crystallize simple ionic solids. However, many multicomponent ionic amorphous materials do exist.

Metallic forces are ordinarily maximized when z is large (typically 12 or 8) and a takes on its optimal value. Once again, there are no strong constraints on the bond angles beyond those with simple geometric origins. Again, this results in the easy crystallization of simple metallic solids. Multicomponent metallic amorphous solids, however, can form rather easily.

In contrast, covalent bonding places strong constraints on all three parameters, z, a and θ. Consequently, it is not difficult to form covalent amorphous structures even from single elements or simple binary compounds. All of these exhibit a great deal of short-range order, as expected. In some covalent amorphous solids, there is evidence for intermediate-range order, in which third and sometimes fourth-neighbor peaks in the correlation function resemble those in the corresponding crystal. For example, this occurs in a-Se [11], which exhibits constraints on the dihedral angles in addition to those on z, a, and θ.

In addition to a determination of the local environments of the constituent atoms, other issues are important in describing the structure of amorphous solids. As indicated previously, a-Si and a-Ge appear to contain significant regions of microvoids and samples may resemble swiss cheese in that respect. In addition to this possibility, many other types of inhomogeneities may exist in particular materials. These may take the form of density fluctuations, of which voids represent an extreme example, or compositional variations, including, e.g., phase separation or more local clustering. When amorphous solids are deposited in thin-film form on a substrate, it should be borne in mind that the region near the substrate may have a different structure than the bulk of the material and that the properties of the interface region can depend strongly on the nature of the substrate. Furthermore, the free (i.e., upper) surfaces of films can exhibit structures which differ from those of the bulk. Finally, the possibility of the incorporation of unintentional impurities in significant concentrations should not be overlooked.

C.    Adiabatic Approximation and Vibrational Structure

Once it is determined that an equilibrium structure exists for the nuclei, we can apply the _adiabatic approximation_ to simplify the quantitative problem enormously [12]. The basis of this approximation is the existence of a small parameter, $m/M$, where m is the electronic mass and M a typical nuclear mass. Since $m/M \sim 10^{-3} - 10^{-5}$, the quantum mechanical problem can be expanded as a function of $(m/M)^{1/4}$, and terms of higher order than the first nonvanishing power can be dropped. Because $m/M$ is so small, the nuclei do not respond rapidly to changes in the electronic state, but the electrons respond immediately to any nuclear motion. This result can be used to separate the remaining problem into two much simpler sub-problems: (1) _vibrational structure_, i.e., determining the normal modes of oscillations of the nuclei about their equilibrium positions under the assumption that the electrons are in the lowest-energy state; and (2) <u>electronic structure</u>, i.e., evaluating the possible states for the electrons under the assumption that the nuclei are in their equilibrium positions. The normal vibrational modes can be thought of in a particle sense, analogous to the photons which can be used to describe the properties of electromagnetic-field oscillations (e.g., light, x-rays, etc.); such particles are called <u>phonons</u>. Another means of describing the adiabatic approximation is that it is equivalent to neglecting any interactions between electrons and phonons to first order.

Given the structure, the density of phonon modes per unit volume per unit frequency, $\rho(\omega)$, can be calculated in principle. In a crystalline solid, all of these modes are extended throughout the material and represent collective oscillations of all of the constituent atoms. The low frequency normal modes can always be modelled as if the solid were on elastic continuum rather than a discrete collection of atoms; these are the modes which represent sound propagation through the material, and consequently the phonons are ordinarily called <u>acoustic</u>. Clearly, for such modes, the amorphous or cryatalline nature of the solid is irrelevant and these phonons are also extended in noncrystalline materials. However, high-frequency modes in amorphous solids can be localized.

As is the case with crystalline materials, anharmonic forces can be present, leading to phonon-phonon coupling. Such anharmonicity has important consequences resulting in,e.g., thermal expansion, absorption of sound, etc.

Even in crystalline solids, accurate calculations of $\rho(\omega)$ from first principles have proved to be quite difficult. Nevertheless, there is experimental evidence that $\rho(\omega)$ in an amorphous solid can have a strong resemblance to that of the corresponding crystalline solid, broadened to take into account the effects of disorder [13]. The existence of disorder-induced localized phonon modes has not yet been demonstrated.

D.     Electronic Structure

After the adiabatic approximation is applied, the electronic part of the problem can be written [12]:

$$\sum_{i=1}^{ZN} \left( -\frac{\hbar^2}{2m} \nabla_i^2 - \frac{1}{2} \sum_{\alpha} \frac{Z_\alpha e^2}{|\underline{r}_i - \underline{R}_{\alpha\circ}|} + \frac{1}{2} \sum_{j=1}^{N}{}' \frac{e^2}{|\underline{r}_i - \underline{r}_j|} \right) \phi_n (\underline{r}_i)$$

$$= E_n \, \phi_n (\underline{r}_i)$$

(4)

where the nuclei are taken as fixed at their equilibrium positions, $R_{\alpha\circ}$. For crystalline solids, the fact that the $R_{\alpha\circ}$ form a periodic array offers a major simplification, and group-theoretical techniques can be used to aid in the solution of the problem. Nevertheless, the complete solution has not yet been attained, because of the mutual electronic repulsions represented by the last term on the left-hand side of Eq. (4). In almost all cases, the one-electron approximation is also applied, in which it is assumed that each electron moves in the average field due to all the others. This approximation neglects electronic correlations, i.e., the possibility that two electrons can correlate their motion in such a manner as to keep away from each other, and thereby minimize their repulsive interaction.

Given the one-electron approximation, we can write the electronic problem as one involving the positions of only a single electron, $\underline{r}_i$:

$$H_i \, \varphi_i (\underline{r}_i) = E_i \, \varphi_i (\underline{r}_i)$$

(5)

where the effects of all the other ZN-1 electrons are included in the operator $H_i$. The solutions to Eq. (5) give the one-electron energies, $(E_i)_n$. The lowest-energy state of the solid is then the one in which the lowest ZN energy states, $(E_i)_n$, are filled and all higher ones are empty. Excited states can be analyzed by transferring electrons from lower energy-filled to higher energy-unfilled states. Since there are so many ($\sim 10^{23} cm^{-3}$) electrons in a typical solid, the energy levels $(E_i)_n$ are closely spaced ($\sim 10^{22} cm^{-3} eV^{-1}$). Consequently, it is preferable to treat them as essentially continuous, replacing the discrete levels $(E_i)_n$ by a function, $g(E)$, which gives the number of one-electron states per unit volume per unit energy; $g(E)$ is called the density of one-electron states, and, if it is known, we can ordinarily deduce the electrical and optical properties of the solid from it. In particular, the Fermi energy, $E_F$, is obtained from the relation:

$$ZN = \int_{-\infty}^{\infty} dE \ g(E) \ f(E) \tag{6}$$

where $f(E)$ is the Fermi-Dirac distribution function. For weakly interacting particles in thermal equilibrium at a temperature, T:

$$f(E) = \frac{1}{\exp[(E-E_F)/kT]+1} \tag{7}$$

where k is Boltzman's constant, equal to $1.38 \times 10^{-23}$ j/K in MKS units. The electrical conductivity can be obtained from the expression:

$$\sigma = \int_{-\infty}^{\infty} dE \ g(E) \ f(E) \ e \ \mu(E) \tag{8}$$

where $\mu(E)$ gives the mobility (velocity attained per unit applied electric field) of an electron in a state at energy E under the condition that the other electrons are in their lowest-energy states. The mobility is controlled by the scattering process that affect the motion of the excited electron as it moves in the presence of the applied field. The scattering can be due to the phonons, the other electrons, or, in some cases, even the ion cores at their equilibrium positions.

The optical absorption also can be obtained from $g(E)$. A photon of frequency w can be absorbed if it can excite an electron from an occupied state of energy E to an unoccupied one of energy $E + \hbar w$. Thus, the absorption coefficient, $\alpha(w)$ can be written:

$$\alpha(w) = \frac{K}{\hbar w} \int_{-\infty}^{\infty} dE \ |M(E)|^2 \ g(E) \ g(E+\hbar w) \ f(E) \ [1-f(E+\hbar w)] \tag{9}$$

where $M(E)$ is the matrix element of the momentum between the states at E and $E + \hbar w$ and K is a constant. If $M(E)$ is not a rapidly-varying function of energy, the spectral variation of $\alpha(w)$ can be deduced just from $g(E)$.

Clearly, this formulation of solid-state theory depends on the validity of both the adiabatic and one-electron approximations. Indeed, in a wide class of materials, the effects phonon coupling and electronic correlations are negligible, and theory and experiment are not widely separated. However, it is important to bear in mind that this is not always the case. For example, consider an ionic solid, in which at least two types of atoms exist, one positively charged and the other negatively charged. There is then a vibrational mode in which the oppositely charged ions move in opposite directions along the axis between them. This mode, called a longitudinal optical phonon, sets up alternating regions of positive and negative charge density in excess of the values when the ions are all in their equilibrium positions. Consider an electron moving in the vicinity of these vibrating ions. Being negatively charged, the electron can lower its energy in the presence of a phonon which brings excess positive charge near its own position and removes negative charge away from it. This is clearly a situation where the electron-phonon interaction is large, and the adiabatic approximation can be quite inaccurate. Analogously, the one-electron approximation can be qualitatively in error in many cases. Consider a collection of N widely-separated hydrogen atoms. In the ground state, each atom has a single electron in the 1s orbital, bound to the proton by the hydrogen ionization potential, 13.6eV. But each 1s orbital has room for two electrons, one with spin up and the other with spin down. If we place the N electrons in 1s orbitals on the N protons at random, N/4 of the protons will have no electrons ($H^+$ ions), N/2 will have one electron (H atom), and N/4 will have two electrons ($H^-$ ions). The second electron on the $H^-$ ions is bound by only the electron affinity of hydrogen, 0.7eV. The electron affinity is lower than the ionization potential by 12.9eV because the two electrons in the 1s orbitals repel each other. The average energy of the collection of N protons and N electrons in the case where the electrons are placed at random is thus -3/4 x (13.6eV) -1/4 x (0.7eV) = -10.4eV. But this is much higher than the average energy of a collection of hydrogen atoms with exactly one electron on each proton, -13.6eV. It is clear that the electrons can correlate their motion to stay out of each other's way and thus lower the total energy. But the one-electron approximation neglects the possibility of such correlations. In solids in which there are many nearly free outer electrons, their rapid motion tends to screen out the effects of electrostatic interactions, and both the correlation energy and the electron-phonon coupling can be reduced to very small values. But, as will be discussed shortly, this is not always the case, not even in perfect crystals.

As indicated previously, the electron-phonon interaction can be very strong in ionic solids. In order to correctly describe the transport in such materials, we must carefully consider the polarization of the ions due to the electronic charge. A reasonable approach is to analyze the motion of an electron always moving together with an associated ionic distortion.

The electron moving with the appropriate numbers of the types of phonons to produce this distortion can be thought of as a quasiparticle, in this case called a polaron [12]. Clearly, a polaron acts as if it has a greater mass than would the electron without the associated ionic distortion. In addition, because of the additional electrostatic attraction which leads to its formation, the polaron has lower energy than the bare electron. In many cases, the effects of the electron-phonon interaction can be taken into account by replacing the one-electron density of states, $g(E)$, calculated with the ion cores at their equilibrium positions (rigid-ion approximation) by a modified $g(E)$, with the bands shifted down in energy and somewhat narrower in width.

If the bandwidth is small within the rigid-ion approximation, a different phenomenon can occur. In this case, an excess electron induces a localized distortion of the neighboring ion cores only. This distortion, in turn, attracts the excess electron, which can form a bound state with the local distortion analogous to a hydrogen atom; this bound state is called a small polaron. Small polarons are stable provided the energy decrease from their formation is larger in magnitude than the energy of delocalization. Since the latter is typically about half the bandwidth [12], electrons in narrow energy bands tend to form small polarons. In periodic crystals, small polarons can overlap and spread into bands, and ordinary transport is possible at low temperatures. However, the bandwidth decreases exponentially with increasing temperature, leading to a rapidly decreasing carrier mobility. Eventually, the mobility becomes sufficiently small that the mean free path between scattering events is less than a nearest-neighbor distance. At this point, bandlike propagation is impossible, and carrier transport occurs only by the hopping of small polarons between localized states. Since this hopping requires the simultaneous transfer of the associated distortion with the carrier, the assistance of phonons is essential for transport. Since there are exponentially more phonons present at high than at low temperatures, the small-polaron mobility is an exponentially increasing function of temperature in the hopping regime. In disordered systems, the small-polaron band regime is not expected to be observable.

Another consequence of narrow bands is the breakdown of the one-electron approximation [12]. As mentioned previously, when bands are narrow, the decrease in energy brought about by electronic delocalization is small. If it is insufficient to overcome the increase in energy arising from the simultaneous presence of two electrons on the same ion core, the electrons will remain localized, even within the rigid ion approximation. Thus, both the adiabatic and the one-electron approximations fail in narrow bandwidth solids.

When the one-electron approximation is invalid, the concept of a density of states, $g(E)$, is inappropriate. Nevertheless, the theory can be extended in certain cases [12]. When correlations are important, the one-electron energies can be taken to be a function of the number of electrons present, in accordance with the previous discussion of the hydrogen atom. But great care must be taken, especially in analyzing the excited states of the system. For example, the thermal equilibrium properties of a correlated system are not properly described by the ordinary Fermi-Dirac distribution function; instead, a modified $f(E)$ must be used [14].

## III.  CRYSTALLINE SOLIDS

### A.  Conventional Viewpoint

The conventional approach to understanding the electronic structure of crystalline solids is well known [15]. The key element is to simplify the problem by making use of the fact that the one-electron potential energy must exhibit the periodicty of the lattice. Group-theoretical methods can then be applied to reduce the problem to one involving only the small number of electrons contained in a single primitive cell (i.e., building block) from which the crystal can be generated. It can easily be shown that all electronic states are extended throughout the solid, in fact having the same probability of being found in each of the primitive cells. The one-electron density of states, $g(E)$, can be calculated by solving the one-electron problem in a single primitive cell and applying periodic boundary conditions. Because of the periodicity, the density of states of any crystalline solid takes the form of alternating regions of energy with large densities, typically $g(E) > 10^{22} cm^{-3} eV^{-1}$, called bands separated from regions where no states are possible, called gaps. As an example, the calculated $g(E)$ for crystalline Si (c-Si) is shown in Fig. 2 [16]. The states with $E < 0$ form the valence band, which is ordinarily filled

at low temperatures. The Fermi energy lies in a gap of about 1.2eV,
which gives c-Si its semiconducting properties. The <u>conduction
band</u>, ordinarily empty at low temperature, lies at higher energies.
The sharp structure that is evident in g(E) at many different energies
arises directly from long-range periodicity and these energies are
called <u>van Hove singularities</u> [17]. It can easily be shown that the
density of states near a van Hove singularity has the behavior,

$$g(E) = A \left| E - E_o \right|^{1/2} \tag{10}$$

either above or below the critical energy, $E_O$. The valence and
conduction band edges, $E_v$ and $E_c$, respectively, are important
examples of van Hove singularities.

Although conventional solid-state theory has had a great
deal of success in explaining the behavior of crystalline solids
and is often in detailed agreement with the results of experiments,
it can be criticized on two major grounds. First, the calculation
of g(E) involving primitive cells and periodic boundary conditions
makes use of the crystalline symmetry but loses sight completely
of the short-range order which reflects the chemistry of the constituent atoms. For example, the symmetry of the primitive cell of

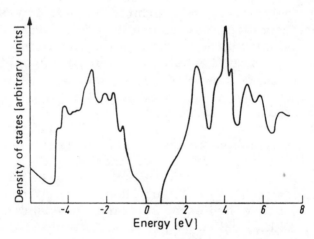

Fig. 2    Density of states as a function of energy for crystalline
silicon. (From ref. 16.)

covalently bonded c–Si is exactly the same as that of metallic Ni and ionic NaCl, despite the very different local coordinations. Second, use of the model fosters the viewpoint that periodicity is vital to the electronic behavior of materials. However, this view is inconsistent with experimental results [18]. If periodicity were essential to the observed electronic transport in crystalline solids, we should expect large changes in conductivity upon melting, at which point the long-range order spontaneously disappears. However, as is clear from Fig. 3, melting has only a very small effect on the conductivity of a large class of materials, including insulators, semiconductors, and metals. Thus, not only must the basic density of states, $g(E)$, be relatively independent of long-range order, but also the mobility, $\mu(E)$, cannot be very sensitive to the crystalline periodicity. Since the Fermi-Dirac distribution function, $f(E)$, exponentially decreases with increasing energy above $E_F$ [see Eq. (7)], the electrical conductivity is dominated

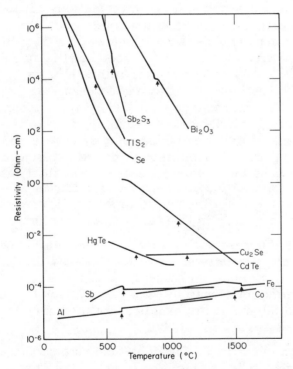

Fig. 3   Resistivity as a function of temperature both below and above the melting point for several materials. The melting point is indicated by an arrow. (From ref. 18.)

by carriers very near the mobility edge. Typical values for the band mobility in crystalline semiconductors are of the order of $1000$ cm$^2$/V-s, which corresponds to carrier mean free paths of about 100 interatomic spacings between scattering events. Since conventional solid-state theory suggests that it is only <u>deviations</u> from crystalline periodicity which scatter free carriers (e.g., phonons or impurities), these long mean free paths are easily understood. However, there is clearly something wrong with this view if they can be maintained in the liquid phase.

B.    Alternative Viewpoint

In view of the problems with the conventional picture, it is worthwhile considering an alternative formulation of solid-state theory, one which emphasizes the chemistry of the constituent atoms. Such formulations exist, including some very sophisticated approaches. In its simplest form, this type of approach employs the <u>tight-binding approximation</u> [19], in which the electronic wave functions are antisymmetric linear combinations of atomic orbitals centered on nearest-neighboring ion cores. Schrödinger's equation is then solved for the one-electron energy levels. Although this method is very crude, it provides a good starting point for analyzing $g(E)$, one which does not lose sight of the chemistry. More recently, the method has been modified and an <u>empirical tight-binding</u> (ETB) approach has become popular [20]. In this technique, the matrix elements of the Hamiltonian and the overlap integrals between electrons centered on neighboring atoms are treated as adjustable parameters to obtain good fits to experimentally measured quantities such as the energy gap. Since the short-range order of corresponding crystalline and amorphous solids is ordinarily identical, it is useful to set the parameters by their crystalline values and keep them the same in their amorphous counterparts.

Perhaps the most sophisticated approach to date is the self-consistent-field $X\alpha$ scattered wave (SCF-$X\alpha$-SW) method [21]. In this technique, a cluster of atoms is analyzed in a self-consistent manner, allowing for atomic relaxations. Correlation effects are included, and the method is excellent for the analysis of optical excitation energies.

In its simplest form, the tight-binding approach can be used in the manner sketched in Fig. 4. In Fig. 4(a), the band structure of Si is estimated. In tetrahedrally bonded Si, the four valence electrons form sp$^3$ hybrids which point in the appropriate tetra-

hedral directions. When four neighboring atoms each form a covalent bond with the atom under consideration, the 8 degenerate
sp³ orbitals split into four lower <u>bonding</u> orbitals and four upper
<u>antibonding</u> orbitals. Introduction of the 12 second-nearest neigh-
bors splits up both the bonding and the antibonding levels. When
all of the rest of the solid is taken into account, the bonding orbitals spread out in energy to form into the valence band and the
antibonding orbitals spread to form the conduction band. Since
each Si atom has four electrons to place into these bands, the
valence band is filled and the conduction band is empty at T = 0.

Analogous considerations hold for the elements from Columns
I-III in the Periodic Table. All N valence electrons, where N is
the appropriate column from which the element comes, can participate in covalent bonding. When N = 1 (the alkalis), a single s
bond can be formed. However, since the s electrons are spherical-

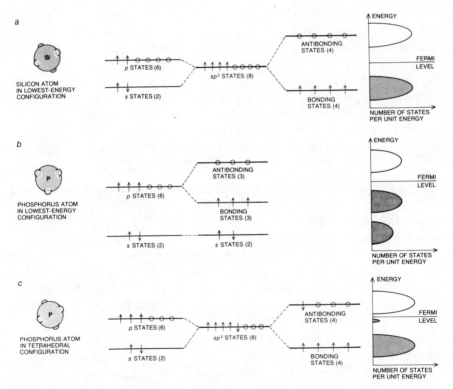

Fig. 4   Tight-binding approximations for the density of states:
(a) a tathogen, such as silicon; (b) a pnictogen, such as
phosphorus; (c) P-doped Si.

ly symmetric, such bonds are weak, and solids containing alkalis typically form either ionic or metallic rather than covalent bonds. The alkaline earths, with $N = 2$, can form two sp hybrids and thus participate in two distinct bonds. Since sp orbitals point in opposite directions, the optimal bond angle is $180^\circ$. A filled bonding band is separated from an empty antibonding band and another empty nonbonding band, which spreads from the remaining two p orbitals. Elements from Column III can form $sp^2$ hybrids and participate in three bonds, lying within a plane and having $120^\circ$ bond angles. A lower $sp^2$ bonding band is filled and upper nonbonding p and antibonding $sp^2$ bands are empty at $T = 0$, again yielding semiconducting behavior.

The case of phosphorus, one of the pnictogen elements, from Column V of the Periodic Table, is sketched in Fig. 4(b). When $N > 4$, the valence electrons cannot all participate in covalent bonding. This comes about because of the general rule that at any interatomic separation, the antibonding orbital is always increased in energy more than the bonding orbital is decreased. Consequently, a pair of electrons in a single orbital always has lower total energy if they remain nonbonded, i.e., a so-called lone pair, than if they form a bonding-antibonding pair. Thus, the optimal covalently bonded state for a pnictogen is one in which the s electrons remain as a low-energy lone pair and the three p electrons form covalent bonds. Pure p-like orbitals form $90^\circ$ bond angles, but some hybridization always takes place and, e.g., the optimal bond angle for phosphorus is in the range $94^\circ - 102^\circ$.

Chalcogen elements, from Column VI, form only two p bonds in their ground states. Not only do they have a low-energy s lone pair, but they also have a p lone pair which is the highest-energy occupied orbital in the ground state. Hybridization effects usually increase the bond angles in chalcogenides to the $98^\circ - 108^\circ$ range. Clearly, the ground state of the halogen elements, from Column VII, is one in which only a single p bond forms. The highest filled orbitals are the two lone-pair p states. Elements from Column VIII do not ordinarily form any covalent bonds at all. Thus, the rules for covalent bonding are that elements from Columns I-IV form N bonds, while those from Columns IV-VIII form 8-N bonds in their ground states. The latter is often called the 8-N rule.

Much of the current semiconductor industry is based on the ability to modulate $E_F$ by doping. The most common example is the introduction of small concentrations of P into c-Si. Because of the constraints imposed by the c-Si lattice periodicity, strains are minimized when the small fraction of P atoms that are present enter the solid _substitutionally_, at a position where a Si atom belongs in the perfect crystal. Thus, the P atoms are tetrahedrally coordinated, and the resulting tight-binding band structure is sketched in Fig. 4(c). The fifth valence electron of the P atom would be expected to enter an antibonding orbital in this configuration. However, its energy would then lie within the conduction band of the doped Si crystal, and it could lower its energy by leaving the vicinity of the P atom and dropping to the bottom of the conduction band in an extended state. But, in fact, it can lower its energy still further by moving in contracted Si antibonding orbitals in the vicinity of the $P^+$ ion from which it came. This effect is completely analogous to a free electron lowering its energy by localizing in the vicinity of a proton, $H^+$, thus forming a hydrogen atom, H. Its kinetic energy is somewhat increased upon localization (from the Uncertainty Principle), but its potential energy decrease from the electrostatic attraction between it and the proton more than makes up for this. Consequently, the excess electron on the tetrahedrally bonded phosphorus atom actually has an energy somewhat below the bottom of the Si conduction band, as shown in Fig. 4(c).

More accurate calculations indicate that the real situation is somewhat more complicated. For example, Fig. 5 shows the results of SCF-Xα-SW calculations on a cluster of five Si atoms (a central atom surrounded by a regular tetrahedron of Si atoms), all of whose valence electrons are bonded either to one of the other Si atoms or 12 saturators (hydrogen atoms placed at the ordinary Si-Si separation of 2.35Å). These results show that the valence band is less hybridized than the tight-binding approximation suggests, the lower part being primarily s-like and the upper part primarily p-like. The conduction band does consist largely of $sp^3$ antibonding orbitals as expected, but the d orbitals also contribute. The photoemission spectrum calculated from the Xα analysis of the simple 17-atom cluster agrees quite well with actual measurements [23].

IV.    EFFECTS OF DISORDER

The considerations of Section III indicate that the density of states, g(E), of an amorphous solid should not be very different from that of its corresponding crystalline solid, provided the short-

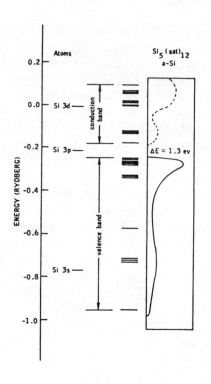

Fig. 5   SCF-Xα-SW molecular orbital energy
levels and optical
density of states of
a 17-atom cluster
representing pure
a-Si.  (From ref.
22.)

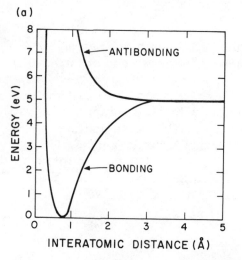

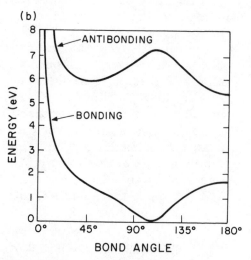

Fig. 6   (a) Energy as a function of interatomic separation for the
lowest-energy bonding and antibonding orbitals of a hydrogen molecule; (b) sketch of the energies of the lowest-
energy bonding and antibonding orbitals as a function of
bond angle for $H_2O$.

range order is identical. As a rough guide, we may again use a tight-bonding approach, now asking how the analysis would be modified if we introduce bond angle and perhaps even bond-length distortions. Figure 6 shows a sketch of the energies of the bonding and antibonding orbitals as a function of bond length and bond angle. It is clear that either the stretching or the bending of a bond tends to increase the energies of the valence-band states and decrease the energies of the conduction-band states. Thus, either bond stretching or bond bending could introduce localized states into the energy gap. This is consistent with the fact that the band edges are just one type of van Hove singularity, which, as we previously discussed, arises only because of long-range periodicity. In the absence of long-range order, we expect these singularities to disappear and be replaced by more gradually decreasing densities of states, as sketched in Fig. 7. These regions of gradually decreasing $g(E)$ are called <u>band tails</u>.

However, the major effect of disorder is not expected to be the replacement of the sharp band edges of crystals by band tails. As discussed previously, another important characteristic of long-range periodicity is the <u>extended</u> nature of the electronic states.

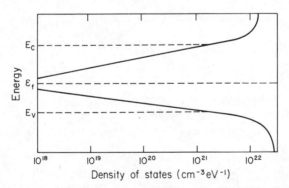

Fig. 7   Density of states as a function of energy for an amorphous semiconductor. $E_V$ and $E_C$ denote the positions of the valence and conduction band mobility edges, respectively; $E_F$ is the zero temperature Fermi energy.

In amorphous solids, the possibility exists that the reverse is true--all states could be localized. But it is not the extent of the states that is of fundamental physical significance; rather, it is quantities such as the mobility, $\mu(E)$, which affects the electrical conductivity via Eq. (8). Even if all states are localized to a distance $r_0$, electrons can still tunnel from an occupied to an unoccupied state at equivalent energies, with a probability proportional to $\exp(-2d/r_0)$, where d is the distance between the two states. Thus, we can write [24]:

$$\mu(E) = \mu_0 \exp(-2d/r_0), \tag{11}$$

where $\mu_0$ is a constant of proportionality. If all states within an energy range $\Delta$ are equivalent, then the average separation between equivalent states is:

$$d = [g(E) \Delta]^{-1/3}. \tag{12}$$

Substitution of Eq. (12) into Eq. (11) yields:

$$\mu(E) = \mu_0 \exp(-2/[g(E) \Delta]^{1/3} r_0). \tag{13}$$

It is not yet clear how $g(E)$ varies in a band tail, and both gaussian and exponential [26] behavior has been inferred theoretically. For the latter, we may write:

$$g(E) = A^3 \exp [-(E_0 - E)/kT_0], \tag{14}$$

where $T_0$ is a parameter describing the extent of the tail. In this case, Eq. (13) yields:

$$\mu(E) = \mu_0 \exp \left\{ -(2/A\, r_0 \Delta^{1/3}) \exp[(E_0 - E)/3kT_0] \right\}. \tag{15}$$

Equation (15) is essentially a step function, in which $\mu(E)$ increases sharply from very small values to $\mu_0$ near the value of E at which point the term in curly brackets is unity. Thus, we can define a critical energy, $E_c'$, by:

$$E_c' = E_0 - kT_0 \ln (A\, r_0\, \Delta^{1/3}/2). \tag{16}$$

$E_c'$ is called the conduction-band <u>mobility edge</u>, and it plays the same role in disordered solids that the band edge, $E_c$, does in crystals. The critical density of states at a mobility edge is:

$$g_c = 8/(\Delta r_o)^3. \tag{17}$$

For example, if we take $\Delta = 0.027eV$ (kT at room temperature) and $r_o = 100\text{Å}$, then $g_c = 3 \times 10^{20}cm^{-3}eV^{-1}$.

If the band tails are gaussians, similar results apply, the mobility edge occurring near the point where the term in parentheses in Eq. (13) is unity. In either case, we should bear in mind that the conductivity is properly an integral over the product $g(E) f(E) \mu(E)$ [see Eq. (8)], and is not simply given by the value of the product at $E_c$. This can be significant when $g(E)$ is such a rapidly varying function of E as is commonly believed [27].

Of course, the valence band should also have a mobility edge, $E_v'$. The difference between $E_c'$ and $E_v'$ is called the mobility gap, $E_g'$:

$$E_g' = E_c' - E_v' . \tag{18}$$

We should expect that the effects of electron-phonon interactions and electronic correlations are small for carriers in extended states. However, this is certainly not the case for carriers in localized states.

Mott [28] was the first to suggest the idea of a mobility edge. Cohen, Fritzsche and Ovshinsky [29] pointed out that the extent of the band tails should be a measure of the amount of disorder. They thus proposed that elemental amorphous solids, e.g., a-Si, a-Se, are characterized by a band structure such as is sketched in Fig. 7, but that a multicomponent amorphous solid such as a-$Te_{40}As_{35}Si_{15}Ge_7P_3$, one of the switching glasses that Ovshinsky [4] had been studying, was more properly represented by the band structure sketched in Fig. 8. In this case, which has come to be called the CFO Model, the valence and conduction band tails are so extensive that they intersect in mid gap. This has several important consequences. First, some normally filled valence-band states have energies in excess of normally empty conduction-band states, resulting in a repopulation to restore equilibrium. Second, there is a finite density of states at the Fermi energy, $g(E_F)$. If $g(E_F) > g_c$, the material is an amorphous metal with high conductivity; on the other hand, if $g(E_F) < g_c$, the solid is still a semiconductor with a mobility gap, $E_g'$, within which the states are localized. However, the finite $g(E_F)$ still would tend to pin the Fermi energy, retarding, e.g., attempts to modulate

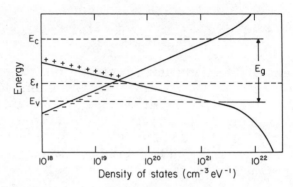

Fig. 8    CFO model for the density of states of a sufficiently dis-
          ordered amorphous semiconductor.  $E_V$ and $E_C$ are the res-
          pective valence and conduction band mobility edges; $E_g$ is
          the mobility gap and $E_F$ is the Fermi energy.  The positions
          of the positively and negatively charged traps at zero
          temperature are schematically noted.

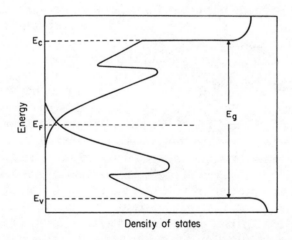

Fig. 9    Effective one-electron optical density of states for a CFO-
          type amorphous semiconductor in which electronic correla-
          tions are important for localized states in the mobility gap.
          (From ref. 18.)

$E_F$ by doping. In addition, the transfer of electrons from localized valence band states to localized conduction band states would create equal concentrations of deep <u>charged</u> traps, the positive ones above $E_F$ and the negative ones below $E_F$. Such traps would be expected to sharply limit the mobility of excess free electrons and holes, respectively.

A major complication with the CFO model is the fact that states near $E_F$ are expected to be strongly localized, and thus characterized by strong electron-phonon interactions and electronic correlations. Because of the correlations, the density of states in the gap depends on their occupation. The likely appearance of a sharp jump in the correlation energy at the valence and conduction band mobility edges should introduce discontinuities in the effective density of one-electron states at $E_v'$ and $E_c'$, respectively [18]. The electron-phonon interaction would tend to reduce the magnitude of the effective correlation energy by stabilizing the charged localized states, but the overall picture would remain basically the same [18]. When these effects are taken into account, the effective density of states is then similar to that sketched in Fig. 9.

If carriers in the localized states in the gap form small polarons, their energy is lowered by the polaron binding energy. Nevertheless, they could contribute to transport via the hopping mechanism discussed previously. In this case, we should expect a mobility of:

$$\mu(E) = \mu_o \exp(-2d/r_o) \exp(-W/kT), \tag{19}$$

where W is about half the polaron binding energy [30]. The extra Boltzmann factor appears because the polarons need the assistance of phonons to hop from site to site. The minimum energy necessary is that when the region around the polaron undistorts somewhat and the region around a nearest-neighbor distorts somwhat until their environments are equivalent; at this point, the carrier can hop between the equivalent sites without any further increase in energy.

It might be expected that the strength of the electron-phonon interaction depends on the degree of localization of the electronic state. Since this degree of localization undergoes a sharp discontinuity near the mobility edge, small polarons could well be stable for would-be carriers within the mobility gap but not beyond.

This would tend to yield more linear behavior on a plot of log $\sigma$ vs $T^{-1}$ than suggested by Eq. (15). Indeed, Cohen et al. [31] showed that as the strength of the electron-phonon interaction increases, $\mu(E)$ changes from a rapidly varying but continuous function of energy to one which is discontinuous at the mobility edge.

Mott [32] pointed out that a parallel conduction mechanism can take place near the Fermi energy if $g(E_F)$ is sufficiently large. In this case, Eq. (19) remains valid, except that W is now identified with the smallest energy difference between an occupied state below $E_F$ and an unoccupied state on one of its nearest neighbors at a distance $d = [g(E_F)\Delta]^{-1/3}$ away. For nearest-neighbor hopping near $E_F$, there is no significant temperature variation in f(E), and a plot of log $\sigma$ vs $T^{-1}$ (often called an Arrhenius plot) should be linear with a slope of W/k. However, as the temperature decreases, exp(-W/kT) becomes very small. Eventually, it may be easier for the electron to hop to a site a distance considerably larger than d from it, provided it can reduce the energy difference. If the states are randomly located in both energy and position and g(E) is not varying rapidly near $E_F$, hopping to a site nd away will reduce W to a value of approximately $W/n^3$ (since there are a factor of $n^3$ more localized states within a sphere of radius nd than in one of radius d). Thus, hopping near $E_F$ should be characterized by a series of parallel conduction mechanisms:

$$\mu(E_F) = \sum_{n=1}^{\infty} \mu_n = \sum_{n=1}^{\infty} \mu_o \exp[-(2nd/r_o + W/n^3 kT)]. \tag{20}$$

The maximum contribution to $\mu(E_F)$ occurs when the exponent is a minimum, at which point:

$$n_o = (3W \, r_o/2d \, kT)^{1/4}. \tag{21}$$

This calculation is valid provided T is sufficiently small that $n_o \gg 1$. In this case, the predominant contribution has the temperature dependence:

$$\mu_{no} = \mu_o \exp[(512 \, W \, d^3/27 \, r_o^3 \, kT)^{1/4}]. \tag{22}$$

Thus, an Arrhenius plot should not yield a straight line at low temperatures. Instead, a plot of log $\sigma$ should be linear as a function of $T^{-1/4}$. When this occurs, the predominant mechanism of conduction is called variable-range hopping.

Mott's model has been justified more rigorously by the analysis of an equivalent network of random resistors by Ambegaokar et al. [33], who then used percolation theory to derive a relationship analogous to Eq. (22) but with a slightly modified coefficient in the exponential; the $T^{-1/4}$ behavior remains valid. An explicit computer simulation [34] has further confirmed the validity of this approach. On the other hand, if $g(E)$ near the Fermi energy is a rapidly varying function of energy, e.g., $g(E) = C(E-E_F)^n$, then $\ln \sigma$ should be linear as a function of $T^{-(1+n)/(4+n)}$ rather than $T^{-1/4}$ [35,36]. Furthermore, the model neglects electronic correlations, which should make the results invalid at very low temperatures, when $kT$ is small compared to the electrostatic attraction between an electron above $E_F$ and a hole below $E_F$. In this case, a bound state between the electron and hole (i.e., an exciton) can form, which is neutral and cannot contribute to conduction. This effect can be looked at as introducing a gap into the effective one-electron density of states near the Fermi energy, even if the valence and conduction band tails overlap [37].

It should also be borne in mind that a linear plot of $\ln \sigma$ as a function of $T^{-1/4}$ is not a proof of the predominance of variable range hopping conduction. It can be shown that the conductivity near a <u>diffuse</u> mobility edge simulates this behavior because of the existence of parallel conduction paths [38]. At low temperatures, relatively low-mobility conduction predominates because of the small carrier concentrations thermally excited to the high-mobility states. In addition, Emin [39] has shown that small-polaron hopping at low temperatures occurs primarily with the assistance of multiple low-energy phonons rather than a single high-energy phonon (which is present in much smaller concentrations), again yielding parallel conduction mechanisms and an approximate $T^{-1/4}$ behavior.

## V.   EXPERIMENTAL TESTS OF THE CFO MODEL

The essential features of the CFO model described in Section IV are: (1) band tails, (2) mobility edges, (3) $g(E_F)$, and (4) local satisfaction of valence requirements. The existence of band tails due to disorder is on firm theoretical grounds. Indeed, since sharp band edges are van Hove singularities (see Section III) and arise only from long-range periodicity, they cannot be present in amorphous solids. However, the shape and extent of the band tails have not yet been calculated with any degree of certainty for any real material. In contrast, even the existence of sharp mobility edges has not yet been demonstrated with any rigor. Most of the theoretical models used to justify their concept emply a periodic lattice and are thus more appropriate to the problem of a disordered alloy [40,41]. In addition, other assumptions have always been invoked.

The density of states at the Fermi energy, $g(E_F)$, can be considered to be simply a parameter. It can arise either from overlap of the valence and conduction band tails, as in Fig. 8, or from localized states in the gap due to impurities or defects. Thus, it may vary enormously depending on the details of the materials processing. In any event, its magnitude is of major importance. If it is small, $g(E_F) < 10^{15} cm^{-3} eV^{-1}$, it has little effect on the electronic properties of the material, and $E_F$ can be easily modulated by, e.g., doping, charge induction, or free-carrier injection. For intermediate values, $10^{15} cm^{-3} eV^{-1} < g(E_F) < 10^{19} cm^{-3} eV^{-1}$, there is an increasing tendency to pin the Fermi energy and thus make it difficult to dope or observe a field effect (i.e., a large change in conductivity via the induction of excess positive or negative charge in the material). For large values, $g(E_F) > 10^{19} cm^{-3} eV^{-1}$, the Fermi energy is strongly pinned, and we should expect to observe variable-range hopping conduction at low temperatures.

The assumption of local satisfaction of valence requirements was implicit in our sketches in Figs. 7 and 8. As indicated previously, localized defect configurations, if present in sufficient concentrations, would yield sharp structure in $g(E)$. For example, the presence of tetrahedrally coordinated P atoms in c-Si:P results in the appearance of a peak below the conduction band, as shown in Fig. 4(c). In general, we would expect that if well-defined defect centers exist in either crystalline or amorphous solids, they should be evident from deviations from the smooth band tails shown in Fig. 8. Cohen et al. [29] made the assumption of local satisfaction of valence requirements on reasonable physical grounds especially for the case in which they were most interested, that of glasses quenched from the liquid phase. In the liquid where atoms are quite mobile, we should expect that strong chemical forces would tend to keep each species in its lowest-energy coordination, obeying the N/8-N rule of covalent bonding. When the temperature is lowered beyond the <u>glass transition temperature</u>, $T_g$, the viscosity of the material increases many orders of magnitude, thus quenching in the local coordination (and the long-range disorder).

We might ask if the essential features of the CFO model can be tested experimentally. Most of these features involve the shape of g(E), particularly within the gap, so that modern spectroscopy techniques should prove to be extremely useful. The most straightforward method for analyzing the shape and extent of the band tails would appear to be direct optical absorption experiments. As is clear from Eq. (9), $\alpha(\omega)$ measures the product of the occupied and unoccupied states separated by an energy $\hbar\omega$. It should be possible, in principle, to deconvolute this joint density of states to yield g(E), particularly if combined with the results of, e.g., photoemission experiments which can provide the density of occupied states directly. Recent advances include the use of photoconductivity [42] to measure $\alpha(\omega)$ in the band-tail region (provided at least one free carrier is excited by the transition) and the development of the technique of <u>photothermal deflection spectroscopy</u> (PDS) [43] to determine $\alpha(\omega)$ even in regions of localized-to-localized excitations. In addition, more refined methods for the analysis of direct optical-absorption experiments are possible [44].

A completely different technique for the determination of g(E) in the band-tail regions is that of transient spectroscopy. In this type of experiment, excess carriers are excited, e.g., by a pulse of light, and the response is measured as a function of time. The induced photoconductivity, optical absorption, and luminescence have been the most common responses investigated. These experiments provide information on the band tails because of the fact that these states are the predominant traps for the excess carriers. Essentially, all of the excess carriers are trapped in times of the order of several picoseconds. It is reasonable to assume [45,46] that the cross section for trapping is the same for all band-tail states, since presumably they all originate from the same types of positional disorder. However, their release times depend exponentially on their energy separation from the mobility edge so that the initial response is due primarily to release of the shallowest trapped carriers. The magnitude of the response at a time t after the pulse probes the density of states near the energy, E, is given by:

$$E = E_c' - kT \ln (\nu t), \tag{23}$$

where $\nu$ is the trap escape frequency, typically $10^{12}$ Hz. Thus, at room temperature, nanosecond response measurements probe localized states about 0.2eV below the conduction-band mobility edge (or above the valence-band mobility edge).

Confirmation of the existence of a sharp mobility edge cannot be accomplished by spectroscopy, since it deals with $\mu(E)$ rather than $g(E)$. Although it is likely that $\mu(E)$ depends on $g(E)$, as discussed in Section IV, some independent transport measurements are necessary to analyze $\mu(E)$. The most straightforward experiment is clearly observation of the electrical conductivity as a function of temperature. If $\mu(E)$ increases sharply from very small values to finite values, Eq. (8) indicates that an Arrhenius plot of $\sigma(T)$ should be quite linear. In contrast, a gradual increase in $\mu(E)$ should yield concave upward plots of $\ln \sigma$ vs $T^{-1}$, since parallel conduction paths are possible. However, the additional possibilities of nearest-neighbor and variable-range hopping discussed previously complicates the analysis. A transition from bandlike propagation to nearest-neighbor hopping as T decreases should yield two distinct linear segments on an Arrhenius plot. However, as mentioned previously, variable-range hopping is very difficult to distinguish experimentally from a gradual increase in mobility with energy. Although, according to Eq. (22), the former should yield a straight line on plot of $\ln \sigma$ vs. $T^{-1/4}$, the small range in $T^{-1/4}$ over which conductivity experiments can ordinarily be carried out (e.g., 77-300K yields a spread of only a factor of 1.4 in $T^{-1/4}$) would tend to straighten out any concave upward Arrhenius plot on a $T^{-1/4}$ graph. In addition, very sharp band tails in a particular material would yield linear Arrhenius plots, even if no sharp mobility edges exist in general; e.g., it is clear that we cannot conclude much about the existence of mobility edges from measurements on single-crystalline semiconductors.

The determination of $g(E_F)$ should be quite straightforward. The simplest means should be by field-effect measurements [47]. Since a known concentration of excess charge can be induced within the sample and the resulting increase in conductivity can be measured easily, the change in $E_F$, and thus $g(E_F)$, are easily determined. The only problem with this experiment is that almost all the excess conduction takes place very near the insulator/

semiconductor interface (an MIS structure is ordinarily used), so the $g(E_F)$ measured represents an interface value. Often the concentration of localized gap states near an insulator/semiconductor interface is many orders of magnitude larger than in the bulk of the semiconductor. Deep-level transient spectroscopy (DLTS) [48] can be used to separate bulk from interface states. However, its conventional use requires the formation of a Schottky barrier, which precludes experiments with many interesting materials. Tunneling spectroscopy [49], isothermal capacitance transient spectroscopy (ICTS) [50], capacitance-voltage measurements [51], transient voltage spectroscopy [52], and space-charge limited-current (SCLC) experiments [53] are among other techniques which have been applied to estimate $g(E_F)$ in amorphous semiconductors.

If $g(E_F)$ is sufficiently large, $E_F$ is pinned and many of the techniques discussed previously cannot be performed or only provide a lower limit to $g(E_F)$ because of an unmeasurable response. In such cases, variable-range hopping could provide some quantitative measure of $g(E_F)$ [54], although some ambiguities are present [55]. In this regime, tunneling experiments have also proved useful [56]. Another test of a large $g(E_F)$ is provided by electron spin resonance (ESR) experiments. If $g(E_F)$ is due to overlap between the valence and conduction band tails, as in Fig. 8, the necessary repopulation of states that must follow should yield large concentrations of unpaired spins. Even if by some diabolical mechanism all the states below $E_F$ at $T = 0$ were spin paired, a large value of $g(E_F)$ would require the unpaired-spin concentration to be a rapidly increasing function of temperature as the states above $E_F$ populate and those below $E_F$ depopulate with thermal excitation.

Many of the techniques that are useful in determining $g(E_F)$ can also be used for a spectroscopy of $g(E)$ near $E_F$. In particular, field-effect and DLTS measurements are useful over rather wide ranges of energy. In order to verify the assumption of local satisfaction of valence requirements, it is necessary to obtain $g(E)$ throughout the gap and show that no sharp structure is present. However, another complication is the maximum resolution in any particular technique. For example, it has been shown that field-effect measurements are ordinarily incapable of resolving bumps in $g(E)$ [57].

Bearing all these cautions in mind, what do the experimental results tell us? Naive application of the CFO model would suggest that simple amorphous semiconductors, e.g., a-Si, a-As, a-Se, should not exhibit extensive band tails. Thus, we might predict that optical absorption experiments should be characterized by sharp absorption edges, electrical conductivity data should be linear on an Arrhenius plot, the Fermi energy should not be pinned, and there should not be a large unpaired-spin concentration. In sharp contrast, we might expect a material like a-$Te_{40}As_{35}Si_{15}Ge_7P_3$ to exhibit a diffuse absorption edge, variable-range hopping conduction, a pinned Fermi energy, and a large ESP signal. Perhaps binary compounds such as a-$As_2Se_3$ and a-GaAs should have intermediate behavior.

The actual results are in sharp disagreement with these predictions. Indeed, two very distinct classes of covalent amorphous semiconductors exist, with a third class exhibiting intermediate behavior. However, the classification is not predicted on the degree of disorder, but rather on the chemistry of the material. Tetrahedrally bonded semiconductors such as a-Si, a-Ge, a-GaAs, and even a-$CdGeAs_2$, fall into one class, while amorphous chalcogenides such as a-Se, a-$As_2Se_3$, and a-$Te_{40}As_{35}Si_{15}Ge_7P_3$ fall into a completely different class. Amorphous pnictides, such as a-P and a-As have intermediate behavior.

For example, pure a-Si and a-Ge ordinarily exhibit none of the predicted behavior. As-deposited films show a very gradual decrease in $\alpha(\omega)$ with photon energy [58,59]. Electrical conductivity data are not linear on an Arrhenius plot, but are quite linear on plots of ln $\sigma$ vs $T^{-1/4}$[60-62]. ESR experiments [63] indicate unpaired-spin concentrations exceeding $10^{20}cm^{-3}$, and no attempts to observe a field effect or to dope these materials have yet been successful. However, more of this behavior appears to be intrinsic. As the films are annealed below the crystallization temperature, the band tail in the optical absorption abates [58], the variable-range hopping conduction decreases [59], and the unpaired-spin concentration goes down sharply [59]. Since the optical absorption edge becomes quite sharp while both variable-range hopping and reasonably large unpaired-spin concentrations persist, we can conclude that the observed $g(E_F)$ is not due to overlapping valence and conduction band tails, but rather must arise from defect states near mid gap. Annealing must reduce both the extent of the band tails and the concentration of these defect states.

All a-Si films are extremely sensitive to both their method of preparation and their thermal history. But even the earliest experiments seemed to indicate that a completely different type of a-Si exists [64]. This material, decomposed from $SiH_4$ gas with the assistance of a plasma discharge, showed exactly the behavior that was predicted from a CFO-type analysis. It has a sharp absorption edge, a linear Arrhenius plot, and a minimal concentration of unpaired spins (as low as $\sim 10^{15} cm^{-3}$), exhibits a large field effect, and can be doped either n type or p type [65]. We now know that these films are not pure a-Si; instead, they contain 5-40% residual hydrogen, and represent an alloy, a-Si:H or hydrogenated amorphous silicon. It is clear that it is the presence of bonded hydrogen, in fact, which is responsible for the relatively very low density of localized states within the gap of this alloy. We shall discuss a-Si:H in more detail in Section VIII.

Although almost all of the a priori predictions about pure a-Si and a-Ge turned out to be incorrect, the experimental results at least had the virtue of <u>consistency</u>. Such did not turn out to be the case for the amorphous chalcogenides. From the analysis based on the CFO model of Fig. 8, we would predict that these materials should be characterized by a strongly pinned Fermi energy and should exhibit a large ESR signal, variable-range hopping conduction, and a diffuse optical absorption edge. In fact, the Fermi energy appears to be very strongly pinned. No dc field effect has ever been reported and the materials are extremely resistant to doping. In sharp contrast, no ESR signal has been observed at any temperature and there is no evidence for variable-range hopping conduction. Although the optical absorption edge is not very sharp, it is typical of those observed in a wide range of crystalline solids and is no more gradual in a-$Te_{40}As_{35}Si_{15}Ge_7P_3$ than in a-Se [66]. Clearly, there is a great deal to explain. The major problem is how $E_F$ can be so strongly pinned without the existence of any unpaired spins or variable-range hopping conduction at any temperature.

Just as turned out to be the resolution of the deviation between theory and experiment in a-Si and a-Ge, a careful consideration of the effects of <u>defects</u> cleared up the puzzle of the amorphous chalcogenides. Consequently, at this point, we shall shift gears and discuss the general structure and behavior of defects in covalent amorphous semiconductors. We shall return to a detailed analysis of chalcogenides in Section VIII.

## VI.    DEFECTS IN AMORPHOUS SOLIDS

In crystalline solids, the equilibrium position of each atom is ordinarily one in which its coordination number, $z$, its bond lengths, $a_i$, and its bond angles, $\theta_{ij}$, are all optimal. In multicomponent alloys, the periodic structure is generally preserved by adjusting the $a_i$ to a value intermediate between those of the constituents (Vegard's Law), since in this manner the overall strain is minimized. Occasionally, small concentrations of specific defect centers such as vacancies, interstitials, dislocations, or grain boundaries are present, sometimes for thermodynamic reasons especially at very high temperatures, at other times due to imperfect growth techniques. In addition, impurity atoms can be incorporated, either by design or because of imperfect purification methods. If these impurities are chemically dissimilar to the host atoms, the periodic constraints usually force them into nonoptimal coordinations, as in the example of c-Si:P discussed previously. Surface and interface states may also be regarded as defects, because of local high-energy environments. In semiconductors, these defects generally yield localized states in the gap, which control the position of $E_F$, and thus the transport properties of the material.

In amorphous solids, no crystalline constraints are present, and the concept of lattice defects such as vacancies, interstitials, dislocations, etc. has little value. However, the same chemical interactions that control the structure of crystals are present, and they provide strong driving forces for optimization of $z$, $a$, and $\theta$, i.e., the short-range order. Furthermore, in multicomponent alloys, a hierarchy of bond strengths can exist, favoring some local environments over others. This suggests a general scheme for the analysis of the electronic structure of an amorphous solid, without reference to any periodic lattice. In this approach, a constituent atom is characterized by its nearest-neighbor environment only. The most important parameter is $z$, the local coordination number. Next, the nature of the nearest-neighboring atoms must be considered, yielding a spectrum of local potential-energy functions for the central atom. The bond lengths, $a_i$, and bond angles (for $z > 1$), $\theta_{ij}$, then pick out one of these potential energies, from which the one-electron energies can be approximated.

For simplicity, we first consider an elemental solid with only one type of atom. It is then straightforward to identify defects, since any nonoptimal value of $z$ yields a local high-energy configuration. For covalent solids with only s and p electrons involved in the bonding, the atom forms a defect center whenever z deviates from N/8-N rule. The energy is still a sensitive function of the bond lengths, but as we shall show shortly, the values of the $a_i$ ordinarily optimize. However, strains can and do yield nonoptimal bond angles in a wide class of materials.

In multicomponent alloys, the situation becomes much more complex. First, there is now a multiplicity of possible bond strengths. If there are n components in the alloy, there are $n(n+1)/2$ different possible two-center bonds, a number which rises rapidly with increasing n. Even if one bond is potentially much stronger than all the others, the stoichiometry might not permit its predominance. A reasonable approach would be to assume that the strongest bond ordinarily forms to the maximum extent allowed by the stoichiometry, the next strongest bond has second priority, etc., until all the atoms are in their optimal coordinations. This then yields the ideal amorphous structure. In this case, we could identify two types of defect center, either a non-optimal coordination or a weak bond not required by the composition. In addition, in order to estimate the energies of the bonding and antibonding orbitals of a heteropolar bond, the difference in electronegativities of the two atoms must be taken into account. Furthermore, other complications are present in multicomponent alloys [67]. For example, dative bonds between a lone pair on one atom and an empty orbital on another can result in an increase in the optimal coordinations of both. An important example of such a bond occurs in III-V compounds, e.g., GaAs. The N/8-N rule suggests that the optimal values for both Ga and As are z = 3, with Ga bonding $sp^2$ and As bonding $p^3$. However, if both atoms form $sp^3$ hybrids, Ga would have an empty orbital while As would have a doubly occupied orbital; thus, a dative bond could form between nearest neighboring pairs, increasing the optimal coordination number to z = 4 for both. Thus, stoichiometric GaAs forms a tetrahedrally bonded structure. In this case, deviations from z = 4 must be identified as defect centers (as well as any Ga-Ga or As-As bonds). In addition, other more exotic chemical environments, including multiple bonding and multicenter bonding, can further complicate the analysis [67]. Nevertheless, the same general principles can be applied: (1) optimize z for each atom, taking all the chemical possibilities into account, (2) optimize the type of bond, taking the composition into account, (3) any deviation from the optimal z or the presence of an unnecessary weak bond

can then by identified as a <u>defect center</u>, and (4) any deviation
from the optimal bond length or bond angle can be classified as a
<u>strained bond</u>. Either defect centers or strained bonds can yield
localized states in the gap of a semiconductor. However, strained
bonds can exhibit a continuum of possible structures and thus
might be expected to induce gradual contributions to g(E). In contrast, well-defined defect centers could produce sharp peaks in
g(E).

It is important to bear in mind that amorphous solids are not
ordinarily the lowest-energy structure for any large collection of
atoms (although for many multicomponent alloys they might well be).
Thus, under ideal preparation conditions, involving, e.g., very
slow cooling, crystalline solids will usually result. Thus, most
amorphous materials are <u>metastable</u> (i.e., locally rather than
globally stable), and they must generally be processed using non-
ideal techniques such as by quenching from the liquid phase or by
direct deposition from a vapor phase onto a relatively cool substrate. The atomic mobility diminishes rapidly with decreasing
temperature below the melting point, and long-range motions
which would normally induce crystallization are retarded. As
mentioned previously, a softening point or <u>glass transition tempera-
ture</u>, $T_g$, usually exists, below which the viscosity of the material increases by many orders of magnitude and the material becomes an amorphous solid. Clearly, the processing techniques
designed to prepare amorphous materials allow short-range atomic
rearrangements, which not only tend to minimize the concentration
of defect centers as defined herein but also tend to produce
longer-range strains. Ovshinsky [68] pointed out that the <u>con-
nectivity</u> of the network is a major factor in controlling the overall
strain in the material, and this in turn is determined by the chemical composition. Let $\bar{z}$ be the average value of the optimal coordination number of the solid; thus, e.g., $\bar{z} = 4$ for Si or GaAs,
$\bar{z} = 3$ for As or GeTe, $\bar{z} = 2.7$ for $GeTe_2$, $\bar{z} = 2.4$ for $As_2Se_3$, and
$\bar{z} = 2$ for Se. It is clear that there are an average of $\bar{z}/2$ (two-
center) bonds per atom, and the strong chemical forces that tend
to optimize the bond lengths introduce $\bar{z}/2$ constraints per atom.
Since each atom is free to move in three dimensions, the optimal
bond lengths can all be achieved provided $\bar{z}/2 \leq 3$ or $\bar{z} \leq 6$. Since
$\bar{z} \leq 4$ for covalent bonding involving only s and p electrons, such
constraints are easily satisfied. However, covalent bonds also
have an optimal bond angle, although, as discussed previously,

the chemical forces constraining them are not so strong as those constraining the bond lengths (see Fig. 6). On the average, there are approximately $\bar{z}(\bar{z}-1)/2$ bond angles per atom [69], so that fixed bond lengths and bond angles can be achieved only if:

$$\bar{z}/2 + \bar{z}(\bar{z}-1)/2 \leq 3$$

or

$$\bar{z} \leq 2.4. \tag{24}$$

Thus, we would expect that a-Se and a-As$_2$Se$_3$ are not ordinarily overconstrained, but a-Si and a-GaAs possess inordinate local strains. Since the extent of the overall strain increases proportional to the deviation of the _square_ of z from 6, even relatively small decreases in z via alloying can be of great help in relieving strains. As we shall discuss in detail in Section VII, this is the major reason why a-Si:H is a much superior semiconductor to pure a-Si. For an overconstrained material, we should expect that most of the strain would be relieved by nonoptimal bond angles. This is why a-Si exhibits a $\pm 10^\circ$ spread in $\theta$ from its optimal $109.5^\circ$ value, as is evident from Fig. 1. However, because of the extent of the overconstrained nature of the network, some concentrations of stretched bonds and undercoordinated atoms are also very likely to be introduced into a-Si films during the preparation process; the latter are defect centers. In contrast to a-Si, a material such as a-Se ($\bar{z} = 2$) is _underconstrained_. Consequently, there is a significant amount of freedom remaining after constraining z, a, and $\theta$, and some dihedral-angle order can be achieved. This is the origin of the intermediate-range order that is observed in a-Se [11].

In addition to those induced by strains, defect centers can also arise from _thermodynamic_ considerations. In this case, the creation energy of the defect, $\Delta E$, is vital. $\Delta E$ is defined as the difference in total energy between the defect center and the optimal configuration. Thermodynamics requires a minimum concentration of such defects, given by:

$$N_d = N_o \exp(-\Delta E/kT_p), \tag{25}$$

where $N_o$ is the density of the material (i.e., the number of atoms per unit volume) and $T_p$ is the temperature at which the solid is prepared; for glasses, $T_p = T_g$, the glass transition temperature. If $\Delta E = 1eV$ and $T_g = 600K$, the thermodynamically required defect concentration, $N_d/N_o$, is about $2 \times 10^{-9}$, a quite small value. Thus, unless $\Delta E$ becomes considerably smaller than 1eV, we should not expect significant concentrations of thermodynamic defects under ordinary conditions.

The creation energy, $\Delta E$, is only one of several important quantities which characterize defect centers in amorphous semiconductors. The effective one-electron energy levels of the resulting localized states which appear within the mobility gap are also of vital significance. These states not only control the position of $E_F$ but also the transport properties of the material. In addition, they act as the traps and recombination centers that determine the kinetics for restoration of equilibrium after any perturbation away from thermal equilibrium. As previously discussed, both the one-electron and the adiabatic approximation are invalid for localized defect states. The one-electron approximation neglects the possibility that two electrons can correlate their motion in order to minimize their mutual electrostatic repulsion. The repulsion between two electrons with opposite spins that are simultaneously present in the same spatial state is usually called the correlation energy, U. If the state is localized, U is typically in the range 0.1 - 1eV. When U is this large, g(E) necessarily depends on the occupation condition of the defect; once an electron of either spin is present on the defect center, the energy of the corresponding state for the electron with opposite spin is increased in energy by U [12].

Consider a localized defect center that is neutral when occupied by a single electron. Let us call the neutral center $D^0$. We set the zero of energy at that of an unoccupied center, $D^+$. A $D^+$ center has two states available at the one-electron energy $E_d$, one with spin up and the other with spin down. Here $E_d$ can be set by the energy it takes to remove the electron from $D^0$ and place it at the conduction band mobility edge, $E_c'$, leaving a $D^+$ defect behind--this energy must be $E_c' - E_d$. If both a spin up and a spin down electron are present, the defect center is negatively charged, $D^-$. Since the two electrons repel each other, the second electron is not at $E_d$ but at $E_d + U$. Only an energy of $E_c' - E_d - U$ is needed to excite it beyond the conduction-band mobility edge. However, we cannot neglect the electron-phonon interactions either. The three states $D^+$, $D^0$, and $D^-$ all possess a minimum-energy local configuration which is necessary distinct.

The correlation energy, U, can be defined as the energy required for the reaction,

$$2D^O \rightarrow D^+ + D^- \tag{26}$$

without any local atomic relaxation. However, in reality, the local environments around both $D^+$ and $D^-$ will relax, lowering the energy of both states. If we take these relaxations into account, the energy required to effect (25) will be lower than U. We define the minimum energy to create an oppositely charged $D^+ - D^-$ pair from 2 $D^O$ centers as the effective correlation energy, $U_{eff}$. The value of $U_{eff}$ is of the utmost importance in characterizing defect centers, as we shall see in Sections VII and VIII.

## VII.   AMORPHOUS SILICON ALLOYS

### A.   Defects

Silicon is a tathogen element, from Column IV in the Periodic Table, and thus ordinarily participates in the maximum number of covalent bonds that can be formed from s and p electrons only. Consequently, almost any defect configuration requires the breaking of some of these bonds, and has a high creation energy, $\Delta E$. The Si-Si bond strength is about 2.4eV, and a-Si crystallizes at a temperature somewhat below 1000K. Clearly, we do not expect any significant concentration of thermodynamically-induced defects in pure a-Si.

However, as discussed in Section V, a-Si films are all characterized by, in fact, large defect concentrations. These abate somewhat with annealing and a great deal with hydrogenation. Clearly, it is vital to uncover the origin and nature of these defects before we can properly analyze the electronic properties of the films.

The key to understanding the electronic structure of a-Si and its related alloys is the fact that it forms an overconstrained network. Strains introduced during deposition lead to both gross structural features such as cracks and microvoids as well as local imperfections such as distorted bonds and defect centers. Clearly, pure a-Si and its alloys with other Column IV elements ($\bar{z} = 4$) are the most overconstrained and thus ordinarily possess the largest defect concentrations. Alloying with any element from one of the other columns in the Periodic Table lowers the value of $\bar{z}$ and thus tends to reduce the overall strains (provided that the spread in optimal bond lengths introduced by the alloying is not too great).

It is simplest to begin the analysis of a-Si alloys by considering pure a-Si. In addition to stretched bonds and distorted bond angles, either of which can, if sufficiently different from the optimal values, yield localized states in the gap, we would expect relatively low-energy defect centers to be present. Since each Si atom in its ground state forms the maximum number of bonds possible involving only s and p electrons, overcoordination is very unlikely to occur. However, several types of undercoordination are possible, in addition to some more exotic chemical behavior. The simplest defect center is a threefold-coordinated Si atom, usually called a dangling bond. The conventional notation [70] is one in which the symbol $A_z$ represents atomic type A (e.g., T for tathogen, P for pnictogen, C for chalcogen, etc.) in z-fold coordination with net charge qe (where e is the magnitude of the electronic charge). Since Si is a tathogen, the ground state is written $T_4^0$, and the neutral dangling bond is $T_3^0$. This notation is incomplete, since it does not specify either the chemical species of the z nearest neighbors or the nature of the bonds (i.e., $sp^3$, $sp^2$, p, etc.), but it is, nevertheless, quite convenient. Tight-binding calculations [71] indicate that $T_3^0$ defects yield two localized states in the gap of a-Si, one filled and one empty. In addition, all $T_3^0$ centers contain an unpaired spin, which can be observed by its characteristic ESR signal (g = 2.0055). Because the ground state of the isolated Si atom is $s^2p^2$, another possible defect that should not be overlooked is the twofold-coordinated Si atom, $T_2^0$ in its neutral state. Such states are completely spin paired, and thus exhibit no ESR signal; so they are difficult to detect. Nevertheless, tight-binding calculations [71] suggest that $T_2^0$ centers may have even lower creation energy than $T_3^0$ centers because sp hybridization is unnecessary. In any event, $T_2$ centers are twice as effective as $T_3$ centers in relieving strains, so that their possible presence in a-Si films should not be overlooked. $T_2^0$ defects can produce up to four localized states in the gap, two filled and two empty [71].

As the Fermi level moves through the gap, charged states of the defect centers could become stable at T = 0. For the dangling bond, the centers $T_3^+$, $T_3^0$, and $T_3^-$ must all be considered. Only the $T_3^0$ state has an unpaired spin and can, thereby, be identified by ESR experiments. For twofold-coordinated Si atoms, $T_2^{2+}$, $T_2^+$

$T_2^0$, $T_2^-$, and $T_2^{2-}$ are all possible centers, and the $T_2^+$ and $T_2^-$ states have unpaired spins. The $T_4^0$ ground state and several of these defect centers are sketched in Fig. 10. Note that the bond angles should ajust with changes in the charge state of the center, particularly for the dangling bond. A positively charged Si atom is isoelectronic to Al, which optimally bonds $sp^2$, in a planar configuration with a $120^0$ bond angle [$\beta$ in Fig. 10(c)]. A negatively charged Si atom is isoelectronic to P, which tends to form predominantly p bonds with $95\text{-}100^0$ bond angles [$\gamma$ in Fig. 10(d)]. For similar reasons, $T_2^{2+}$ very likely bonds in an sp configuration with a $180^0$ bond angle.

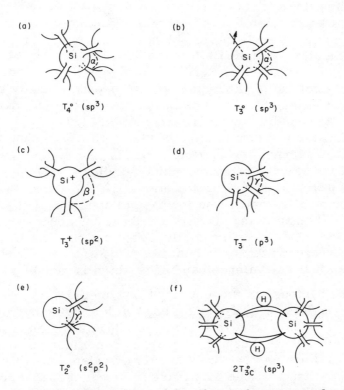

Fig. 10 Sketches of the optimal local coordinations of a Si atom in several different configurations: (a) ground state, $T_4^0$; (b) neutral dangling bond, $T_3^0$; (c) positively charged dangling bond, $T_3^+$; (d) negatively charged dangling bond, $T_3^-$; (e) twofold-coordinated Si atom, $T_2^0$; (f) complex consisting of two three-center bonds with bridging H atoms. The bond angles identified are $\alpha = 109.5^0$, $\beta = 120^0$, $\gamma \simeq 95^0$. (From ref. 83.)

One other possibility that should not be overlooked is the possibility of Si-Si double bonds. Although rare, such Si double bonds are known in chemical molecules and there is reason to believe that they could well be stable in a strained amorphous network.

Hydrogenated amorphous silicon, a-Si$_{1-x}$H$_x$ has an average coordination number z = 4-3x, so that a typical concentration of 20% hydrogen decreases z from 4 to 3.4, thus inducing a considerable reduction in the overall strain. This reduction manifests itself not only by minimization of the large-scale imperfections, i.e., the cracks, microvoids, etc., but also by a sharp decrease in local distorted bonds and defects. In addition, the Si-H bond strength is approximately 3.4eV, about 40% greater than the Si-Si bond strength. Since H is more electronegative than Si, tight-binding calculations indicate that the Si-H bonding states are deep in the a-Si:H valence band, while the Si-H antibonding orbitals are not too far from the conduction-band mobility edge. Sketches of the tight-binding estimates for the configurations Si$_5$, Si$_4$H, and the dangling bond Si$_4$ are shown in Fig. 11. Note that these estimates indicate that the band gap of a-Si:H should increase with hydrogen concentration, by shifting states from near the top of the valence band to considerably lower energies. The results of more accurate X$\alpha$ calculations [22] are shown in Fig. 12. These results confirm the increase of band gap, but suggest that the Si-H antibonding orbitals are about 3eV above the conduction band edge. Experimentally, there seems to be little doubt that E$_g$ increases with H concentration [72], and some evidence that indeed it is the valence band edge which is receding [73].

Once hydrogen is present, another type of chemical bond becomes possible, a three-center bond with bridging hydrogen atoms [67]. Such a bond is sketched in Fig. 10(f). X$\alpha$ calculations [74] indicate that hydrogen atoms can bridge between two Si atoms as close as 2.5Å apart and remove states from the gap of a-Si:H. We can conclude that stretched bonds probably do not contribute to the band tail of a-Si:H, which more likely arises from distorted bond angles. Indeed, there is strong experimental evidence for rotational modes of H$_2$ molecular units in a-Si:H [75-77], which do not effuse readily. The two bridging hydrogen atoms in Fig. 10(f) are free to rotate in the plane perpendicular to the line joining the two Si atoms, and this could easily account for the data. (Molecular H$_2$ trapped in microvoids provides an alternative explanation; however, if it is not bound to the network, it is puzzling why it does not effuse below 700K.)

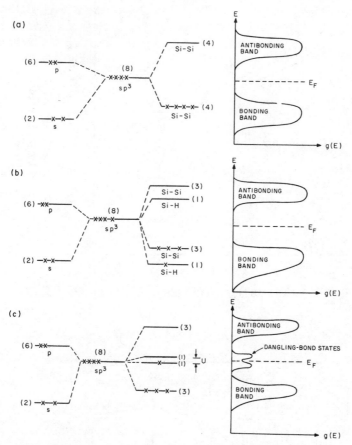

Fig. 11   Tight-binding approach to the band structure of hydrogenated amorphous silicon films: (a) A central Si atom
tetrahedrally coordinated by four neighboring Si atoms;
the eight sp$^3$ hybridized orbitals are split into four bonding and four antibonding orbitals, which respectively
spread into the valence and conduction bands of the solid.
(b) A central Si atom tetrahedrally coordinated by three
neighboring Si atoms and one H atom; the Si-H bonding
orbital falls deep within the valence band, while the Si-H
antibonding orbital lies near the lower edge of the conduction band. (c) A central Si atom surrounded by only three
neighboring Si atoms forming hybridized sp$^3$ bonds.  The
fourth electron on the central atom occupies a nonbonding
orbital (a dangling bond), separated from its unoccupied
partner by the correlation energy U; both nonbonding
states lie within the gap.  (From ref. 83.)

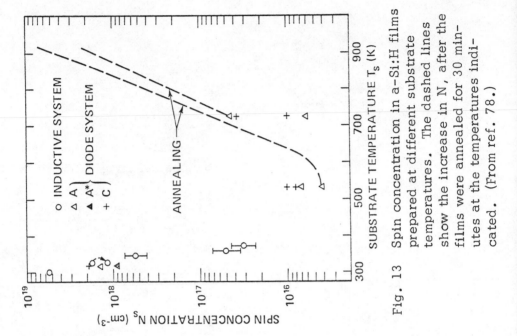

Fig. 13  Spin concentration in a-Si:H films prepared at different substrate temperatures. The dashed lines show the increase in N, after the films were annealed for 30 minutes at the temperatures indicated. (From ref. 78.)

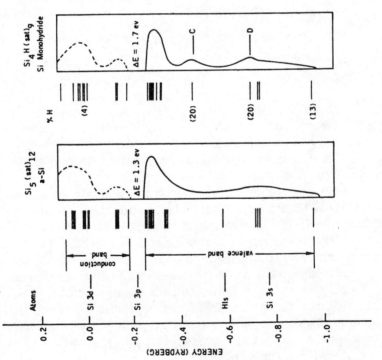

Fig. 12  SCF-Xα-SW molecular-orbital energy levels and optical densities of states of saturated clusters representing pure a-Si and a-Si:H. (From ref. 22.)

Although a-Si:H ordinarily possesses fewer $T_3^0$ centers than pure a-Si by a factor of $10^3 - 10^5$, there is always a residual concentration [78]. As shown in Fig. 13, the unpaired-spin density, $N_S$, exhibits a broad minimum as a function of substrate temperature or annealing temperature. This minimum must be the result of two competing effects. The decrease of $N_S$ with increasing T up to about 450K most likely represents the effects of local atomic relaxations, which extend over larger volumes as the temperature is raised. It was originally believed that the increase of $N_S$ with T above about 550K was due to the effusion of hydrogen, which begins to occur at approximately that temperature and is complete by the crystallization temperature. However, recent evidence suggests that other effects may be related more to the film morphology than to the H concentration [79].

We might ask if any defect centers other than dangling bonds are present in either a-Si or a-Si:H films in significant concentrations. In fact, although there is no direct evidence, it presently appears very likely that the predominant defect centers are spinless [80]. For example, field-effect [81] experiments generally indicate that a much larger density of mid-gap states is present than is suggested by the unpaired-spin concentration, while DLTS [82] measurements have been interpreted to show a great deal of structure at several different energies within the mobility gap. There are a number of candidates among the centers already discussed for these spinless defect states. One is the $T_2^0$ center, which appears to have a relatively low creation energy. Divalent Si is common in many chemical compounds, notably $SiH_2$, and $T_2^0$ centers lower the value of z twice as effectively as $T_3^0$ centers and so are more efficient in relieving strains in a-Si:H films. There is some evidence from ESR and light-induced ESR (LESR) experiments for the presence of twofold-coordinated Si centers in a-Si:H films [80].

A second possibility for spinless defects in a-Si:H films is $T_3^+ - T_3^-$ pairs [83]. These pairs can exist in significant concentrations only if the effective correlation energy, $U_{eff}$, is small or negative.

Unfortunately, in an overconstrained material such as a-Si:H, there may not be a unique value for $U_{eff}$ for each defect type. This comes about because the atomic relaxations necessary to attain the minimum-energy configurations for the positively- and negatively-charged centers might be retarded by the local strains. Such effects tend to increase $U_{eff}$ and lower the relative concentration of the charged states. Tight-binding estimates [71] suggest that $U_{eff}$ could well be negative for the $T_3$ defect in a-Si:H. The physical origin of this result can be extracted from Fig. 10. If an electron is removed from a $T_3^0$ center, it converts to $T_3^+$ [Fig. 10(c)], which optimally has bond angles of $120^\circ$. Thus, the local distortion involved tends to move the central Si atom into a position within the plane of the three neighbors to which it bonds. Analogously, a $T_3^-$ center optimally has bond angles in the range $95^\circ$ – $100^\circ$ [Fig. 10(d)], induced by a relaxation in which the central Si atom moves away from the plane of the three neighbors to which it bonds. Since threefold coordination represents the ground-state configurations of both $Si^+$ and $Si^-$ ions but not of neutral Si atoms, it is not unreasonable to conclude that $U_{eff}$ could be negative. Somewhat more sophisticated calculations [84] have indicated that $U_{eff}$ is positive, but complete relaxations were not taken into account. In any event, since there are always some unpaired spins in a-Si:H films, and their concentration is not a strong function of temperature, we must conclude that $U_{eff}$ is positive for at least some of the $T_3$ defects. On the other hand, there is some evidence [80] that large concentrations of $T_3^+$ and $T_3^-$ centers are also present in a-Si:H films. A possible explanation [85] is that $U_{eff} < 0$ when complete relaxations around both the $T_3^+$ and $T_3^-$ centers can take place, but that such relaxations are retarded in particularly strained regions of the film. In that case, isolated $T_3^0$ centers would coexist with $T_3^+$ - $T_3^-$ pairs. We shall return to this possibility later.

The electronic structure of the $T_2$ centers is quite complex, since four states could appear in the gap [83]. However, there is no question that $U_{eff}$ is positive for $T_2^0$ defects [71]. There is some evidence from ESR experiments on P-doped films that two ordinarily unoccupied states of the $T_2$ centers are located in the mobility gap below $E_C'$, and from similar experiments on B-doped films that two ordinarily occupied states are located above $E_V'$ [80].

When oppositely charged pairs such as $T_3^+$ and $T_3^-$ are present in any material, it is clear that their total energy can be lowered by reducing their spatial separation [86]. The maximum energy decrease occurs when the oppositely charged centers are nearest neighbors, in which case the complex is called an intimate charge-transfer defect (ICTD) [87]. In general, the total energy of an oppositely charged pair separated by a distance R is reduced by $e^2/\kappa R$, where $\kappa$ is the effective dielectric constant. Thus, an

ICTD is stabilized by an energy $M = e^2/\kappa R_0$, where $R_0$ is the nearest-neighbor Si-Si separation, approximately 2.35Å. If we use the long-range dielectric constant of a-Si:H, $\kappa = 12$, we can estimate that $M \simeq 0.5eV$. However, the actual value for M is probably considerably larger, since $\kappa$ is likely to be effectively reduced at short distances.

When spatially correlated pairs are present, the electronic structure of the defects is complicated considerably, because of the many possible conditions of occupancy and the fact that different charge states necessarily have different electrostatic interactions [70]. For an ICTD, the two levels associated with an isolated center are each split, so that a total of four levels are possible for each value of R. Because of the change in electrostatic interaction with occupancy, the Fermi energy unpins to the extent of $2M$ [87]. If we assume that the concentration of defects is given by the law of mass action, despite the fact that they are strain-induced rather than thermodynamically required, then we should expect an approximately exponential distribution of intrapair separations. For undoped films characterized only by ICTDs, the effective density of states would be as is sketched in Fig. 14.

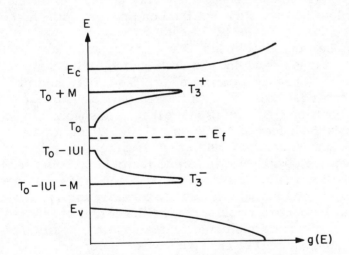

Fig. 14    Effective one-electron density of states for a tetrahedrally bonded amorphous semiconductor containing a distribution of spatially correlated dangling-bond defects with $U_{eff} < 0$. $T_0$ represents the energy of the nonbonded electron on a neutral dangling bond and M is the electrostatic attraction between a $T_3^+$ center and a $T_3^-$ center located on nearest-neighboring sites. (From ref. 83.)

Figure 14 assumes that the sign of $U_{eff}$ is negative. If $U_{eff}$ is actually positive, the overall picture remains the same, but the two defect bands overlap in mid gap, and a small concentration of unpaired spins results even at low temperatures. In this case, the Fermi energy is pinned only to the extent of $g(E_F)$. Furthermore, if $M - U_{eff}/2$ is 0.9eV or larger, the high-density portions of the defect bands merge with the valence and conduction bands. In this case, the defect bands are indistinguishable from band tails, and the overall density of states is much the same as that predicted by the CFO model for conditions of gross disorder (see Fig. 8). Such a situation may be the real reason why pure a-Si films behave exactly as predicted by the CFO model for highly disordered materials (recall Section V).

The introduction of dopants or the incorporation of unintentional impurities can lead to many new defect types. Some possibilities that involve P and B atoms, the most common dopants, are sketched in Fig. 15. It is very important to analyze these centers, in order to understand the doping mechanism, its efficiency, and its consequences on the electronic structure. Since the strength of the Si-P bond is comparable to that of the Si-H bond and P has an electronegativity only slightly lower than that of H, it might be expected that the bonding and antibonding orbitals arising from covalent Si-P bonds have roughly the same energies as those from Si-H bonds, i.e., the bonding orbitals lie well within the valence band and the antibonding orbitals in the vicinity of the conduction-band mobility edge. Normal structural bonding for phosphorus is threefold-coordinated (predominantly $p^3$) bonding, $P_3^0$ in our notation [Fig. 15(b)]. If all the phosphorus atoms introduced into a-Si:H formed $P_3^0$ centers, P would not act as a substitutional donor; rather, it would serve as a network relaxer, lowering the average coordination, $\bar{z}$, and it would probably tend to open up the band gap somewhat. In addition, it is also possible that the Si-P antibonding orbitals are actually located just below $E_c'$, so that $P_3^0$ centers could increase the extent of the conduction-band tail. This could then tend to decrease the position of the conduction-band mobility edge, $E_c'$, with increasing phosphorus concentration, thus lowering the electrical activation energy, $E_c' - E_F$, and providing a quasi-doping.

Experimentally, $E_c' - E_F$ decreases with phosphorus incorporation from about 0.9eV to 0.1eV, so that it is likely that a significant fraction of the P atoms enters the network in a tetrahedral ($sp^3$) configuration. Since this is not the lowest-energy configuration for phosphorus, there are no crystalline constraints, and fourfold coordination introduces more network strain than threefold coordination, the appearance of tetrahedrally-coordinated phosphorus is a major puzzle. The probable resolution [71] is the fact that $P_4^+$ - $T_3^-$ pairs have low energy, perhaps even lower energy than the apparent $P_3^o$ - $T_4^o$ ground-state configuration. The total number of bonds is seven for both pairs, but the $P_4^+$ - $T_3^-$ configuration contains an additional Si-P in place of an Si-Si bond, and the former is stronger than the latter. If this analysis is

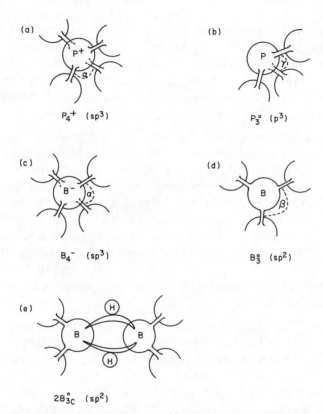

(a)                                    (b)

$P_4^+$ (sp$^3$)                        $P_3^o$ (p$^3$)

(c)                                    (d)

$B_4^-$ (sp$^3$)                        $B_3^o$ (sp$^2$)

(e)

$2B_{3C}^o$ (sp$^2$)

Fig. 15   Sketches of the optimal local coordinations of dopant atoms in several different configurations: (a) positively charged tetrahedrally coordinated P atom, $P_4^+$; (b) neutral threefold-coordinated P atom, $P_3^o$; (c) negatively charged tetrahedrally coordinated B atom, $B_4^-$; (d) neutral threefold-coordinated B atom, $B_3^o$; (e) complex consisting of two B three-center bonds with bridging H atoms. The bond angles are the same as in Fig. 10: $\alpha = 109.5^o$, $\beta = 120^o$, $\gamma \simeq 95^o$. (From ref. 83.)

correct, the $P_3^O$ centers should be considered as defects, intro duced to relieve strains in the network.

In principle, the centers and electronic structure arising from the introduction of boron into a-Si:H should be analogous to those arising from phosphorus. However, because of its complex chemistry, the analysis of the effects of boron is much more difficult.

As is the case with P, normal structural bonding for boron is threefold coordinated, $B_3^O$. This bonding, predominantly $sp^2$, is sketched in Fig. 15(d). In contrast to the case of P, the Si-B bond appears to be weaker than the Si-Si bond, an effect that favors $B_3^O$ centers over $T_3^+ - B_4^-$ pairs. Since B is more electro- negative than Si and $B_3^O$ centers have two empty nonbonding p or- bitals, the presence of $B_3^O$ centers may introduce states in the gap of a-Si:H:B. In fact, if these empty states are sufficiently low in energy, $B_3^O$ centers can act as a p-type dopant in a-Si:H.

Boron chemistry allows for several new possibilities for local structure in a-Si:H:B, the most likely being the formation of three-center bonds with bridging hydrogen atoms [see Fig. 15(e)] [67]. Such three-center bonds are the chemical basis of the $B_2H_6$ molecule, the gas of which is ordinarily used to dope a-Si:H with boron. The electronic structure of the bond is such that empty nonbonding orbitals are introduced into the gap, another potential p-doping mechanism. It should be noted that the strength of the B-B bond appears to be somewhat greater than that of the Si-B bond, suggesting the possibility of boron clustering in a-Si:H:B films.

Other defect centers can be expected to arise from the presence of oxygen (a chalcogen element with very large electro- negativity), nitrogen (a pnictogen element like P, but much more electronegative and with a predilection for hybridized $sp^2$ bonding), and carbon (another tathogen element like Si itself, but one which easily forms multiple bonds) [83]. It would be extremely useful if the electronic structure of all of the impurity-related centers were calculated accurately, but at this point we have to rely pri- marily on tight-binding estimates.

B.    Ground-State Electronic Structure

The main features of the ground state of any amorphous semi- conductor can be divided into three regions: (1) the structure of the valence and conduction bands, and the magnitude of the mobility gap, (2) the nature of the valence and conduction band tails, and (3) the density of deep localized states in the gap. In the case of pure a-Si, the defect concentration is so large that even the magnitude of the mobility gap is unknown at present. There is large optical absorption down to low values of photon energy and dc conduction is dominated by hopping through localized states. However, there is some evidence from extrapolation of the optical gaps of a-Si:H alloys to zero hydrogen concentration that the gap of pure a-Si is in the 1.4 - 1.6eV range [88]. This value is larger than the 1.1eV gap of c-Si, a fact which appears to be due to the presence of fivefold rings in a-Si (whereas c-Si only contains six-fold rings) [89]. A more important difference between the optical properties of a-Si and c-Si is the relative sharpness of the absorption edges in the two materials. In c-Si, $\alpha$ rises slowly with increasing $\omega$ above the gap, because the minimum-energy optical transition across the gap is <u>indirect</u> (i.e., it does not conserve crystal momentum); since photons do not carry sufficient momentum, optical transitions in the energy range 1.1 - 3.4eV require the assistance of phonons, and $\alpha$ is correspondingly reduced by about a factor of 100. In amorphous solids, on the other hand, crystal momentum is not conserved and all optical transitions are necessarily direct.

As hydrogen is incorporated into a-Si, the optical gap increases [88]. However, because of variations in the extent of the disorder, which changes the shape of the band tails, it is difficult to determine a precise value for the optical gap from measurements of $\alpha(\omega)$. It is clear from Eq. (9) that $\alpha(\omega)$ measures the joint density of states, independent of the spatial extent of these states except via the matrix elements, $M(E)$. Consequently, the optical absorption edge need not be directly related to the mobility gap, $E_g'$. Indeed, there is reason to believe that transitions between localized and extended states should have matrix elements of the same order of magnitude as those connecting two extended states [66], so that no discontinuity in $\alpha(\omega)$ can be expected at the mobility gap. Even localized-to-localized transitions can have large matrix elements if the initial and final states are localized in the same region of space. The usual procedure [90] in analyzing $\alpha(\omega)$ is to assume that the densities of states of both the valence and conduction bands in the vicinity of the mobility edges are given by:

$$g_V(E) = A_V(E_V' - E)^{1/2} \tag{27}$$

and

$$g_C(E) = A_C(E - E_C')^{1/2} \tag{28}$$

respectively. Such expressions are rigorous only for systems with long-range periodicity [see Eq. (10)], and so their use in amorphous solids is dubious, to say the least. However, the substitution of Eqs. (27) and (28) into Eq. (9) yields:

$$\alpha(\omega) = \frac{B}{\hbar\omega} \ (\hbar\omega - E_g')^2, \tag{29}$$

where $E_g' = E_C' - E_V'$ and B is a constant. Thus, this model suggests that a plot of $(\alpha\hbar\omega)^{1/2}$ as a function of $\hbar\omega$ should be linear above $E_g'$, at which point $\alpha = 0$. When this procedure is used, the zero absorption extrapolated value of $\hbar\omega$ is ordinarily called the optical gap, $E_{opt}$. If the densities of states given by Eqs. (27) and (28) are correct near both mobility edges, $E_{opt}$ is actually the mobility gap. However, there is absolutely no theoretical justification for this association. In reality, there is no unique linear region of $(\alpha\hbar\omega)^{1/2}$ and $E_{opt}$ is somewhat arbitrary. Often, the linear region is taken to necessarily include the absorption in the vicinity of $\alpha = 10^4$ cm$^{-1}$ to reduce the degree of arbitrariness in $E_{opt}$. An alternative procedure [66] is to assume that $g_V(E) = C_V(E_V' - E)$ and $g_C(E) = C_C(E - E_C')$ when disorder is present, suggesting that the optical gap is better obtained from extrapolations of $(\alpha\hbar\omega)^{1/3}$ vs $\hbar\omega$ to zero absorption. There is some evidence [91] that the latter plots are linear over wider frequency ranges than square-root plots, not surprising considering the additional compression of the data, but once again any theoretical justification for linear densities of states extrapolating to the mobility edges is completely lacking. Thus, the physical significance of $E_{opt}$ obtained by either procedure is nonexistent, except for comparative purposes. In most cases, the cube-root plots tend to give smaller values for $E_{opt}$ than the square-root plots.

In any event, $E_{opt}$ for a-Si:H as determined from Eq. (29) increases with increasing hydrogen concentration from approximately 1.6eV to about 2.4eV, the latter value being characteristic of polysilane [i.e., $(SiH_2)_n$]. The highest-quality films appear to have optical gaps in the 1.8 - 2.0eV range.

Photoemission experiments [92] have suggested that the increase in optical gap with hydrogen is primarily the result of a

recession of the valence band edge, in accordance with the simple
model discussed previously (see Fig. 11) as well as with more
sophisticated calculations [93,94]. Photoemission results [95]
further indicate that the sharp valence band structure of c-Si is
not present in a-Si, but that the addition of hydrogen introduces
two new peaks about 5eV and 9eV below the valence band maximum.
These peaks are very likely to be the Si(3p)-H(1s) and Si(3s)-H(1s)
bonding states (peaks C and D in Fig. 12).

There is more ambiguity with regard to the band tails, due
to their sensitivity to the details of film preparation, the difficulties involved in measuring low absorption in thin films, and their
possible masking by exciton effects. Optical-absorption measurements [96] indicate an exponential decrease in $\alpha$ with decreasing
photon energy, $\hbar\omega$, below the optical gap for a wide array of
a-Si:H films. In this region, typical films were found to obey relations such as:

$$\alpha = \alpha_0 \exp(E - E_0)/kT_0 , \tag{30}$$

where $\alpha_0 \simeq 10^6 cm^{-1}$ and $E_0 \simeq 2.2eV$ for many different values of
$T_0$. This behavior, plotted in Fig. 16, is reminiscent of the Urbach
edge of many diverse semiconductors, crystalline as well as amorphous [97]. It is ordinarily believed to arise from excitonic effects
in conjunction with internal electric fields. This could well account
for the focus in Fig. 16, provided the different films were characterized by different strengths of the internal fields. Such fields
could be provided by $T_3^+ - T_3^-$ pairs in a-Si:H. If the density of
these ICTDs is larger in the more ordered and thus less strained
films, in accordance with the idea that complete local relaxations
could induce negative values of $U_{eff}$, then $T_0$ would be a measure
of disorder as observed. In this case, we should identify $E_0$ with
the optical gap of a-Si:H. Alternatively, if this region of absorption is due to transitions between localized band-tail states and
the extended states in the other band, it may be indicative of an
exponential band tail. It is clear that the predominant contribution
would be from the broader band tail, usually assumed to be the
valence band tail in a-Si:H. This interpretation leads to the conclusion that disorder not only increases $T_0$ but decreases $E_{opt}$,
which is related to $T_0$ by [96]:

$$E_{opt} = 2.0eV - 6.2kT_0 .$$

This result is consistent with a recent model [98], in which a-Si:H is approximated by a virtual crystal with varying site disorder.

In principle, the problem of the origin of the exponential region of absorption should be resolvable by transient photoconductivity or time-of-flight studies of the band tails themselves, independent of excitonic effects. However, such experiments are usually carried out for times greater than ~ 10ns, so that they probe only the regions beyond 0.2eV from the mobility edges. There is some evidence from time-of-flight measurements that both the valence and conduction band tails are purely exponential,

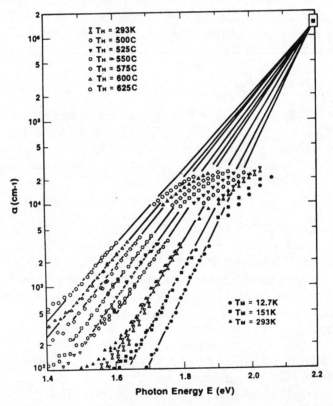

Fig. 16    Optical absorption coefficient, $\alpha$, as a function of photon energy. The solid symbols refer to data obtained at different measurement temperatures, $T_M$. The open symbols refer to a film that has been isochronally heated at temperature $T_H$. (From ref. 96.)

with $T_O \simeq 500K$ for the valence band and $T_O \simeq 325K$ for the conduction band [99]. In contrast, transient photoconductivity experiments [100] appear to yield more gradual behavior than exponential near the mobility edges, in agreement with field-effect
results [101].

There have been many attempts to measure the density of
localized states deep in the gap. Clearly, this is a difficult
task, not only because of the wide dispersion in properties among
films deposited by different preparation techniques and deposition
parameters, but also due to the effects of unintentional impurities,
interface states, surface absorption, and even incident light. In
addition, the previously-discussed ambiguities due to electron-
phonon coupling and electronic correlations are particularly important for deep localized states.

Optical-absorption experiments can be analyzed to yield information about localized gap states [44], as can photoconductivity [102], and photothermal deflection stectrocopy [103]. Typical
results [96] are shown for both undoped and phosphorus-doped
films in Fig. 17, and for boron-doped films in Fig. 18. It is clear

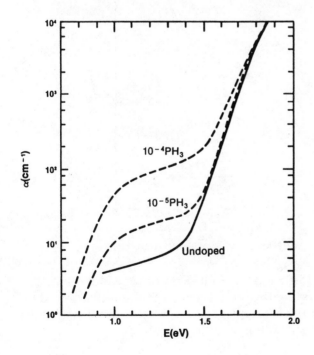

Fig. 17 Absorption edge of lightly phosphorus doped a-Si:H$_x$. The indicated concentrations refer to gas phase below $\alpha = 10^2$ cm$^{-1}$. The absorption edge was determined by photoconductivity below $\alpha = 10^2$ cm$^{-1}$ and above this value by optical transmission. (From ref. 96.)

that phosphorus and boron doping do not affect the valence and conduction bands very much, but both yield large densities of localized states deep in the gap. Deconvolutions of the absorption spectra cannot be trusted because of the lack of knowledge about the dependence of the matrix elements in Eq. (9) on energy, but it appears that the data can be explained by the assumption that phosphorus introduces a localized state about 1.1eV below $E_C$ [96], while boron introduces a localized state about 1.3eV above $E_V$. If, as discussed previously, phosphorus dopes primarily by the creation of $P_4^+ - T_3^-$ pairs and boron by the creation of $T_3^+ - B_4^-$ pairs, then the effective correlation energy of the dangling bond defect can be formed from the relation [104]:

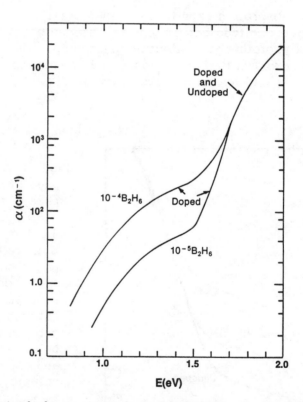

Fig. 18   Optical absorption edge of lightly boron doped a-Si:H$_x$. The indicated concentrations refer to the gas phase. Below $\alpha = 4 \times 10^2$ the absorption edge was determined by photoconductivity. (From ref. 96.)

$$U_{eff} = E_{opt} - E_P - E_B,$$  (31)

where $E_P$ is the energy of the phosphorus-induced peak (i.e., 1.1eV) and $E_B$ is that of the boron-induced peak. Equation (31) is easily derived by noting that $E_P = E_C' - E(T_3^-)$ and $E_B = E(T_3^+) - E_v'$. Since $U_{eff} = E(T_3^-) - E(T_3^+)$ and $E_{opt} = E_C' - E_v'$, Eq. (31) follows. But Eq. (30) shows that $E_{opt}$ is less than 2.0eV, so that the present evidence is in favor of a negative $U_{eff}$ for the vast majority of the dangling bonds in a-Si:H. Further evidence that the shoulders in $\alpha$ induced by boron and phosphorus are actually $T_3$ defects is contained in the optical-absorption results on undoped films, Fig. 19 [96]. It is clear from these data that shoulders in $\alpha$ in the 1.1 - 1.3eV range appear in films annealed above 500°C. But

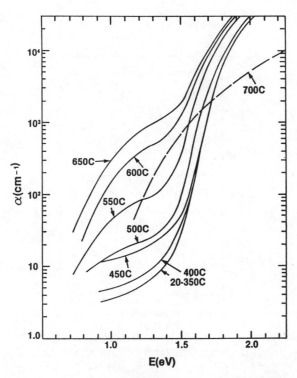

Fig. 19   Absorption edge for a glow discharge film of a-Si:H$_x$ prepared at a substrate temperature of $T_s$ = 100C and annealed at the indicated temperatures $T_A$. The measured spin density is indicated in the parenthesis. (From ref. 96.)

this is just the regime where hydrogen effuses and the dangling bond concentration increases [78]. However, it is important to note that only about one additional unpaired spin is created for each 100 H atoms that effuse [105]. On the other hand, the shoulder in $\alpha$ induced by H effusion appears to be sufficiently in-

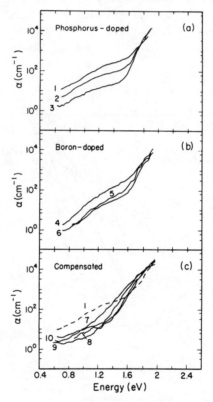

Fig. 20   Absorption coefficient vs energy for various dopants: (a) $PH_3$-doping concentration of the films is 1:1 x $10^{-3}$, 2:3 x $10^{-4}$, and 3:1 x $10^{-5}$; (b) $B_2H_6$-doping concentration is 4:$10^{-3}$, 5:3 x $10^{-4}$, and 6:$10^{-4}$; (c) compensated samples, all have $10^{-3}$ $PH_3$ and the $B_2H_6$ concentrations are 1:0, 7:2 x $10^{-4}$, 8:4 x $10^{-4}$, 9:2 x $10^{-3}$, and 10:4 x $10^{-3}$. All concentrations refer to the relative concentration of the dopant in the gas phase $T_s$ = 230°C and rf power is 2 W. (From ref. 106.)

tense to be brought about by excitations of the same order of magnitude as the number of effusing H atoms. But if $U_{eff}$ is negative, the effusing hydrogen would tend to leave primarily $T_3^+$ and $T_3^-$ centers behind, rather than $T_3^0$ centers, leading to large increases in mid-gap absorption, but not in $N_s$.

Photothermal deflection spectroscopy yields results such as those shown in Fig. 20 [106]. Although there is a great deal of noise in the data, the results appear to be consistent with the previous discussion. It is interesting that compensated films do not exhibit quite so large an absorption shoulder in the 1.0 - 1.2eV region as do singly doped films, indicating that the formation of $P_3^0$ and $B_3^0$ centers or perhaps $P_4^+ - B_4^-$ pairs predominate in compensated samples. In fact, since both $P_3^0$ and $B_3^0$ centers reduce z, and thus relieve strains, compensated samples could have even fewer dangling bonds than do undoped films.

The density of deep gap states can also be determined by specialized techniques such as DLTS, photoacoustic spectroscopy, capacitance, or field-effect measurements. The results are not always in agreement. Some of this dispersion is clear from Fig. 21, which shows several experimental determinations of g(E) in the upper half of the gap [107]. The DLTS results (curve E) indicate a much cleaner gap than do the other experiments, but the latter may be dominated by interface states, which DLTS measurements can separate out. A typical DLTS result for a P-doped sample is shown in Fig 22 [48]. It suggests that there is a peak in g(E) about 1.0eV below $E_c'$, consistent with the optical absorption results. There is also evidence for a large density of localized states in the lower part of the gap, but the errors in this region are large.

## C.    Transport Properties

In most semiconductors, the dc transport properties are very well understood. Electrical conductivity, Hall effect, and thermoelectric power (or thermopower) measurements as functions of temperature can be interpreted to yield the position of the Fermi energy, the effective mass of the carriers, the band mobility, and the predominant carrier scattering mechanism. Intrinsic and extrinsic regions can be distinguished and investigated separately. Small-polaron formation and hopping transport can be identified, if present. Even ambipolar conduction can be analyzed quantitatively.

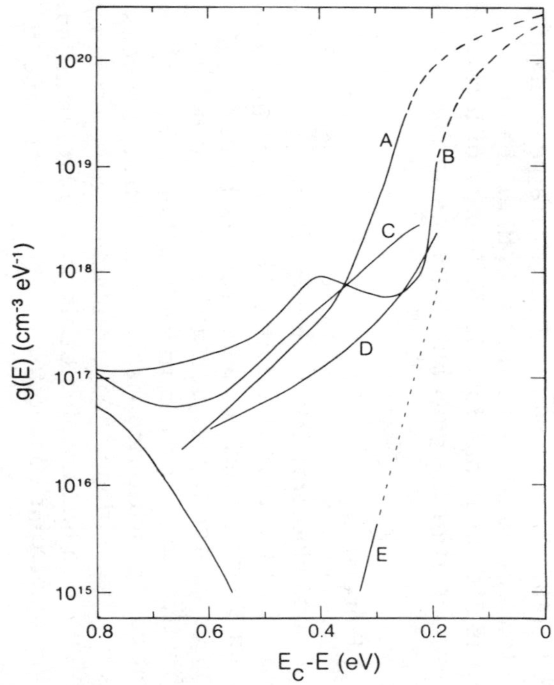

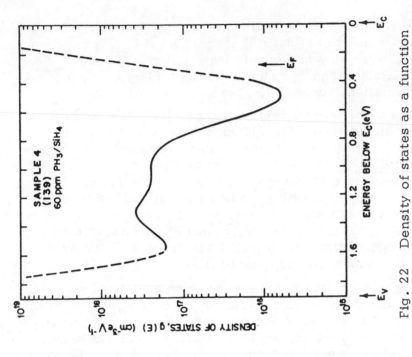

Fig. 22  Density of states as a function of energy for the best fit to the DLTS spectra of an a-Si:H film doped with 60 ppm $PH_3$ in the gas phase.  (From ref.[3] 48.)

Fig. 21  Density of states, $g(E)$, as a function of energy, $E_C - E$, for a series of experiments based on (A) transient and steady-state photoconductivity, (B) field-effect, (C) field-effect, (D) capacitance, and (E) DLTS measurements.  (From ref. 107.)

Unfortunately, much less can be expected from transport studies of a-Si:H. First, there are large sample-to-sample variations in composition, not only in hydrogen content but also in unintentional impurity concentrations. Second, the presence of strains during deposition yields different defect concentrations in different films. Third, there is the strong possibility of both compositional and structural inhomogeneities; e.g., microvoids or hydrogen-rich and hydrogen-poor regions have been inferred from structural studies. Fourth, surface and interface regions of major importance to transport in thin-film geometries appear to be very different from the bulk structure and composition, and space-charge regions are routinely observed. Fifth, metastable states with major changes in transport properties can be induced by absorption of light, injection of charge, or rapid temperature variations. Sixth, in the absence of crystalline constraints, dopants need not enter the network in a substitutional manner, nor even in a consistent manner. Seventh, the defect states and the band-tail states can vary greatly with temperature, due to annealing effects at low and moderate temperatures. In addition, at sufficiently high temperatures (T > 600K), hydrogen begins to effuse, irreversibly affecting the material. Eighth, changes in the degree of disorder for any reason can induce Fermi-energy shifts, which, due to correlation effects, can shift the effective one-electron density of states. Clearly, interpretation of transport data in a-Si:H films is fraught with danger. This exhortation is borne out by the actual transport results, which are extremely puzzling in many respects.

One of the most peculiar results in a-Si:H is the variation of the electrical conductivity as a function of temperature. Typical behavior is sketched in Fig. 23. For a wide class of samples, the conductivity, $\sigma$, obeys the relation:

$$\sigma = \sigma_{00} \left[\exp(-E_A/kT)\right]^{1-\alpha}, \tag{32}$$

where $\alpha = T/T_0$ for an array of different activation energies. Typical values are $\sigma_{00} = 10^{-1}\,cm^{-1}$ and $T_0 = 700K$. This puzzling behavior, in which the $\sigma(T)$ curves tend to converge to a single point at $T = T_0$, is observed in many different types of semiconductors from organics to transition-metal oxides. It is often called the Meyer-Neldel rule [109]. This rule has not yet been adequately explained, although it could arise from a disorder-induced shift of $E_F$ [110].

Hall-effect experiments have been carried out, but they have not elucidated the transport mechanisms. A major complication is that B-doped samples exhibit an n-type Hall effect, while P-doped samples exhibit a p-type Hall effect, the opposite sign from what logic and thermopower measurements dictate [111, 112]. This so-called Hall-thermopower sign anomaly can be explained if the carriers form small polarons [113], or if ambipolar conduction exists in films characterized by large concentrations of negatively-correlated defects [114], but its real origin still remains a mystery.

The thermopower, S, for electrons moving above the conduction-band mobility edge, $E_c'$, is given by:

$$S = - \frac{k}{e} \left[ \frac{E_c' - E_F}{kT} + a \right],$$
                                                                        (33)

where a is a measure of the average kinetic energy of the carriers and is typically in the range 2-4. For holes moving below $E_v'$, the analogous expression is:

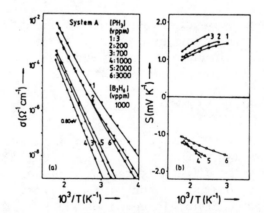

Fig. 23   (a) conductivity σ and (b) thermopower S as a function of 1/T of a-Si:H films doped with 1000 ppm $B_2H_6$ and various concentrations of $PH_3$. (From ref. 108.)

$$S = \frac{k}{e} \left[ \frac{E_F - E_v'}{kT} + a' \right]. \tag{34}$$

Since $(E_c' - E_F)/kT$ is the exponent in the Boltzmann factor, $\exp[-(E_c' - E_F)/kT]$, which controls the electrical conductivity, $\sigma$, we might expect a plot of S vs $T^{-1}$ to have an identical slope with the Arrhenius plot, $\log \sigma$ vs $T^{-1}$. In fact, this is not the case experimentally, the activation energy of S being generally less than that of $\sigma$ by about 0.2eV [115]. This could represent an activation energy in the mobility, $\mu(T)$, which enters the expression for $\sigma$ [Eq. (8)] but not for S. In that case, conduction could be due to small-polaron hopping. However, the conductivity data appear to require sufficiently large mobilities that this mechanism is highly unlikely [110]. Typical results for the thermopower are also shown in Fig. 23.

At the present time, none of the three standard transport properties, $\sigma$, $R_H$, and S can be considered to be understood in a-Si:H. It is clear that $\sigma$ can be modulated over many orders of magnitude by both n-type and p-type doping, but a quantitative analysis in terms of a well-defined g(E) and a particular carrier scattering mechanism is lacking.

D.   Nonequilibrium Transport

Considering that the transport properties of a-Si:H films near equilibrium (i.e., at low values of applied fields) are so poorly understood, it is not surprising that nonequilibrium transport is also a subject of much controversy. The main experiments investigated are steady-state and transient photoconductivity, steady-state and transient photoluminescence, and photo-induced absorption. In addition to the problems enumerated in the last subsection, when the sample is far from thermal equilibrium, the distribution function, f(E), which gives the actual occupancy of the one-electron states is extremely difficult to achieve. This adds still another major complication to an already intractable analysis.

For example, consider a steady-state photoconductivity experiment. The exciting light, even if monochromatic, is absorbed nonuniformly through the sample. When initially absorbed, light creates electron-hole pairs. But in an amorphous semiconductor, these electrons and holes are not necessarily free to move, since either or both types of photo-induced charge can be in localized states. Even when both are free, they initially move in each other's field, and if they do not possess sufficient kinetic energy, they can recombine with each other, a process called geminate recombination [116]. If they avoid geminate recombination, they can then move by hopping through localized states. The mobile carriers can also be trapped in localized states. Once trapped, they can be rereleased into the band or they can recombine via a trapping of the opposite type of carrier. Even a free electron and hole can directly recombine. When either type of carrier reaches a contact, it can either leave the semiconductor or accumulate near the contact. When one carrier exists at a contact, either another can enter at the opposite contact or the semiconductor can become depleted of that type of carrier. In order to analyze the transport, we need to know the complete one-electron density of states, $g(E)$, the associated carrier mobilities, $\mu(E)$, the charge condition of the localized states, the cross-sections of each for trapping carriers of either type, the escape frequencies, and the details of the internal-field distribution, especially near the contacts. The problem is often simplified by assuming that carriers within each band quickly thermalize, i.e., approach a Fermi-Dirac distribution.

$$f_i(E) = \{\exp[(E - E_{Fi})/kT] + 1\}^{-1} , \tag{35}$$

where i is the carrier type and $E_{Fi}$ is an associated quasi-Fermi energy which is fixed by the condition that the appropriate total number of excited electrons and holes (i.e., the equilibrium concentration plus the photogenerated carriers) is correct [117]. Another simplification is that a demarcation level for each carrier type exists, above which the carrier is more likely to be released than to trap a carrier of the opposite type and below which the reverse is true. Thus, the demarcation level separates the localized states which act primarily as traps from those which act as recombination centers.

It is clear that the cross-section for trapping of, e.g., electrons, is much larger for positively-charged centers such as $T_3^+$ than for neutral centers such as $T_3^0$. Thus, a knowledge of the sign $U_{eff}$ for the dangling bond is vital for any analysis of the nonequilibrium transport. This relates to a major present subject of controversy, the magnitude of the carrier mobility beyond the mobility edge. It is generally agreed that the directly measurable quantity, the drift mobility, $\mu_d$, is relatively small, of the order of $1 cm^2/V$-s for electrons at room temperature [118]. But the drift mobility is only a fraction of the band mobility, $\mu_o$, since it is

limited by the trapping of the photogenerated carriers. The relationship can be expressed [119]:

$$\mu_d = \mu_o \frac{n_o}{n_o + n_t} \, , \tag{36}$$

where $n_o$ and $n_t$ are the concentrations of free and trapped carriers,
respectively. The details of extracting $\mu_o$ from measurements of $\mu_d$
are very dependent on the position of $E_c'$ and the shape of the density of states, and $\mu_o$ has been estimated to be $\sim 10 cm^2/V\text{-}s$ [99]
or $\sim 800 cm^2/V\text{-}s$ [120], depending upon the inferred $g(E)$. There is
an equivalent spread in estimating $\mu_o$ from transient measurements [121, 122]. In fact, there are now two diverse views of
transport in a-Si:H, one in which conduction beyond the mobility
edges proceeds with scattering events every second or third neighbor, the other in which the band mobility is of the same order as
that of c-Si, but large densities of charged traps greatly reduce
the drift mobility. This puzzle must be considered as unresolved
at the present time. However, the apparent complete absence
of geminate recombination [123,124], indicating extremely rapid
separation of photogenerated electrons and holes, and recent
work on multilayer structures [125], suggesting delocalized wave
functions with coherence lengths in excess of 30Å, are strong
evidence in favor of the higher values of band mobilities.

E.    Metastable Effects

Staebler and Wronski [126] observed that a flux of photons
with energies greater than 1.6eV induce metastable changes in
a-Si:H films. They called the fully annealed (i.e., stable) phase
A and the light soaked (i.e., metastable) phase B. The A → B
transition can be induced by any process that creates excess free
electrons and holes, photogeneration, double injection, electron
or ion bombardment, etc. In contrast, the B → A transition is
induced only by thermal annealing.

There are major differences in the electronic structures of
states A and B [85]. In state B, the Fermi energy lies closer to
mid-gap than it does in state A, resulting in a many orders-of-
magnitude decrease in room-temperature conductivity upon light
soaking (except in the regime where $E_F$ is already near mid-gap
in the annealed phase) [127]. The photoconductivity is generally
lower in state B, which also is characterized by a factor of 2 - 4
increase in the concentration of $T_3^o$ centers [128]. The photogenerated centers have separations in excess of 10Å. The two

transitions, $A \rightarrow B$ and $B \rightarrow A$, appear to both be thermally activated with activation energies in the 1eV range [129], but a dispersion in activation energies may be present. However, both processes are completely reversible. The optical edge does not appear to be affected significantly, but there are shifts of localized states within the gap. Finally, films with relatively large defect densities do not exhibit any measurable $A \rightarrow B$ transition.

It is clear that the $A \rightarrow B$ transition is induced by a metastable trapping of the excess electrons and holes. The resulting local relaxations evidently stabilize the trapped carriers, which must then produce the excess $T_3^O$ centers. Since the metastable states are local potential minima after the relaxations, a potential barrier must exist retarding both the $A \rightarrow B$ and the $B \rightarrow A$ transitions. Since the electrons and holes are trapped in different regions of space, the light-induced $T_3^O$ centers are then more than 10Å apart, as observed.

Several specific defect models which can produce metastable $T_3^O$ centers have been proposed. The simplest would have a high probability if $U_{eff}$ is negative for the $T_3$ defect in a-Si:H [85]. This would then yield significant concentrations of $T_3^+$ and $T_3^-$ centers in the less-strained regions of the film, as discussed previously. The more-strained regions would possess neutral $T_3^O$ centers, which account for the observed unpaired-spin density. The effective density of states for this state, phase A, is sketched in Fig. 24(a). Upon the trapping of excess electrons by $T_3^+$ centers and holes by $T_3^-$ centers, additional $T_3^O$ defects are stabilized after the appropriate relaxations. This yields phase B, for which g(E) is as sketched in Fig. 24(b). The Fermi energy tends to move to a point half-way between the two $T_3^O$ bands, presumably near mid gap. If there were no significant potential barrier retarding the re-equilibration process, $2T_3^O \rightarrow T_3^+ + T_3^-$, then phase B would be unstable. However, this process requires two bond-angle distortions, effectively changing the $109.5^O$ angles that characterize the $T_3^O$ centers to one of $120^O$ ($T_3^+$) and one near $95^O$ ($T_3^-$). A possible configuration coordinate diagram for the re-equilibration process is given in Fig. 25. Large values for the activation barriers for the charge-transfer processes would be expected with this model.

If $U_{eff} > 0$, other possibilities must be invoked. There is some evidence that oxygen incorporation in the film enhances the photo-induced effects [130,131]. This suggests that charged defect centers in the vicinity of O impurities can trap photoexcited carriers metastably and produce excess $T_3^0$ centers. The possibility of relatively weak hydrogen bonding in the film suggests a model such as the one sketched in Fig. 26 [131]. In this case, both electrons and holes again yield excess $T_3^0$ defects. However, recent work [132] indicates that even films with very low concentrations of oxygen exhinit photo-induced effects, so that an intrinsic mechanism is still required.

Xα calculations [133] suggest that it is difficult to stabilize localized charge near strained bonds in a-Si:H. However, the possibility of Si double bonds together with the inferred existence of both five-membered and seven-membered rings in a-Si:H has led to the novel suggestion [133] that structures such as those sketched in Fig. 27 could be the origin of the photo-induced $T_3^0$ centers. Figure 27(a) shows the silicon analogue of carbonium, which can stabilize positive charge, while (b) is the analogue of a carbanion, known to stabilize negative charge. Appropriate excess-carrier trapping once again leads to metastable $T_3^0$ centers. Clearly, other possibilities can also be envisioned.

F.     Effects of Fluorine

Ovshinsky and Madan [134] prepared an amorphous alloy incorporating F as well as Si and H. Fluorine is the most electronegative element of all, and the Si-F bond is ionic and extremely strong. Consequently, F should be efficient in both relieving strains and removing potential dangling bond states from the gap. One way of looking at this process is to consider an Si-F bond as an intimate $T_3^+ - F^-$ pair. The $F^-$ state is very low in the valence band and the neighboring $T_3^+$ state is probably within the conduction band. With this perspective, it is clear that non-intimate pairs are also possible, so that F can passivate dangling bonds from some distance away, albeit at a cost in energy. Such a situation would yield exponential tails of $F^-$ and $T_3^+$ states analogous to the case sketched in Fig. 14. Although the $F^-$ states are likely to be located completely within the valence band, the $T_3^+$ states could contribute to the conduction band tail. The suppression of the $T_3^-$ states when F is added should make a-Si:F:H alloys less susceptible to the photo-induced changes that characterize a-Si:H. This, indeed, is borne out by the experimental results [135]. In addition to its electronic effects, the fact that fluorine does not effuse as easily as hydrogen makes a-Si:F:H stable to a higher temperature than a-Si:H, while the introduction of ionic bonding results in a-Si:F:H being mechanically harder than a-Si:H [135].

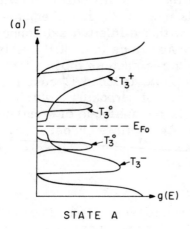

(a)

STATE A

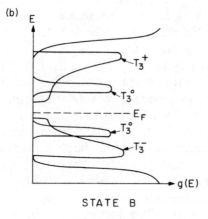

(b)

STATE B

Fig. 24  Effective one-
electron density of states
for a–Si:H films under the
assumption that dangling-
bond defects are charac-
terized by $U_{eff} < 0$ if
complete atomic relaxa-
tions are possible, but
that such relaxations are
retarded in certain
strained regions, result-
ing in the presence of
stable $T_3^0$ centers; (a)
equilibrium state (A);
(b) light–soaked state (B).
(From ref. 83.)

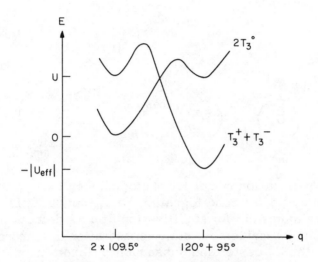

Fig. 25 Total energy as a function of a configuration coordinate
q that represents the variations in both of the sets of
bond angles around a pair of dangling-bond defects. Two
charge configurations are possible for the pair, $2T_3^0$ and
$T_3^+ + T_3^-$, and the energies of both are plotted. The two
curves shown will interact, resulting in a potential
barrier between the lowest-energy states of the possible
charge configurations. (From ref. 83.)

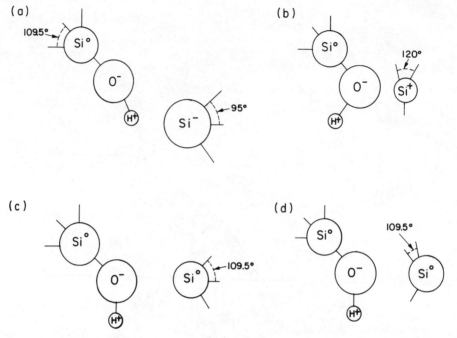

Fig. 26   Isomerization model for photo-induced effects in a-Si:H:O.
(a) hydrogen-bond formation stabilizing a $T_3^-$ Si defect
configuration; (b) stabilization of a $T_3^+$ defect configuration; (c) metastable state formed after trapping of a hole
by the $T_3^-$ center and subsequent relaxation; (d) metastable
state formed after trapping of an electron by the $T_3^+$ center
and subsequent relaxation.   (From ref. 131.)

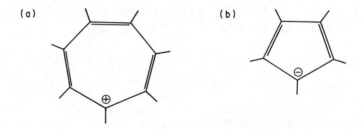

Fig. 27   (a) Si analogue of carbonium ion, capable of stabilizing
positive charge; (b) Si analogue of a carbanion, capable
of stabilizing negatige charge.   (From ref. 133.)

## VIII.  AMORPHOUS CHALCOGENIDE ALLOYS

### A.    Defects

In general, amorphous chalcogens and chalcogenide alloys are characterized by $\bar{z} < 2.4$, and so are not ordinarily overconstrained. Thus, we would not expect the existence of significant concentrations of strain-related defects, as is the case with amorphous tathogens. On the other hand, many experiments suggest that well-defined defects do exist in chalcogenides [136]. Therefore, it is extremely likely that these defects are thermodynamically required (recall the discussion in Section VI). We are thus faced with the necessity of identifying a defect with low creation energy, $\Delta E$.

It is clear that no neutral center has low creation energy in chalcogenides [86]. A neutral dangling bond, $C_1^0$, costs a bond strength, $E_b$, to create, typically $2 - 6eV$. However, in elements other than tathogens, it is also possible to overcoordinate. For example, a threefold-coordinated chalcogen center can be formed by taking the high-energy lone pair that characterizes an atom in its ground-state configuration (i.e., $C_2^0$) and converting it into a bonding-antibonding pair. This always costs a positive energy, $\Delta$ (recall Section IIIB), typically $1 - 4eV$. Thus, the lowest-energy neutral chalcogen center is $C_3^0$ if $\Delta < E_b$ or $C_1^0$ if $E_b < \Delta$.

The key to understanding the unusual properties of amorphous chalcogenide alloys is to consider the possibility of charged defect pairs [86]. A $C_3^+$ center has very low energy because it forms three bonds. (Note that a positively-charged chalcogen ion is isoelectronic to a pnictogen atom, which optimally forms three bonds.) However, for charge neutrality, this must be compensated by negatively-charged centers, which automatically cost at least the correlation energy, $U$, to form. The lowest-energy negatively charged chalcogen center is $C_1^-$. (Note that a negatively charged chalcogen ion is isoelectronic to a halogen atom, which forms one bond in its ground-state configuration.) Tight-binding estimates, sketched in Fig. 28, show that the creation energy of a $C_3^+ - C_1^-$ pair is $U$, typically $0.5 - 1.5eV$. Since this is the creation energy of a _pair_ of defects, the defect concentration is given by:

$$N_d = N_o \exp[-U/2kT_g],  \tag{37}$$

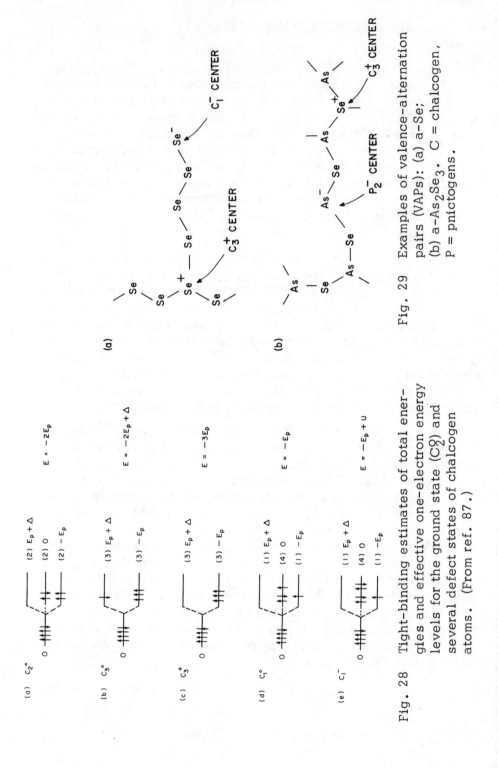

Fig. 28  Tight-binding estimates of total energies and effective one-electron energy
levels for the ground state ($C_2^0$) and
several defect states of chalcogen
atoms.  (From ref. 87.)

Fig. 29  Examples of valence-alternation
pairs (VAPs): (a) a-Se;
(b) a-As$_2$Se$_3$.  C = chalcogen,
P = pnictogens.

where $N_o$ is the total atomic concentration and $T_g$ is the glass transition temperature, typically 300 - 900K. For example, if $N_o = 10^{23} \text{cm}^{-3}$, $U = 1.0 \text{eV}$, and $T_g = 600K$, $N_d = 5 \times 10^{18} \text{cm}^{-3}$. In the extensively-studied amorphous arsenic-chalcogenide compounds, e.g., a-As$_2$Se$_3$ or a-As$_2$Te$_3$, the lowest-energy defect is very likely a $C_3^+ - P_2^-$ pair [70]. Kastner et al. [86] called such defects valence alternation pairs (VAPs). Two examples are sketched in Fig. 29.

The reason for the low creation energy of VAPs is that the total number of bonds is the same as when both atoms are in their ground-state configurations; e.g., a $C_3^+ - P_2^-$ pair has five bonds, the same as a $C_2^0 - P_3^0$ pair. Furthermore, VAPs are spinless defects, since both centers are optimally coordinated for their particular electronic configurations. But the most interesting property of VAPs is that they are characterized by a negative $U_{eff}$, as can be inferred from Fig. 29. Consider, for example, the injection of electrons into the conduction band of a-As$_2$Se$_3$. We would expect that the first electron is most likely to be trapped by a $C_3^+$ center, since it has a large cross section for electron trapping because of its negative charge. This reaction,

$$C_3^+ + e^- \rightarrow C_3^0, \tag{38}$$

results in the electron being localized in an antibonding orbital, with relatively high energy (probably in the upper part of the mobility gap). This localized electron represents an unpaired spin, which should be evident in an ESR experiment. Once a $C_3^0$ is created, however, another method for electron trapping becomes possible--mobile electrons can be trapped on a nearest neighbor of the $C_3^0$ center, provided that the bond between that As atom and the $C_3^0$ center breaks. This bond-breaking can be represented by:

$$C_3^0 + P_3^0 \rightarrow C_2^0 + P_2^0. \tag{39}$$

Reaction (39) essentially converts an overcoordinated chalcogen center to a pnictogen dangling bond. Since the latter also has an unpaired spin, it can be identified by its ESR signature. But since a $P_2^0$ center has an unoccupied nonbonding orbital, it can trap an electron in a relatively low-energy state:

$$P_2^o + e^- \rightarrow P_2^-. \tag{40}$$

Since this state is very likely to be in the lower part of the mobility gap, the second electron is trapped in a lower-energy state than the first. But the difference in energy between trapping a second and a first electron in the vicinity of the same defect is just $U_{eff}$; thus, $U_{eff}$ is ordinarily <u>negative</u> for VAPs.

The VAP model clears up the major puzzles of amorphous chalcogenides, viz. how $E_F$ can be strongly pinned without any evidence for unpaired spins or variable-range hopping. Since the VAP creation energy is so small, large concentrations are frozen in at $T_g$. The positively-charged defects, e.g., $C_3^+$, have one-electron energies in the upper part of the gap, and act like ionized donors; the negatively charged defects, e.g., $P_2^-$, have one-electron energies in the lower part of the gap, and act like ionized acceptors. Neither center has any unpaired spins. When excess positive or negative charge is injected or induced, holes or electrons are trapped in pairs, the net effect of which is to interconvert type of center to the other. For example, two excess electrons are accommodated by the net effects of reactions (38)-(40):

$$C_3^+ + 2e^- \rightarrow P_2^-. \tag{41}$$

The Fermi energy is the average energy necessary to add a few electrons to the material. If, for example, $C_3^+$ - $P_2^-$ VAPs are present, adding two electrons simply converts a $C_3^+$ center to a $P_2^-$ center, and this procedure can take up all excess electrons until no $C_3^+$ centers remain. The process occurs in three steps, given by (38)-(40). The average energy to add the two electrons is:

$$E_F = \frac{1}{2}\left[E(C_3^+) + E(P_2^-)\right], \tag{42}$$

where $E(C_3^+)$ and $E(P_2^-)$ are the donor and acceptor energies, respectively. Thus, $E_F$ is pinned near mid gap, despite the fact that there are no one-electron states in that region of energy. Variable-range hopping would require two electrons to move between defect centers, a process which necessitates both a bond breaking, reaction (39), and a bond formation, represented by:

$$C_2^o + P_2^o \rightarrow C_3^o + P_3^o. \tag{43}$$

The net effect is the motion of two electrons with an accompanying bond-breaking, which can be thought of as a quasi-particle, often called a bipolaron. It is clear that the binding energy of the bipolaron is large, because of the bond-breaking distortion. Thus, the hopping activation energy, approximately half the binding energy [30], is also large and hopping conduction is ordinarily unobservable.

Just as in the case of charge-transfer defects in a-Si:H, discussed in the previous section, intimate VAPs (IVAPs) are energetically favorable. Since VAPs are thermodynamically induced defects, we would expect the law of mass action to apply. Then, the concentration of IVAPs in which the separation between the positively and negatively charged centers is R is given by:

$$N_d(R) = N_d \exp(e^2/2\epsilon RkT_g),\qquad(44)$$

where $N_d$ is the concentration of non-intimate pairs (NVAPSs), given by Eq. (37). In this case, the effective density of states is just the same as the one for ICTDs given in Fig. 14. When IVAPs are present, $E_F$ is pinned only to the extent of the non-intimate VAP concentration, $N_d$. When sufficient additional charge is introduced to convert all of one type of the spatially separated pairs to the other, $E_F$ will begin to move. This arises because, e.g., additional electrons which convert $C_3^+$ to $P_2^-$ center concomitantly change the $C_3^+$ - $P_2^-$ coulomb attraction to a $P_2^-$ - $P_2^-$ repulsion, an effect which costs an energy equal to $(e^2/\epsilon R) + (e^2/\epsilon R')$, where $R'$ is the $P_2^-$ - $P_2^-$ separation (somewhat larger than R since the new $P_2^-$ is a neighbor of the original $C_3^+$).

There are three types of VAP which have low creation energies and tend to pin $E_F$. In addition to the $C_3^+$ - $C_1^-$ and $C_3^+$ - $P_2^-$ pairs already discussed, $C_3^+$ - $T_3^-$ pairs are possible in amorphous IV-VI alloys. When tathogens are involved, however, the local relaxations necessary to provide the negative $U_{eff}$ may not always be possible. For example, when two excess free holes are introduced into a IV-VI alloy, the trapping reactions induced are:

$$T_3^- + h^+ \rightarrow T_3^0 \qquad(45)$$

$$T_3^0 + C_2^0 \rightarrow T_4^0 + C_3^0 \qquad(46)$$

$$C_3^0 + h^+ \rightarrow C_3^+ . \tag{47}$$

But (46) requires the formation of an extra bond, with optimal values for the interatomic separation and the bond angles. If $\bar{z} > 2.4$, the entire network may be overconstrained, and it might not be possible to locally optimize the structure. For example, the stoichiometric compound, $GeSe_2$, has $\bar{z} = 2.7$ and is somewhat overconstrained.

Other defects can exist in multicomponent alloys, particularly involving weak bonds. For example, a Se-Se bond is considerably weaker than an As-Se bond, and the former could introduce localized gap states in amorphous As-Se alloys. Nonstoichiometric alloys, in fact, require some weak bonding. Even though the 8-N rule is obeyed, when weak bonds yield states in the gap, we must consider the center a defect. Such defects could also be characterized by negative values of $U_{eff}$.

B.     Ground-State Electronic Structure

One of the simplifying features in understanding the ground-state properties of amorphous chalcogenide alloys is the strong similarities between corresponding crystalline and amorphous materials [136]. For example, the x-ray photoemission spectroscopy results on crystalline and amorphous samples of $As_2S_3$, $As_2Se_3$ and $As_2Te_3$ are compared in Fig. 30 [138]. The great similarity between the valence-band structure of the ordered and disordered phases is remarkable. This result may follow primarily from the fact that the valence band consists of nonbonding chalcogen orbitals. In any event, however, it makes the study of these solids relatively straightforward, since crystalline band-structure calculations on the stoichiometric compound can be carried out rather easily.

Even the value of the optical gap is similar in corresponding crystalline and amorphous solids. The optical absorption edge of the two phases of c-Se are compared with that of a-Se in Fig. 31. while Fig. 32 shows the spectra of $c-As_2Se_3$ and $a-As_2Se_3$. The optical absorption of trigonal and amorphous Se below the edge appear to be virtually identical, suggesting similar types and concentrations of defects in both phases. In $As_2Se_3$, the above-gap absorptions and the values of $E_{opt}$ are also very similar in the two phases, but the band tail appears to be more extensive in the amorphous sample.

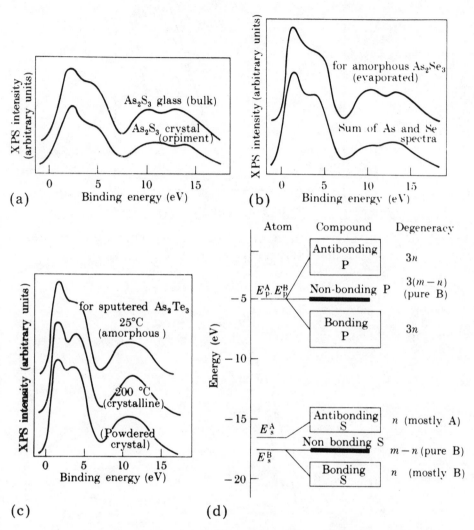

Fig. 30     Monochromatized X-ray photoemission spectroscopy
intensities (hν = 1486.6eV) for the density of valence
band states for (a) crystalline and amorphous $As_2S_3$,
(b) amorphous $As_2Se_3$ and (c) crystalline and amorphous
$As_2Te_3$; (d) bonding in arsenic chalcogenides. (From
refs. 66 and 138.)

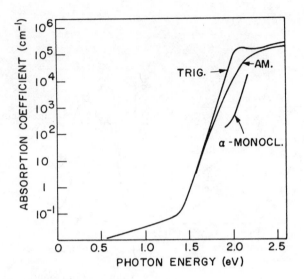

Fig. 31  Optical absorption coefficients in amorphous and crystalline selenium.  (From ref. 66.)

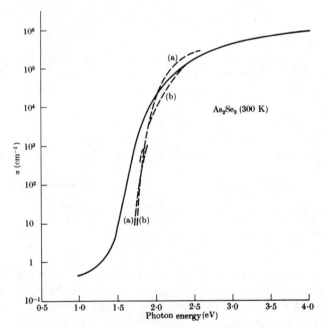

Fig. 32  Optical absorption coefficients at room temperature (300K)
for amorphous (——) and crystalline (---) $As_2Se_3$: curves
a, for the crystal the electric field vector is parallel to
the a axis; curves b, for the crystal the electric field
vector is perpendicular to the a axis.  (From ref. 66.)

## C.  Transport Properties

The most remarkable feature of the electrical-conductivity data in amorphous chalcogenide alloys is the strong pinning of the Fermi energy. This yields very linear Arrhenius plots, as is evident from Fig. 33. The electrical conductivity of all eight glasses maintains a constant activation energy over a region of up to ten decades. No extrinsic conduction or transition to hopping transport is evident for any sample. Even when large impurity concentrations are added, no change in activation energy appears at any temperature. Note also the Meyer-Neldel behavior of the different glasses, a focus in the Arrhenius plots existing near 1500K.

The Hall effect in many but not all chalcogenides is negative and opposite to the sign of the thermopower [139]. The activation energy of the thermopower is less than that of the conductivity [140], similar to the data on amorphous silicon alloys. Although this be-

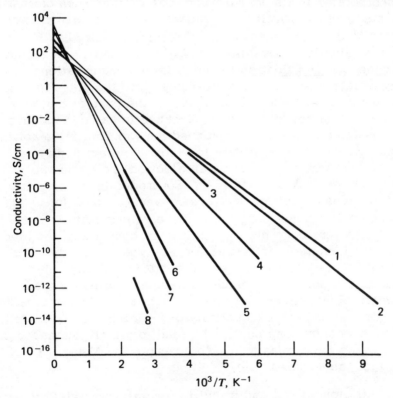

Fig. 33   Electrical conductivity as a function of temperature for several amorphous chalcogenide alloys (S/cm = $\Omega^{-1}cm^{-1}$). (From ref. 66.)

havior is consistent with small-polaron formation [141], it can
also be a consequence of ambipolar conduction in the presence of
negatively correlated defects [114]. Since there is evidence from
switching experiments on the same glasses that the free-carrier
mobility is of the order of 10 $cm^2/V-S$ [142], it appears unlikely
that small polarons form in these materials.

Because of the presence of large VAP concentrations, conventional doping is ordinarily impossible in amorphous chalcogenides. However, if a sufficient concentration of additional
electrons or holes could be chemically provided to eliminate all
of the positively or negatively charged centers, $E_F$ could be
moved. Of course, this requires the introduction of the order of
0.1% or more electrically active impurities, i.e., atoms in configurations in which electrons are present with energies above $E_F$,
or empty states are present with energies below $E_F$. An impurity
atom incorporated in its normal structural configuration ordinarily
will be electrically inactive. Because of this, simple addition of
impurities to the melt during glass formation usually has no major
effect on conduction. Ovshinsky [143] developed a new technique
called chemical modification in which large concentrations of
suitably chosen impurities are introduced into the glass in a nonequilibrium manner. A simple example is the post-diffusion of
indium into a chalcogenide alloy. Because In has odd valence, it
cannot easily enter the already-formed network in a fully-bonded
position. However, a relatively low-energy defect configuration
for In is one in which it gives up an electron and forms two
ordinary and two dative bonds [67]. For moderate In concentra-
tions, the excess electrons in pairs convert $C_3^+$ to $T_3^-$ (or $P_2^-$ or $C_1^-$,
depending on the glass) centers, and the conductivity is unaf-
fected. However, once the $C_3^+$ centers are completely depleted,
$E_F$ begins to increase. Initially, for a p-type glass, the electrical activation energy is increased and $\sigma$ is lowered. But
eventually, the conduction becomes n-type, and increases by
many orders of magnitude. Typical results are shown in Fig. 34.
With 1% or less In, the conductivity of a-GeTeSe is unchanged.
However, concentrations beyond 1% lead to greatly increased conduction, and a p $\rightarrow$ n transition. Again, note the Meyer-Neldel
behavior for the samples with 0 - 4% In.

Alkali atoms [144] and transition-metal atoms [145] can also
be used as chemical modifiers, and co-sputtering and ion-
implantation techniques can be effective under certain condi-

tions [146]. Alkalis such as Li have a high-energy occupied s
orbital which often lies above $E_F$, and is thus electronically active
in nonbonded configurations. Transition-metal atoms have high-
energy occupied d orbitals, and often can multiply ionize in a
semiconductor matrix [67]. Because of the nonequilibrium preparation techniques, however, the chemistry of modified chalcogenides
can be extremely complex.

## D. Nonequilibrium Transport

In addition to their effects on the ground-state and near-
equilibrium transport, we might anticipate that the negatively-
correlated defects also control the nonequilibrium transport properties of chalcogenide glasses. The VAPs introduce large concentrations of positively charged electron traps located above $E_F$ and

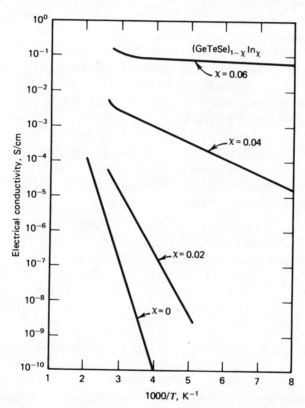

Fig. 34 Electrical conductivity as a function of temperature for
pure a-GeTeSe and films chemically modified with
In (S/cm = $\Omega^{-1}$cm$^{-1}$).

negatively charged hole traps below $E_F$. Since both types of center should have large cross-sections as well as concentrations, we should expect that they trap excess free carriers very quickly. However, another possible array of trapping centers exist, viz. localized states in band tails that arise from the disorder. These could be present in even larger concentrations than the VAPs, although they are neutral rather than charged.

Dispersive transport was first identified in chalcogenide glasses [147], and was originally analyzed in terms of hopping conduction among a distribution of spatially separated localized states. More recently, Orenstein and Kastner [45] showed from transient photoconductivity and photo-induced absorption experiments on a-$As_2Se_3$ that the re-equilibration kinetics are controlled by multiple trapping in an exponential distribution of traps. This exponential distribution, in fact, is characterized by a characteristic temperature, $T_O$ = 550K, not too far from the glass transition temperature of a-$As_2Se_3$, $T_g$ = 450K. Although, we would expect $T_O \simeq T_g$ for IVAPs frozen in at $T_g$ [see Eq. (44)], this need not be true in general since the law of mass action (which governs the defect concentrations) involves total energies, while g(E) involves effective one-electron energies which could be very different. Indeed, in chemically modified samples of a-$As_2Se_3$ and in a-$As_2S_3$, the correlation between $T_O$ and $T_g$ is not clear. In any event, the conclusion that excess carriers in a-$As_2Se_3$ are multiply trapped in an exponential density of states with $T_O$ = 550K also follows from time-of-flight experiments [148], and must be considered to be on solid ground.

The situation is not so clear with regard to the photoluminescence. The present weight of evidence is that radiative recombination occurs between carriers trapped in centers that are charged when unoccupied [149], although alternative explanations have been proposed. The charged traps could well be the IVAPs, independent of whether they or the band tails are responsible for the transient photoconductivity results.

A recent study of transient photoconductivity in a-$As_2Se_3$ beyond the glass transition temperature by Thio et al. [150] has confirmed the predictions of the VAP model and enabled a determination of the magnitudes of both $\Delta E$ and $U_{eff}$. The results have also shed some light on the origin of the exponential trap distributions. Using low-intensity excitation, Thio et al. were able to measure the monomolecular recombination time, $\tau$, as a function of temperature in the range 295 - 580K. The existence of the large exponential distribution of traps protects the carriers from recombination for quite long times, and serves to increase $\tau$ to the order of minutes at room temperature. The fact that dispersive transport continues beyond these times indicates that the exponential trap densities extend all the way to the mid-gap region. The recombination time was then related to the recombination rate, $b_r$, and the concentration of recombination centers, $N_r$. With the reasonable assumption that $b_r$ is temperature independent, it was found that an Arrhenius plot of $N_r$ has two linear regions, one below $T_g$ and one above, with different activation energies. The model of Kastner et al. [86] predicts that the concentrations of VAPs above $T_g$ should be given by:

$$N_d = N_o \exp(-\Delta E/2kT), \tag{48}$$

since the defects are thermodynamically induced. In contrast, if the VAPs are frozen in at $T_g$, below that temperature they should obey the relation [see Eq. (37)]:

$$N_d = N_e \exp(-\Delta E/2kT_g). \tag{49}$$

But it is the neutral defects, e.g., $C_3^o$ and $P_2^o$, which are expected to act as the monomolecular recombination centers, since they are attractive to the opposite type of carrier after a trapping event. Since neutral defects require an energy of $U_{eff}$, the density of recombination centers above $T_g$ is given by:

$$N_r = N_o \exp[-(\Delta E - U_{eff})/2kT], \tag{50}$$

but below $T_g$, it is instead given by:

$$N_r = N_o \exp(-\Delta E/2kT_g) \exp(U_{eff}/2kT). \tag{51}$$

Using this analysis, Thio et al. found $\Delta E = 0.9$eV and $U_{eff} = -0.7$eV for a-$As_2Se_3$. Thus, about $4 \times 10^{17} cm^{-3}$ VAPs are frozen in at $T_g$.

If $U_{eff} = -0.7$eV, there must be a mid-gap region this wide containing no one-electron states, with $E_F$ pinned in the center of this region. Since the optical gap of a-$As_2Se_3$ is about 1.8eV and the electrical activation energy is 0.9eV, $E_F$ is very near the center of the mobility gap. The non-intimate VAPs are located at

$E_F = |U_{eff}|/2$. If IVAPs are present in concentrations given by
Eq. (44), an exponential distribution of positively charged traps
begins at $E_F + |U_{eff}|/2$ and ends at $E_F + |U_{eff}|/2 + M$; in addition,
a mirror-image distribution of negatively charged states extends
from $E_F - |U_{eff}|/2 - M$. If M is sufficiently large, these states
merge with the conduction and valence bands, respectively. Because of the Coulomb attraction, the activation energy to create
neutral defects from IVAPs is larger than that from NVAPs,
guaranteeing that NVAPs control the recombination kinetics. However, the IVAPs could still control the multiple trapping.

A surprising feature of the data of Thio et al. [150] is that
the dispersive transport at room temperature appears to persist for
more than $10^3$s. This suggests that the exponential density of
traps continues at least through 0.8eV above $E_v'$. Since the activation energy for conduction is about 0.9eV, the exponential g(E)
appears to extend beyond the point where the isolated $P_2^-$ one-
electron levels are located. If this is correct, then we must conclude that the relevant traps are the band tail states and, more
surprisingly, the neutral traps have a larger cross-section for hole
capture than the negatively charged $P_2^-$ centers. This is reasonable
only if a potential barrier which retards hole trapping is present.
Such a barrier would create a bottleneck in the re-equilibration
kinetics that would retard not only hole trapping but also the effects of the negative $U_{eff}$. If no such bottleneck existed, e.g.,
excess electrons would be trapped in pairs via the processes
(38) - (40), and the quasi-Fermi energies would quickly return to
$E_F$, thus restoring thermal equilibrium.

The concept of a bottleneck to the re-equilibration kinetics
first arose during the attempt to understand the results of field-
effect experiments in arsenic-tellurium glasses. In a field-effect
configuration, a step voltage is applied to a so-called gate electrode, which is separated from the semiconductor being investigated by an insulating oxide (MOS geometry). This induces excess
positive or negative charge in the semiconductor, decreasing or
increasing $E_F$, respectively, and thus changing the electrical
conductivity. From this conductivity change, $\Delta E_F$ can be obtained,
and this, together with the known concentration of induced charge,
should yield $g(E_F)$. However, for a material with negatively
correlated defects, the Fermi energy remains pinned even in field-
effect experiments. For example, if negative charge is induced
in amorphous arsenic chalcogenides, excess electrons should be

trapped via the processes (38)-(40), ensuring that $E_F$ remains pinned and no field effect is observed. Indeed, no field effect has been reported in a-$As_2Se_3$. However, several groups [151,152] have measured a field effect in glasses in the arsenic-tellurium system, including a-$As_2Te_3$.

The breakthrough in understanding the observation of a field effect in a system characterized by negatively correlated defects was the discovery that the effect is transient [153,154]. In fact, the decay curve contains several segments, representing the combined effects of dispersive transport and collapse of the field toward the gate oxide at early times and much longer decays (of the order of hours) later [155].

The long decay itself has two components, both of which are activated in temperature although with some dispersion. It cannot be due to multiple trapping, because a-$As_2Te_3$ has a gap of only 0.9eV, and all trapped carriers must thus be thermalized after $10^4$s. A straightforward explanation [156] of the long-time decay is the existence of a potential barrier retarding neutral defect interconversion, e.g., (39) or (43). Since it would be expected that each neutral defect is at a local minimum in potential energy, then the necessary barrier between these minima would retard defect interconversion and mask the effects of the negative $U_{eff}$. A barrier of about 0.7eV is sufficient to account for the long transients observed in As-Te glasses. Since the two neutral defects have different total energies, two barrier energies enter, $\Delta V_1$ and $\Delta V_2$, depending on whether bond breaking, e.g., (39), or bond formation, e.g., (43), is involved. These energies define two characteristic times:

$$\tau_1 = \nu_0^{-1} \exp(\Delta V_1/kT) \tag{52}$$

and

$$\tau_2 = \nu_0^{-1} \exp(\Delta V_2/kT), \tag{53}$$

where $\nu_0$ is a typical phonon frequency. At times short compared to the smaller of these characteristic times, the glass resembles a compensated semiconductor. The Fermi energy is unpinned, and field effect is observable. In contrast, at times long compared to the larger of the characteristic times, the effect of the negative $U_{eff}$ becomes apparent and $E_F$ is strongly pinned. In this regime, the field effect has decayed to very small values.

Transient photoconductivity is analogous to the field effect, except for the field redistribution in the latter and the additional requirement of charge neutrality in the former [157]. A key factor is the relative energy of the two neutral defects. If the donor-like defect (e.g., $C_3^0$) has lower energy than the acceptor-like defect (e.g., $P_2^0$), then electrons are more strongly trapped than holes. Since the trapping is by charged centers with large cross-sections, the short-term effect is to suppress the net increase of free electrons. Because the trapped charge is primarily negative, charge neutrality requires the concentration of free holes to rise faster than that of free electrons, thus leading to a high initial photoconductivity in the p-type chalcogenide glasses. The con-comitant excess of trapped holes then drives acceptor-to-donor (e.g., $P_2^0 \rightarrow C_3^0$) interconversions to restore thermal equilibrium. If a potential barrier exists, the interconversion is retarded. How-ever, eventually the requisite concentration does transform, and most of the resultant neutral donors quickly thereafter become ionized. This final component of positive trapped charge offsets the negative charge arising from the excess trapped electrons, thus reducing the photoconductivity.

The results of Thio et al. [150] now suggest that a modifi-cation of these ideas may be required, at least for a-As$_2$Se$_3$. In a-As$_2$Se$_3$, the free holes appear to be much more mobile than the free electrons. A possible explanation for this would be an asym-metry in the trapping kinetics. If processes (38) – (40) were not retarded by any potential barriers, the trapping of excess electrons would proceed very rapidly. In contrast, holes are trapped by the inverse processes:

$$P_2^- + h^+ \rightarrow P_2^0 \tag{54}$$

$$P_2^0 + C_2^0 \rightarrow P_3^0 + C_3^0 \tag{55}$$

$$C_3^0 + h^+ \rightarrow C_3^+ . \tag{56}$$

Although these processes are analogous to (38) - (40), they are intrinsically asymmetric in the sense that (39) represents a bond-breaking, which can always take place relatively easily, while (55) represents a new bond formation, which requires the presence of a chalcogen atom near the $P_2^0$ defect. Thus, the electron quasi-Fermi energy is more easily pinned than the hole quasi-Fermi energy in agreement with observations.

But, in addition, the results of Thio et al. [150] suggest that a potential barrier may also retard (54). It is certainly true that hole capture by a $P_2^-$ center will be accompanied by some local relaxation, since the p lone pair is broken up upon trapping. If both the configurations are local potential minima, a barrier must be present. Of course, the existence of a barrier to (54) also implies one which retards the inverse process, (40), although the latter will likely be smaller in magnitude.

It is also important to explain the photoluminescence, a fundamentally more complex phenomenon because of the existence of both radiative and nonradiative recombination branches. Clearly, the photoluminescence cannot occur after one of the two neutral centers has converted to the other, since both defects then have the same energy. Thus, interconversion represents a nonradiative recombination branch which should be very temperature-dependent. Indeed, the photoluminescence is strongly quenched upon increasing the temperature [158]. In fact, the detailed variation does not represent a simple activated process but rather a more complex one which suggests a distribution of IVAPs. Even at very low temperatures, the photoluminescence fatigues with time [159]. This fatigue could be the result of defect interconversions driven by the imbalance between positive and negative trapped charge discussed previously. The fatigue correlates with the growth of photo-induced unpaired-spin concentrations [160], presumably due to the now metastable neutral defects.

E.    Switching

Amorphous chalcogenide alloys were among the first and certainly have been the most investigated materials which exhibit the phenomenon of threshold switching discussed by Ovshinsky in his landmark paper [4]. When electric fields in excess of about $10^5$V/cm are applied to these materials, a metastable state of high conductance appears, in which of the order of $10^{19}$cm$^{-3}$ free electrons move with mobilities of approximately 10cm$^2$/V-s [161]. When the current is reduced below a critical value, the material returns to its original low-conductance state.

The detailed experimental observations of switching in chalcogenides can be explained by the VAP model [162]. Near equilibrium, the charged centers, e.g., $C_3^+$ and $P_2^-$, act as efficient traps for field-generated, as well as photogenerated, carriers, and the trapping time is considerably shorter than the transit time.

However, beyond a critical value of the applied field, sufficient
free-carrier generation takes place so that the charged traps are
all occupied and thus neutral. Since the concentrations of the
positively and negatively-charged centers were originally equal,
the material remains neutral after charged trap saturation. Only
neutral traps remain, so that the trapping time increases sharply.
If it becomes large compared to the transit time, the current rises
dramatically, initiating the switching. Note that it is essential
that a sufficiently high barrier to neutral defect interconversion
exists to retard this possibility over the time necessary for the
switching transition, since such interconversion would tend to
pin the quasi-Fermi energies. When the current is reduced below
the values necessary to sustain sufficient carrier concentrations to
keep the charged traps filled, the material quickly transforms to
the nonconducting state.

## F.    Metastable Effects

Many varieties of metastable effects, initiated by, e.g.,
applied electric fields or light, have been discovered in amorphous
chalcogenide alloys. Two of the more interesting are photodarkening and amorphous-to-crystalline transitions. For example, in
$As_2S_3$, prolonged exposure to light of energy in excess of the
optical gap induces a darkening of the material, which appears to
be either a decrease in the gap or an increased Urbach
edge [163,164]. These are accompanied by a build-up of large
($\sim 10^{19} cm^{-3}$) concentrations of new defect centers which exhibit
photo-induced unpaired spins and quench the photoluminesc-
cence [165]. There are several possible mechanisms for this
photodarkening. Since the energy from the light is transmitted to
the amorphous network only upon trapping or recombination of
the photoexcited carriers, the new defects must be created by
energies of the order of 1eV. Of course, this is no problem in
amorphous chalcogenides, in which, as has been discussed previously, VAPs can be created with energies of just this magnitude.
The additional internal electric fields from the photo-induced
VAPs could easily cause an enhancement in the Urbach edge and
thus the photodarkening. Alternatively, the photodarkening could
be due to direct transitions involving localized states near the
band edges arising from the photo-induced defects. In either
case, annealing of the material removes the new defects and
restores thermodynamic equilibrium.

In underconstrained amorphous chalcogenides (e.g., $Te_{83}Ge_{17}$), under highly nonequilibrium conditions, nucleation and growth of crystallized material is often observed [4]. This has been observed both after threshold switching has been sustained for times of the order of milliseconds [166], or after the absorption of intense light [167]. Trapping and recombination of the excess carriers created by the applied field or by the light heats the material beyond $T_g$, allowing for cooperative atomic motions on a relatively large scale. Crystallization can occur between $T_g$ and $T_m$, the melting temperature. It appears that the crystallization rate is enhanced by the bond-breaking which necessarily occurs upon creation of excess carriers. It is possible that the new VAPs which are induced further enhance the rate of crystallization. Once crystallization occurs, it is necessary to heat the material above $T_m$ and cool rapidly to restore the glass. However, this can be accomplished either electronically or optically in a straightforward manner [166].

## IX. CONCLUSIONS

It is clear from a comparison of papers reviewing the field of amorphous semiconductors written several years apart, even by the same author, that the theory is in a constant state of flux. Our perspective changes every few years as new insights are obtained. Consequently, it is very presumptuous to try to summarize the present situation without the exhortation that the next review may recant everything. Our current viewpoint is that the foremost consideration in any analysis of the properties of a particular material is the composition, which determines the chemistry and the average coordination and, thus, the local structure. Next, we must focus on the preparation conditions, which primarily affect the morphology but can also influence the local chemical bonding. We must pay careful attention to the creation energy and effective correlation energy of the low-energy defects, involving not only the nominal constituents but also the unintentional impurities. In addition, it is vital to analyze the surface and interface states as well as the space-charge conditions. Trapping and recombination kinetics far from equilibrium can be extremely subtle, and long-time transients and metastable phases are commonplace.

Despite all these problems, steady progress has been made. It took many years to realize the sharp distinctions between amor-

phous tathogens and chalcogenides. Our present perspective is
that the electronic properties of the former are controlled by strain-
induced defects and the latter by thermodynamically induced
defects. Because strain-relieving elements such as H and F can
be introduced in amorphous silicon networks, and careful processing procedures have been developed, the defect concentrations
can be reduced to small values and these materials can be doped.
In contrast, there are almost always very large concentrations of
negatively correlated defects required by thermodynamics in
chalcogenides, thus precluding conventional doping. Nevertheless, conductivity can be controlled by chemical-modification
techniques and even clever alloying. In recent years, the sharp
distinctions between the two classes discussed above have begun
to become more hazy. Both tathogens and chalcogenides exhibit
similar transient behavior, including dispersive transport, photoluminescence, photo-induced absorption, metastable phases, and
photo-induced ESR. It is somewhat ironic that many of these
phenomena are now being observed and studied in crystalline semiconductors, after having been first understood in amorphous
phases. Although we do not know what new insights will emerge
over the next few years, we can be confident that the study of amorphous materials will take its proper place at the forefront of solid-
state physics in the near future. After all, even professors cannot
fool all of the students all of the time.

## ACKNOWLEDGMENTS

I have benefitted from many long and fruitful discussions
with my colleagues. Among the most helpful have been Marc Kastner,
Hellmut Fritzsche, Nevill Mott, Marvin Silver, and, above all,
Stan Ovshinsky, who continues to introduce insightful, profound,
and imaginative ideas to a field too often characterized by regur-
gitation.

## REFERENCES

1.    S.R. Ovshinsky, J. de Phys. 42, C4-1095(1981).
2.    F. Bloch, Z. Physik 52, 555(1928).
3.    A.H. Wilson, Theory of Metals, Cambridge University Press,
         N.Y., 1936.
4.    S.R. Ovshinsky, Phys. Rev. Lett. 21, 1450(1968).

5.    S.C. Moss and J.F. Graczyk, in Proceedings of the Tenth
      International Conference on the Physics of Semiconductors,
      edited by S.P. Keller, J.C. Hensel, and F. Stern, U.S.
      Atomic Energy Commission, Washington, D.C., 1970, p.
      658.
6.    W.H. Zachariasen, J. Am. Chem. Soc. 54, 3841(1932).
7.    D.E. Polk, J. Noncryst. Solids 5, 365(1971).
8.    See, e.g., H. Koizumi and T. Ninomiya, J. Phys. Soc.
      Japan 44, 898(1978).
9.    G. Etherington, A.C. Wright, J.T. Wenzel, J.C. Dore,
      J.H. Clarke, and R.N. Sinclair, J. Noncryst. Solids 48,
      265(1982).
10.   See, e.g., D. Beeman and B.L. Bobbs, Phys. Rev. B 12,
      1399(1975).
11.   T. Takahashi, K. Ohno, and Y. Harada, Phys. Rev. B 21,
      3399(1980).
12.   D. Adler, in Handbook on Semiconductors, edited by T.S. Moss,
      North-Holland, N.Y., 1982, volume 1, p. 805.
13.   M.H. Brodsky and M. Cardona, J. Noncryst. Solids 31,
      81(1978).
14.   E.J. Yoffa and D. Adler, Phys. Rev. B 12, 2260(1975).
15.   See, e.g., O. Madelung, Introduction to Solid-State Theory,
      Springer-Verlag, Berlin, 1978.
16.   B. Kramer, Phys. Stat. Sol. 47b, 501(1971).
17.   L. van Hove, Phys. Rev. 89, 1189(1953).
18.   D. Adler, Amorphous Semiconductors, CRC Press, Cleveland,
      1971.
19.   J.R. Reitz, Solid State Phys. 1, 1(1955).
20.   V. Heine, Solid State Phys. 35, 1(1980).
21.   J.C. Slater and K.H. Johnson, Phys. Rev. B 5, 844(1972).
22    K.H. Johnson, H.J. Kolair, J.P. deNeufville and D.L. Morel,
      Phys. Rev. B 21, 643(1980).
23.   L. Ley, S. Kowalczyk, R. Pollak and D.A. Shirley, Phys.
      Rev. Lett. 29, 1088(1972).
24.   D. Adler, Solar Energy Mats. 8, 53(1982).
25.   T.P. Eggarter and M.H. Cohen, Phys. Rev. Lett. 25, 807
      (1970).
26.   C.M. Soukoulis and M.H. Cohen, J. Noncryst. Solids 66,
      279(1984).
27.   F.R. Shapiro and D. Adler, J. Noncryst. Solids 66, 303
      (1984).
28.   N.F. Mott, Adv. Phys. 16, 49(1976).

29.     M.H. Cohen, H. Fritzsche and S.R. Ovshinsky, Phys.
        Rev. Lett. 22, 1065(1969).
30.     T. Holstein, Ann. Phys. (N.Y.) 8, 343(1959).
31.     M.H. Cohen, E.N. Economou and C.M. Soukoulis, Phys.
        Rev. Lett. 51, 1202(1983).
32.     N.F. Mott, Phil. Mag. 19, 835(1969).
33.     V. Ambegaokar, B.I. Halperin and J.S. Langer, Phys. Rev.
        B 4, 2612(1971).
34.     G.E. Pike and C.H. Seager, in Amorphous and Liquid Semi-
        conductors, edited by J. Stuke and W. Brenig, Taylor and
        Francis, London, 1974, p. 147.
35.     M. Pollak, J. Noncryst. Solids 11, 1(1972).
36.     E.M. Hamilton, Phil. Mag. 26, 1043(1972).
37.     V. Nguyen, B. Shklovskii and A. Efros, Sov. Phys. Semi-
        conductors 13, 11(1979).
38.     D. Adler, L.P. Flora and S.D. Senturia, Solid State
        Commun. 12, 9(1973).
39.     D. Emin, Phys. Rev. Lett. 32, 303(1974).
40.     E.N. Economou, S. Kirkpatrick, M.H. Cohen and T.P.
        Eggarter, Phys. Rev. Lett. 25, 520(1970).
41.     E.N. Economou and M.H. Cohen, Phys. Rev. Lett. 25
        1445(1970).
42.     T.D. Moustakas, Solid State Commun. 35, 745(1980).
43.     W.B. Jackson, N.M. Amer, A.C. Boccara and D. Fournier,
        Appl. Opt. 20, 1333(1981).
44.     C.F. Fuleihan and D. Adler, Bull. Am. Phys. Soc. 27,
        873(1982).
45.     J. Orenstein and M. Kastner, Phys. Rev. Lett. 46, 1421
        (1981).
46.     T. Tiedje and A. Rose, Solid State Commun. 37, 49(1981).
47.     A. Madan, Solar Cells 2, 277(1980).
48.     D.V. Lang, J.D. Cohen and J.P. Harbison, Phys. Rev. B
        25, 5285(1982).
49.     I. Balberg, Phys. Rev. B 22, 3853(1980).
50.     H. Okushi, Y. Tokumaru, S. Yamasaki, H. Oheda and
        K. Tanaka, Phys. Rev. B 25, 4313(1982).
51.     I. Balberg, E. Gal and B. Pratt, J. Noncryst. Solids 59-60,
        277(1983).
52.     N.M. Johnson, J. Noncryst. Solids 59-60, 265(1983).
53.     K.P. Chik, C.K. Yu, P.K. Lim, B.Y. Tong, S.K. Wong and
        P.K. John, J. Noncryst. Solids 59-60, 285(1983).
54.     M.L. Knotek, M. Pollak, T.M. Donovan and H. Kurtzman,
        Phys. Rev. Lett. 30, 853(1973).

55. P. Viscor and A.D. Yoffe, J. Noncryst. Solids 35-36, 409 (1980).
56. J.A. Sauvage, C.J. Mogab and D. Adler, Phil. Mag. 25, 1305(1972).
57. N.B. Goodman and H. Fritzsche, Phil. Mag. B 42, 149 (1980).
58. M.L. Theye, Mat. Res. Bull. 6, 103(1971).
59. M.H. Brodsky, R.S. Title, K. Weiser and G.D. Pettit, Phys. Rev. B 1, 2632(1970).
60. P.A. Walley and A.K. Jonscher, Thin Solid Films 1, 367 (1967).
61. K.L. Chopra and S.K. Bahl, Phys. Rev. B 1, 2545(1970).
62. M.L. Knotek, Solid State Commun. 17, 1431(1975).
63. M.H. Brodsky and R.S. Title, Phys. Rev. Lett. 23, 581 (1969).
64. R.C. Chittick, J.H. Alexander and H.F. Sterling, J. Electrochem. Soc. 116, 77(1969).
65. W.E. Spear and P.G. LeComber, Adv. Phys. 26, 811(1977).
66. N.F. Mott and E. A. Davis, Electronic Processes in Non-Crystalline Materials, Second Edition, Clarendon Press, Oxford, 1979.
67. S.R. Ovshinsky and D. Adler, Contemp. Phys. 19, 109 (1978).
68. S.R. Ovshinsky, in Structure and Excitations of Amorphous Solids, edited by G. Lucovsky and F. Galeener, A.I.P., N.Y., 1976, p. 31.
69. J.C. Phillips, J. Noncryst. Solids 34, 153(1979).
70. D. Adler, J. Noncryst. Solids 35-36, 819(1980).
71. D. Adler, Phys. Rev. Lett. 41, 1755(1978).
72. G.D. Cody, C.R. Wronski, B. Abeles, R.B. Stephens and B. Brooks, Solar Cells 2, 227(1980).
73. B. von Roedern, L. Ley, M. Cardona and F.W. Smith, Phil. Mag. B 40, 433(1980).
74. M.E. Eberhart, K.H. Johnson and D. Adler, Phys. Rev. B 26, 3138(1982).
75. W.E. Carlos and P.C. Taylor, Phys. Rev. B 25, 1435(1982).
76. H. N. Lohneysen, H.J. Schink and W. Beyer, Phys. Rev. Lett. 52, 549(1984).
77. J.E. Graebner, B. Golding, L.C. Allen, D.K. Biegelsen and M. Stutzmann, Phys. Rev. Lett. 52, 553(1984).
78. H. Fritzsche, C.C. Tsai and P. Persans, Solid State Technol. 21, 55(1978).

79.  R. Tsu, personal communication, 1983.

80.  D. Adler, J. de Phys. 42, C4-3(1981).

81.  A. Madan, P.G. LeComber and W.E. Spear, J. Noncryst.
     Solids 20, 239(1976).

82.  J.D. Cohen, D.V. Lang and J.P. Harbison, Phys. Rev.
     Lett. 45, 197(1980).

83.  D. Adler, in Semiconductor and Semimetals, edited by
     R.K. Willardson and A.C. Beer, Academic Press, N.Y.,
     1984, Vol. 21 A, p. 291.

84.  D.C. Allen and J.D. Joannopoulos, Phys. Rev. Lett. 44,
     43(1980).

85.  D. Adler, Solar Cells 9, 133(1983).

86.  M. Kastner, D. Adler and H. Fritzsche, Phys. Rev. Lett.
     37, 1504(1976).

87.  D. Adler and E.J. Yoffa, Canad. J. Chem. 55, 1920(1977).

88.  A. Kasdan, D.P. Goshom and W.A. Lanford, J. Noncryst.
     Solids, in press.

89.  J.D. Joannopoulos and M.L. Cohen, Solid State Phys. 31,
     71(1976).

90.  J. Tauc, in Optical Properties of Solids, edited by F. Abels,
     North-Holland, Amsterdam, 1974, p. 279.

91.  V. Vorlicek, M. Zavetova, S.K. Pavlov and L. Pajasova,
     J. Noncryst. Solids 45, 289(1981).

92.  B. von Roedern, L. Ley and M. Cardona, Phys. Rev. Lett.
     39, 1576(1977).

93.  W.E. Pickett, Phys. Rev. B 23, 6603(1981).

94.  W.Y. Ching, D.J. Lam and C.C. Lin, Phys. Rev. Lett.
     42, 805(1979).

95.  L. Ley, in Semiconductor and Semimetals, edited by
     R.K. Willardson and A.C. Beer, Academic Press, N.Y.,
     Vol 21 B, in press.

96.  G.D. Cody in Semiconductors and Semimetals, edited by
     R.K. Willardson and A.C. Beer, Academic Press, N.Y.,
     1984, Vol. 21 B, in press.

97.  J.P. Dow and D. Redfield, Phys. Rev. B 5, 594(1972).

98.  S. Abe and Y. Toyazawa, J. Phys. Soc. Japan 50, 2185
     (1981).

99.  T. Tiedje in Semiconductors and Semimetals, edited by
     R.K. Willardson and A.C. Beer, Academic Press, N.Y.,
     1984, Vol. 21 B, in press.

100. C. -Y. Huang, S. Guha  and S.J. Hudgens, Phys. Rev.
     B 27, 7460(1983).

101. W.E. Spear, J. Noncryst. Solids 59-60, 1(1983).

102.    C.R. Wronski, B. Abeles, T. Tiedje and G.D. Cody,
        Solid State Commun. 44, 1423(1982).
103.    W.B. Jackson, N.M. Amer, A.C. Boccara and D. Fournier,
        Appl. Opt. 20, 1333(1981).
104.    D. Adler, A.I.P. Conf. Proc., in press.
105.    D.K. Biegelsen, R.A. Street, C.C. Tsai and J.C. Knights,
        Phys. Rev. B 20, 4839(1979).
106.    W.B. Jackson and N.M. Amer, Phys. Rev. B. 25, 5559
        (1982).
107.    C.-Y. Huang, S. Guha and S.J. Hudgens, Phys. Rev. B
        27, 7460(1983).
108.    W. Beyer, H. Mell and H. Overhof, J. de Phys. 42,
        C4-103(1981).
109.    W. Meyer and H. Neldel, Z. Tech. Phys. 18, 588(1937).
110.    F.R. Shapiro and D. Adler, J. Noncryst. Solids 66, 303
        (1984).
111.    P.G. LeComber, D.I. Jones and W.E. Spear, Phil. Mag.
        35, 1173(1977).
112.    M. Roilos, Phil. Mag. B 38, 477(1978).
113.    D. Emin, Phil. Mag. 35, 1189(1977).
114.    E.J. Yoffa and D. Adler, Phys. Rev. B 15, 2311(1977).
115.    D.I. Jones, P.G. LeComber and W.E. Spear, Phil. Mag.
        36, 541(1977).
116.    L. Onsager, Phys. Rev. 54, 554(1938).
117.    A. Rose, Concepts in Photoconductivity and Allied Problems,
        Interscience, N.Y. 1963.
118.    T. Tiedje, J.M. Cebulka, D.L. Morel and B. Abeles, Phys.
        Rev. Lett. 46, 1425(1981).
119.    W.E. Spear, J. Noncryst. Solids 1, 197(1969).
120.    M. Silver, E. Snow, and D. Adler, Solid State Comm.
        51, 581(1984).
121.    W.E. Spear and H. Steemers, Phil. Mag. B 47, L77(1983).
122.    M. Silver, E. Snow, M. Aiga, V. Cannella, R. Ross,
        Z. Yaniv, M. Shaw and D. Adler, J. Noncryst. Solids
        59-60, 445(1983).
123.    D. Adler, M. Silver, A. Madan and W. Czubatyj, J. Appl.
        Phys. 51, 6429(1980).
124.    J. Tauc, in Semiconductors and Semimetals, edited by
        R.K. Willardson and A.C. Beer, Academic Press, N.Y.,
        1984, Vol. 21 A, p. 299.
125.    B. Abeles and T. Tiedje, Phys. Rev. Lett. 51, 2003(1983).
126.    D.L. Staebler and C.R. Wronski, J. Appl. Phys. 51,
        3262(1980).

127.   M. Tanielian, Phil. Mag. B 45, 435(1982).

128.   H. Dersch, J. Stuke and J. Beichler, Phys. Stat. Sol. B 105, 265(1981).

129.   R.S. Crandall, J. de Phys. 42, C4-413(1981).

130.   D.E. Carlson, A. Moore and A. Catalano, J. Noncryst. Solids 66, 59(1984).

131.   S. Yamazaki, T. Shiraishi and D. Adler, J. Noncryst. Solids, in press.

132.   C.C. Tsai, J.C. Knights and M.J. Thompson, J. Noncryst. Solids 66, 45(1984).

133.   D. Adler, M.E. Eberhart, K.H. Johnson and S.A. Zygmunt, J. Noncryst. Solids 66, 273(1984).

134.   S.R. Ovshinsky and A. Madan, Nature 276, 482(1978).

135.   A. Madan, S.R. Ovshinsky and E. Benn, Phil. Mag. 40, 259(1979).

136.   D. Adler, Solar Cells 2, 199(1980).

137.   D. Adler and E.J. Yoffa, Phys. Rev. Lett. 36, 1197(1976).

138.   S.G. Bishop and N.F. Shevchik, Phys. Rev. B 12, 1567 (1975).

139.   P. Nagels, R. Callaerts and M. Denayer, in Amorphous and Liquid Semiconductors, edited by J. Stuke and W. Brenig, Taylor and Francis, London, 1974, p. 867.

140.   C.H. Seager, D. Emin and R.K. Quinn, Phys. Rev. B 8, 4746(1973).

141.   D. Emin, Adv. Phys. 22, 57(1973).

142.   D. Adler, H.K. Henisch and N.F. Mott, Rev. Mod. Phys. 50, 209(1978).

143.   S.R. Ovshinsky, in Amorphous and Liquid Semiconductors, edited by W.E. Spear, C.I.C.L., University of Edinburgh, Scotland, 1977, p. 519.

144.   B.A. Khan and D. Adler, J. Noncryst. Solids 66, 35(1984).

145.   A.A. Andreev, Z.v. Borisova, E.A. Bishkov and Ym. G. Vlasov, J. Noncryst. Solids 35-36, 901(1980).

146.   R. Flasck, M. Izu, K. Sapru, T. Anderson, S.R. Ovshinsky and H. Fritzsche, in Amorphous and Liquid Semiconductors, edited by W.E. Spear, C.I.C.L., University of Edinburgh, Scotland, 1977, p. 524.

147.   G. Pfister and H. Scher, Adv. Phys. 27, 74(1978).

148.   B.A. Khan, M. Kastner and D. Adler, Solid State Commun. 45, 187(1983).

149.   G.S. Higashi and M. Kastner, Phys. Rev. Lett. 47, 124 (1981).

150.   T. Thio, D. Monroe and M. Kastner, Phys. Rev. Lett. $\underline{52}$, 667(1984).

151.   J.M. Marshall and A.E. Owen, Phil. Mag. $\underline{33}$, 457(1976).

152.   J.E. Mahan and R.H. Bube, J. Noncryst. Solids $\underline{24}$, 29 (1977).

153.   N.A. Rajdy and M. Green, Phil. Mag. $\underline{41}$, 497(1980).

154.   R.C. Frye and D. Adler, Phys. Rev. Lett. $\underline{46}$, 1027(1981).

155.   B.A. Khan, P. Bai and D. Adler, J. Noncryst. Solids $\underline{66}$, 321(1984).

156.   R.C. Frye and D. Adler, Phys. Rev. B $\underline{24}$, 5812(1981).

157.   R.C. Frye and D. Adler, Phys. Rev. B $\underline{24}$, 4855(1981).

158.   R.A. Street, Adv. Phys. $\underline{25}$, 397(1976).

159.   F. Mollot, J. Cernogora and C. Benoit a la Guillaume, Phys. Stat. Sol. A $\underline{17}$, 521(1974).

160.   S.G. Bishop, U. Strom and P.C. Taylor, Phys. Rev. Lett. $\underline{34}$, 1346(1975).

161.   D. Adler, in Amorphous and Liquid Semiconductors, edited by W.E. Spear, C.I.C.L., University of Edinburgh, Scotland, 1977, p. 695.

162.   D. Adler, M.S. Shur, M. Silver and S.R. Ovshinsky, J. Appl. Phys. $\underline{51}$, 3289(1980).

163.   J.P. deNeufville, R. Seguin, S.C. Moss and S.R. Ovshinsky, in Amorphous and Liquid Semiconductors, edited by J. Stuke and W. Brenig, Taylor and Francis, London, 1974, p. 737.

164.   K. Tanaka, in Structure and Excitation of Amorphous Solids, edited by G. Lucovsky and F. Galeener, A.I.P. Conf. Proc. No. 31, A.I.P., N.Y. 1976, p. 148.

165.   D.K. Biegelsen and R.A. Street, Phys. Rev. Lett. $\underline{44}$, 803 (1980).

166.   D. Adler and S.C. Moss, J. Vac. Sci. Tech. $\underline{9}$, 1182(1972).

167.   D. Adler and J. Feinleib, in Physics of Opto-Electronic Materials, edited by W.A. Albers, Jr., Plenum Press, N.Y., 1971, p. 233.

# FUNDAMENTALS OF AMORPHOUS MATERIALS

Stanford R. Ovshinsky

Energy Conversion Devices, Inc.
1675 West Maple Road
Troy, Michigan 48084

## I.    INTRODUCTION

When I first began to study amorphous materials in the mid 1950's, the field appeared to be as mysterious as hieroglyphics had been to renaissance scholars. While it was taken for granted that amorphous materials had no real significance scientifically or technologically, it was clear to me, even then, that this was a rich, unexplored, and important area of science [for early references see 1-5]. Until then its major thrust was in the ancient art of glass making, and glass meetings devoted inordinate amounts of time to discussing "What is glass?"

Just as there was a Rosetta Stone which allowed the deciphering of the hieroglyphics of ancient civilizations, the following is the key I provided to make the nature of amorphous materials clear and to understand their physical properties.

It is the purpose of this paper to discuss how we broke the code, and how we have applied this insight to the development of an array of new devices, several of which will be described in detail. Such an understanding of our field is not yet widespread. For example, I recently received a book from Professors Yonezawa and Ninomiya [6], both fine scientists. In discussing topologically disordered systems, i.e., amorphous materials, they state, "In this kind of disordered system, long-range order in the atomic distribution is completely broken while the short-range order (... referred to as SRO), is maintained in the sense that the coordination number of each atom remains the same as in the case of a corresponding ordered crystal, although bond lengths and angles in a disordered system fluctuate."

That statement is insufficient and can be misleading since the characteristics of amorphous materials are controlled not only by the fluctuations of bond lengths and bond angles with the consequent loss of periodicity and the establishment of chemical short-range order [7], but also by the following interrelated factors which make up the Rosetta Stone for understanding amorphous materials. First, there is an average coordination number which defines the structural integrity of the material and its gap and is determined only by the chemistry of the constituent atoms; I have called this its normal structural bonding (NSB). Second, it is the deviations from the optimal coordination number, the deviant electronic configurations (DECs), that are essential to the understanding of the important phenomena in amorphous materials [8,9]. It is these DECs which determine the transport properties of amorphous materials and are responsible for the states in the gap. Third, there need not be "corresponding crystal structures," the central dogma of many working in the amorphous field, a leftover from crystalline physics with its inherent dependence on a lattice structure. The ability to design and synthesize a great variety of amorphous materials depends on the fact that many do not have corresponding crystal structures.

There is a subtle but important insight which should be kept in mind. It is that while short-range order and deviant electronic bonding represent distinct configurations whose total energy can be calculated, there is another distinction that reflects a localized region, the total interactive environment (TIE). This TIE depends on a number of factors of which the nearest-neighbor bonding is but one; others include the effects of nearby chemical forces and of electrical charge distribution which are reflected in the overall three-dimensional topology and in the character of the states in the gap. Perturbations of the TIE can occur by excitational processes [10].

It is difficult to understand now, but the absolutist belief of physicists in the dogma of the crystalline lattice as the basis of semiconductor science can be appreciated by tracing the attitude of Ziman, one of the leading figures in solid-state theory. In 1965 he wrote in his well-known introductory book on solid-state physics [11], "A theory of the physical properties of solids would

be practically impossible if the most stable structure for most
solids were not a regular crystal lattice." Later, in 1969, at the
Third International Conference on Amorphous and Liquid Semiconductors, he delivered a paper [12] entitled "How Is It Possible To
Have An Amorphous Semiconductor?" In this talk he proved that,
since there is no regular lattice in amorphous materials, there can
be no band gap, and therefore these materials cannot be semiconductors. Of course, this misses the whole point of the CFO model
with its concept of a mobility gap [13], illustrated in Fig. 1.
Finally, indicating how science progresses, or better, how
scientists progress, Ziman later published another book, Models
of Disorder [14], in which he states, "Condensed-matter physics
has expanded in recent years and shifted its centre of interest to

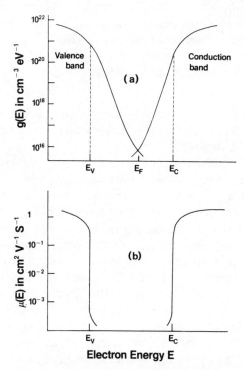

Fig. 1   Sketch of the densities of states of the valence and con-
         duction bands and corresponding electron and hole mobi-
         lities. The magnitude of the mobilities should be regarded
         as approximate because no quantitative calculations have
         been made. States which are neutral when occupied are
         associated with the valence band, those neutral when
         empty with the conduction band; they overlap in the mobi-
         lity gap. (Ref. 13.)

encompass a whole new range of materials and phenomena. Fundamental investigations on the molecular structure of liquids, on
amorphous semiconductors, on polymer solutions, on magnetic
phase transitions, on the electrical and optical properties of liquid
metals, on the glassy state, on metal ammonia solutions, on disordered alloys, on metallic vapours--and many other interesting
systems--now constitute a significant proportion of the activity of
innumerable physical and chemical laboratories around the world."
He continues, "This research is not purely academic: disordered
phases of condensed matter--steel and glass, earth and water, if
not fire and air--are far more abundant, and of no less technological value, than the idealized single crystals that used to be the
sole object of study of 'solid state physics.'" These contradictory
quotes [15] suggest the climate in which we were living when I
first discussed amorphous materials at scientific meetings. While
the situation is much better these days, there are still remaining
misconceptions which this paper will attempt to clarify. In so
doing, we will address the fundamental principles of amorphous
materials.

Kuhn's book [16] should be required reading for the historical and philosophical understanding of how scientific paradigms
are developed. He discusses how anomalies appear in physical
understanding, how the new solutions based upon original thinking
unacceptable to the status quo physicists emerge, and then after
a period of struggle, how a new mind set is generated and a new
field is created. "More clearly than most other episodes in the
history of at least the physical sciences, these display what all
scientific revolutions are about. Each of them necessitated the
community's rejection of one time-honored scientific theory in
favor of another incompatible with it. Each produced a consequent
shift in the problems available for scientific scrutiny and in the
standards by which the profession determined what should count as
an admissible problem or as a legitimate problem-solution. And
each transformed the scientific imagination in ways that we shall
ultimately need to describe as a transformation of the world within
which scientific work was done. Such changes, together with
controversies that almost always accompany them, are the defining
characteristics of scientific revolutions." It is amazing how
Kuhn's description applies to the development of amorphous materials in contradiction to the accepted and dogmatically defended
crystalline approach. I was particularly struck by his description
of how scientists attempt to explain new phenomena by seeking to

extend their conventional approaches to the point of irrationality. This can have unfortunate consequences. For example, what a waste of time it was for us to have to prove over and over that threshold switching was really electronic rather than thermal [17-32]! Shaw [33], who for a time embraced the thermal theory, has recently put the final nail in its coffin.

## II.    CHEMICAL CONSIDERATIONS

It is important to discuss some of the still-remaining misconceptions. It appears puzzling to many that materials composed of exactly the same elements can have completely different structural and electronic properties, depending upon how they are processed. The reason that many amorphous materials are preparation-dependent is that the same elements can combine with each other in a number of different and distinct configurations. The local order actually chosen depends on the nature of the chemical bonding, which in turn is predicated on several factors, including dynamic considerations, for ours is not a chemistry of equilibrium states. The possibility of steric isomerism results in the same elements in different configurations displaying very different chemical reactivities and electronic properties. The internal freedom for placement of atoms in three-dimensional space without long-range order allows for new design possibilities not found in crystals. Indeed, it was stereo- and polymer chemistry that was my guide from the beginning.

Instead of a lattice of repetitive atoms, amorphous solids form a matrix where bonding and nonbonding orbitals with different energies interact in three-dimensional space, sometimes yielding charged centers and thus internal electric fields. The particular bonding option chosen by an atom as it seeks out an equilibrium position on the surface of a growing film is dictated by the kinetics, the orbital directionality, the state of excitation of the relevant atoms, and the temperature distribution during the deposition process. With the constraints of crystalline symmetry and lattice specificity lifted, new internal configurations can and do develop. Rheology plays a role, disclosing important differences between amorphous and crystalline materials, with the former exhibiting unique electron-phonon relaxation processes and pseudo-equilibria.

The chemical foundation of amorphous materials can be clarified by considering, e.g., why and how the carbon atom forms the basis of organic chemistry. For just as the varied bonding possibilities of carbon can generate many different configurations [9,34], even though the same elements are involved, the multi-orbital choices of the elements in an amorphous material can lead to differing configurations. This is the basis for my broadly classifying our materials as synthetic. The difference between amorphous inorganic materials and synthetic organic materials is qualitatively important since we can make not only high temperature, chemically stable, passive materials which in themselves outperform plastics, but we can also make amorphous solids that are electronically active and can be used as switches, memories, transducers, photovoltaic cells, batteries, catalysts, superconductors, etc. Be reminded that there was no such list in 1960.

Directionality of bonding, multi-valences, and varied coordination possibilities, all of which are involved in the offering of multi-orbital choices, become the building blocks for amorphicity. It should not be a surprise that the temperature of the substrate, the state of excitation of the atoms, and the reequilibration kinetics all affect how orbital relationships are formed and how atoms select one another to make up a desired material. Therefore, substrate temperature, orbital directionality, multi-atomic interactions, sticking coefficients, free radical chemistry, and diffusion coefficients are important considerations in how atoms in amorphous materials relate to other atoms and build up their local geometries. These controllable parameters are important assets for they permit us to engineer many new and useful materials as well as being of great scientific value.

It is often asked if there is a fundamental difference between glasses and amorphous materials. The difference is simply that scientists who prepare materials by quenching from the melt make use of a longer time scale than those who deposit atoms on surfaces directly from vapor or plasma phases. Therefore, more equilibrium structures can be expected in glasses. The time and energy required for two atoms to bond to one another can be considered to be design parameters. For example, if there are four outer p electrons, as in a chalcogenide material, but only two bond in the NSB configuration, it is easier to prepare an amorphous material than if one has to bond four outer $sp^3$ electrons, as in elemental amorphous silicon. We can chemically aid the process by making it easier for atoms to bond to each other. How do we accomplish this? It is exceptionally difficult, if not impossible, to form amorphous silicon from the melt (except under laser energization), but it is easy to form amorphous selenium in this way. In the former case, the liquid is not tetrahedrally coordinated

and quickly crystallizes upon quenching. One has to add an inter-
fering additive to prevent this crystallization and, more importantly, one has to bond all four outer electrons to obtain the tetrahedral structure. The rigidity of that structure can be understood by
anyone who has tried to fit four surfaces together. In mechanics,
one must insert a shim or a gib to do so. In stereo- and polymer
chemistry, a "fitting link," either a crosslink or a bridge, is
needed. In elemental amorphous silicon, it costs too much strain
energy to try to bond all four orbitals [35-38] when all must be
distorted to fit the local geometry. The result is that there are
many strained bonds, dangling bonds are prevalent, and voids are
formed in the solid. One would expect this from free energy considerations.

In contrast, in chalcogen elements, only two of the outer
electrons need to be utilized for structural bonding. The remaining lone pair can assume a spectrum of nonbonding or bonding relationships [39]. Consequently, more flexible chain and ring
structures result in the chalcogenides, more rigid structures in the
tetrahedral materials. In both cases, I utilized stereo- and polymer chemistry concepts to control rigidity. In the tetrahedral materials, additional alloying elements are needed to reduce the strain
and lower the average coordination of the structure. They can also act in a bridging manner, like oxygen in fused silica. In the
chalcogenide materials, alloying elements should be preferentially
those that effectively crosslink the material, thus increasing the
average coordination [17,35,36,38-40], and making for more stable
structures; i.e., they should add rigidity. If these alloying and
bridging rules are not followed, then the rapid quench rate that one
achieves, e.g., by sputtering, only leads to the freezing in of
local atomic mismatches and strains. Wherever there are strains
or, more importantly, wherever there are bonding options, DECs
are ordinarily created yielding large densities of localized states
whose origin and significance need to be understood, especially if
one wants to control or eliminate them. For example, the DECs in
elemental silicon are generated by undercoordination; the DECs in
chalcogenide materials arise from the various lone-pair configurations [35,36,39]. We can control them in the former by compensating the dangling bonds, e.g., with fluorine and hydrogen [41],
and in the latter by interacting the lone pairs with modifying elements [8,38,42,43].

Chemical understanding must be translated into specific
topological configurations, since the local geometries reflect the
appropriate chemistry in amorphous materials and structure and

function are indivisible. Grigorovici [44] was early interested in
the structural configurations and internal topology of amorphous
materials. Our work emphasizes the correlation of internal geometries with electronic properties. Surface topology in periodic
materials is related to the lifting of restrictions in the free space
above the surface. Therefore, the study of crystalline surfaces
can be a useful first step in the study of bulk amorphous mater-
rials [45,46]. In fact, the unusual back-bondings at crystalline
surfaces can provide clues of internal bulk configurations of amorphous materials. The TIE is different on the surface than in the
bulk, for the third dimension in the bulk sets up its own chemical
and electrical constraints.

Understanding that the types of defects available in amorphous materials are intimately related to the internal degrees of
freedom unique to noncrystalline solids, one can appreciate that
the defects are really part of the total interactive environment and
part of the energy considerations therein. Defects need not be
only dangling bonds, but can be very similar to the unusual bonding configurations that occur in amorphous chalcogenides or variations of the back-bondings that occur at surfaces. In the same
amorphous material, there can be a whole spectrum of bonds including metallic, covalent, ionic and coordinate [2]. Whether they
appear as defects or not depends upon the particular design of the
material.

## III.    THERMODYNAMIC CONSIDERATIONS

Thermodynamically, if we have a system that has several
possible configurations with essentially equal bulk energies open
to it, depending, e.g., on the temperature distribution, we may
well ask how the atoms developing into a solid choose between
them. I would like to briefly discuss the meaning of metastability
in amorphous solids. How often have we heard that amorphous
materials are metastable? We should bear in mind that so is
diamond! Should we consider tectites as unstable? Amorphous
materials can be very stable indeed. When we want to utilize
their metastability, we do so by design. The understanding of
energy barriers on an atomic scale, as well as on a more macroscopic scale, is a crucial point. As we have shown, the barrier
between the amorphous and crystalline phases can be controlled,
as is sketched in Fig. 2. It is adjustable by altering the bond
strengths of the atoms involved, and it can be lowered or overcome

by external energy sources. For crystallization to occur, there
must be a cooperative action of a large cluster of atoms, but many
subtle changes can occur first. Far more subtle barriers exist than
the one between the amorphous and crystalline phases. Slight
differences in energy can have important influences on the various
conformations and configurations that are inherent in amorphous
materials and the transformations available to them. One internal
structure can be converted into another without affecting important
properties of the material or, for that matter, without even breaking
bonds, as indicated in Fig. 3. (However, the TIEs would be affected.) The closeness of energy of the various conformations and
configurations can be masked by thermal vibrations (phonons) down
to very low temperatures [35]. I interpret the so-called universal, low-temperature, two-level atomic tunneling systems seen in

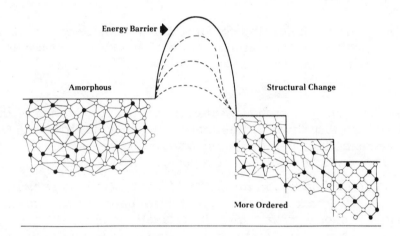

Energy barrier can be reduced by any of the
following-applied singly or in combination:

• Light
• Heat
• Electric field
• Chemical catalyst
• Stress-tension pressure

Transformations in amorphous materials
produce changes in:

• Resistance
• Capacitance
• Dielectric constant
• Charge retention
• Index of refraction
• Surface reflection
• Light absorption,
  transmission and
  scattering
• Differential wetting
  and sorption
• Others, including
  Magnetic
  Susceptibility

Fig. 2    Information storage/retrieval and display by structural
         transformation. (Ref. 50.)

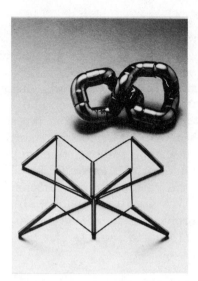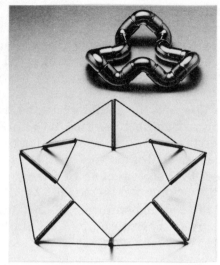

Fig. 3   Models illustrating conformational changes without bond
          breaking--the interconversion of one structural configura-
          tion into another.


glasses and many amorphous materials as direct evidence of the
multi-equilibrium possibilities that I have been describing.  Other
such evidence includes the photostructural changes that characterize both chalcogenide and tetrahedral alloys.

     Rather than postulate that only bond switching is the source
of the specific-heat anomalies which have been viewed as atomic
tunneling phenomena, larger-scale relaxations unique to the disordered and amorphous state could be the most accurate explanation, especially since these represent the conformational changes
discussed in this paper.  Such changes are directly related to variations in the TIE which reposition atoms, ions, and charged as well
as neutral defects, to new positions related to the rest of their environment; i.e., there can be a new TIE as one changes the phonon
concentration.  This is reflected in the character and number of
states in the gap.  From the very beginning of my work in this field,
I have been emphasizing that the coupling between electrons and
phonons is, and must be, basically different in amorphous and crystalline materials.  It is in pursuit of a direct demonstration of this
concept that I have been actively working on superconductors since
the early 1970's [47,48].  I am certain that investigations of the
phonon spectra of amorphous material will some day be one of the
most exciting new areas of scientific research.

Our concept of metastability begins on an atomic level, or, because atoms are not isolated in amorphous materials, rather on a molecular one. Let us assume that a local atomic cluster has been excited by inducing a transition from a low-energy molecular orbital to a higher-energy one, and ask what happens to the TIE? It must change, but how? It will change transiently if the local environment absorbs the excitation energy as it does in the Ovonic Threshold Switch; but if the added energy is dissipated through structural interactions that cannot contain the local conformational changes, as in the Ovonic Memory Switch, then the surrounding structure will disperse the energy in a manner which not only reshapes the conformation with an attendant redistribution of charge but also results in a configurational change, i.e., a breaking of bonds. These configurational changes can be designed to be reversible. There can be a whole spectrum of such changes, including the formation of crystallites. Whether the process is reversible or irreversible is basically a matter of the bond strengths, the size of the crystallites, and the topological and chemical environment. There are not only energy barriers in amorphous materials inhibiting crystallization but also many more subtle barriers involved with atomic and molecular scale changes which are part of the relaxation process unique in amorphous materials.

Reversible amorphization can be pictured as the dissolving of the periodic structure into the surrounding matrix [49,50]. This solute-solvent concept is an apt analogy since it conjures up the picture of precipitating under certain sets of conditions and dissolving under others. Unlike the absorption of energy in a crystal which then propagates throughout the entire lattice, such events in amorphous materials can be very localized. That is why recombination of carriers has more important consequences in amorphous materials than in crystalline. A knowledge of the principles and processes of relaxation, nucleation, and of catalytic effects is necessary for the understanding of crystallization mechanisms in amorphous materials.

Not satisfied with the conventional wisdom that one had to have melting, i.e., a transition to the liquid phase, in order to reach the amorphous state, I proposed the concept of "amorphization" to describe the process of going from an ordered to a disordered system, and placed emphasis on this process occurring from chemical interactions without the temperature having to exceed the melting point although, of course, it may [51]. My theory, which has now been vindicated by many experiments, was that there is a dynamic chemical force tending to bring about the amorphous state, which can represent a configuration equally as attractive as the crystalline one under certain conditions. As an example of what we might call an "anticrystalline" configuration, let us consider a tellurium atom which is initially part of a chain as in crystalline tellurium but is also near an arsenic atom. If energy is supplied to the vicinity of this local area, the resultant displacement of the tellurium atom under consideration can cause it to align its orbitals within the chemical field of the arsenic atom and form a tellurium-arsenic bond. Since this is more stable than the tellurium-tellurium bond which it replaced, absorption of the energy in this case has led to a destruction of the crystal structure. The tellurium-arsenic configuration and, even more, the selenium-arsenic configuration are crosslinked, disordered ones, and can be thought of as anticrystalline. We have shown that there is an analog of the amorphization process in the mechanism of crystallization. If many free carriers are generated by light or electric field, then the relaxation processes favor ordering without the need for melting.

IV.     CHEMISTRY AS A DESIGN TOOL

Right from the beginning of my work in amorphous materials, I have used a chemical approach as a basic design tool. The Periodic Chart of the Elements shown in Fig. 4 [52] has been for me

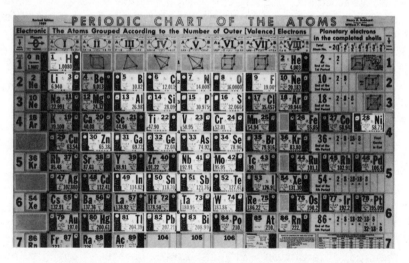

Fig. 4  Periodic Chart of the Elements with examples of the various elements that can be utilized to fabricate amorphous materials or to modify or dope them.

primarily a means of deciding which elements could bond to each other in such a way as to control not only the shape and magnitude of the mobility gap but also the density of localized states in the gap. Since many of our materials are multi-component alloys, this concept can be illustrated by examples which will be detailed subsequently, but whose simple premises follow.

I have emphasized that it is not only the bond strengths but the type and number of crosslinks which control the barrier to crystallization. One can frustrate crystallization by <u>steric hindrances</u>. For a <u>unistable</u> material, we utilize maximum numbers of strongly bonded atoms and crosslinks, e.g., silicon, germanium, arsenic, and oxygen, as the crosslinks for a tellurium-based alloy. For a <u>bistable</u> material, we reduce the bond strengths of the alloying elements and also reduce their concentration, e.g., some or all of the arsenic can be replaced by antimony, which forms weaker bonds, and some or all of the silicon can be replaced by germanium, or even by tin or lead, for the same reasons. A glance at Fig. 4 shows the chemical logic in this method. It also follows that as we reduce the bond strengths, we concomitantly reduce the band gap of the material. The lone pairs in chalcogenides and the various configurations that they enter into control the transport properties of these materials [39]. We utilize small amounts of additional elements in our multi-component materials not only for their spatial, structural and chemical effects, but also for the influence they have on the electronic activity.

From the above, we can see that mere atomic displacements or simple distortions of a crystalline structure do not do the concept of amorphization justice. The two phases of our bistable materials can co-exist at room temperature. The balance can be shifted, from one state to the other. For example, excitation in a memory material can result not only in crystallization but also in a tendency toward the amorphous state because of the chemical forces discussed above. We showed, e.g., that excitation could either inhibit crystallization or expedite it [53]. The outer electron lone pairs of the chalcogens are analogous to the double bonds of carbon in that their possible conversion into bonding-antibonding pairs opens up a host of different configurational structures with nearly the same energy. These new configurations fall into the category of DECs.

Fundamental to my way of thinking relative to the Periodic Chart has been the fact that low average coordination ordinarily favors the amorphous state. In multi-component alloys, the additional elements aid in assuring optimal coordination [9,35]. The balance between adding constraints, completing structures, and assuring rigidity becomes a chemical design parameter as will be seen from our subsequent discussion of tetrahedral materials.

Chemistry and structure are related through the concept of connectivity [38], for lattice constraints limit the ways in which atoms connect to each other, but the different possibilities in three-dimensional space in amorphous materials allow many new geometric configurations. The consequence of this concept, together with free energy considerations, is that there is not just one equilibrium but various equilibria. Structure and function are connected. If one wants to design and define a local order, then the entire local environment, the TIE, must be taken into account. Selective excitation can, in fact, add an important dimension in designing new configurations that would otherwise not be available through the usual thermodynamic considerations. We can also use such electronic pathways and their recombination events in amorphous materials for various memory and photographic applications [2,45,46,49-51,55,56].

By perturbing the TIE, one also perturbs the density of states. It is no wonder that the Staebler-Wronski effect [54] can be understood as an example of a photostructural change [41] instead of appearing as some new esoteric phenomenon. If one excites carriers and recombination events occur in a material which has several different structural relaxations available to it, one can forget the conventional picture of a well-defined density of states [55,57]; one can readily see that there would be a redistribution of the localized states as a consequence of the redistribution of atomic configurations in three-dimensional space. From the beginning of our work we have used electro- and photostructural effects constructively for device applications. As has been shown in our laboratory by Guha et al. [58], recombination is the mechanism which explains the worrisome Staebler-Wronski effect in hydrogenated amorphous silicon alloys.

Fritzsche's "hills and valleys" model [59] puts into perspective some of the consequences of the atomic fluctuations related to the density of states. The charge density of fluctuations in such situations is connected with positional relationships. Therefore, while local chemical bonding is of great importance because it has calculable bond strengths, and therefore short-range order, it does not adequately reflect the true spatial state of affairs of an amorphous solid. The overall positional charge-density

fluctuations in three-dimensional space as well as the nature of the
chemical bonds are an integral part of the total interactive environment.

It is important to reemphasize that in amorphous materials
we do not use the concept of lattice but of matrix [60]. The normal
structural bonding (NSB) makes up the great majority of bonding
configurations, and therefore is responsible for the cohesive energy,
the structural integrity, and the optical energy gap of the material.
This gap, as we have discussed, can be adjusted by alloying, and
is related to the bond strengths of the elements involved. Compositional, positional, and translational disorder inherent in amorphous materials are reflected in the shape and sharpness of the
mobility edge and the density of states in its vicinity. The origin
of the density of states in the gap as well as its control are also
now quite clear. More subtle effects related to states near the
mobility edge itself are still interesting areas of investigation, for
these can act as traps and thus can have important device consequences. I am sure that as research progresses, we will be finding
fine structure near and in the edge itself. We have already been
successful in affecting the sharpness and steepness of the mobility
edge by the choice of materials, the control of impurities and the
generation of intermediate order.

The use of the term "disorder" is unfortunate since it ordinarily means deviations from periodic reference points, but if periodicity is not dominant, then we must substitute our own basic and
specific noncrystalline principles. As pointed out previously, one
can tailor the optical gap by the use of different covalently bonding
elements which also affect the cohesive energy of the material.
The alloying elements can further act as structural crosslinks, assuring amorphicity. Following our rules, one can very specifically
design materials: e.g., to increase the band gap of a tellurium
alloy, add germanium; to increase it further, add stronger-bonding
silicon. Similar increase of the band gap occurs if one substitutes
arsenic for antimony, or adds selenium, sulfur, or oxygen. It is
not unusual for amorphous materials to be multi-component alloys,
with four or more elements. The bond strengths of all the elements
affect and determine the overall gap. In terms of defects, DECs
are generated by the three-dimensional spatial freedom of individual atoms counterbalanced by the chemical and electrical forces
surrounding them, i.e., their environment. Therefore, a silicon
alloy is primarily tetrahedral but its electronic properties, i.e.,
its transport properties, are controlled by the deviations from the

NSB. These DECs are primarily responsible for the deep states in the gap of amorphous materials, and, depending upon their position in energy, can also play a role in the aforementioned shape of the mobility edge. The matrix that we are discussing not only has relaxation modes which are different from a lattice structure, but has a degree of elasticity which becomes an exceedingly important parameter in material design [35,36,38,39,61].

V.      MECHANICAL PROPERTIES

The concept of elasticity is a common theme throughout this paper. For simplicity, consider the fact that as one changes the average coordination by replacing divalent materials in Group VI by tetrahedral materials in Group IV, e.g., silicon, the elasticity decreases. In order to attain necessary elasticity to make useful materials, atoms of lower valence are utilized. In contrast, if we start with divalent materials, we must add crosslinks to assure and control rigidity and stability. If we begin with tetrahedral materials, we add monovalent atoms such as hydrogen and fluorine to decrease the rigidity and to control and assure the tetrahedral structure of the Group IV atoms.

In order to understand how one goes from a flexible to a rigid structure, I proposed that the controlling influence was the network connectivity, which is characterized by a single parameter, the average coordination number C [38]. This average coordination number is related, of course, to the NSB. As one goes from primarily divalent materials, which have the greatest tendency for flexibility and the formation of glass, to tetrahedral materials with the greatest rigidity, the alloying and crosslinking elements that are added accomplish two purposes. They not only play a structural role, e.g., as can be seen in Fig. 5 [39] where the nonchalcogenide elements add rigidity to the solid, but they also provide an increase of the average coordination number. As the average coordination number is increased, the freedom in three-dimensional space is limited by placing a greater number of constraints on each atom; however, just as important, the freedom of chain and ring folding and twisting is also controlled and inhibited. We need not go to more tetrahedral materials to increase the coordination of tellurium; oxygen and/or arsenic can increase both the average coordination and the size of the gap. In tetrahedral materials, there is much strain added as the bonding orbitals seek to complete their configurations. To relieve the strain, one alloys with atoms of a lower valency or with those which tend to form ionic bonds. As coordination is increased, e.g., in elemental amorphous silicon, the alloying atoms play the role of permitting completion of the tetrahedral structure by providing flexibility and electronic compensation. If they did not, DECs would be induced by virtue of the resulting undercoordination, and dangling bonds would be formed. Therefore, elasticity is intimately connected with coordination number: in chalcogenides, crosslinks and bridges play an important role; in a material such as amorphous silicon, alloying reduces the coordination. However, the average coordination of the <u>silicon</u> atoms themselves is increased, e.g., by the addition of fluorine, carbon, oxygen, nitrogen, etc. due to the reduction of the concentration of dangling bonds.

## VI.    PHONONS

I wish to emphasize that phonon activity in amorphous materials differs basically from that in crystalline materials, although there is a wide spectrum and in some materials similarities can

Fig. 5    Model of an Ovonic Threshold Switch illustrating a large amount of strongly-bonded crosslinks assuring stability. The dark balls are Ge, Si, and As atoms. The light balls are Te atoms. (Ref. 39.)

exist. One should start with the simple premise that although crystalline solids only exhibit extended phonons, both localized and extended phonons characterize amorphous solids. There is a tendency for strong localized coupling in the flexible materials and weaker coupling in the tetrahedral materials. The matrix mediates the orbital energies in the divalent materials. The resulting spin pairing usually produces completely diamagnetic material. When the matrix is not deformable enough, such as in an as-deposited elemental amorphous silicon material, the electron-phonon interactions cannot provide the necessary pairing. The relaxations that are inherent in amorphous materials are, therefore, different from those of crystalline materials.

## VII.    MATERIALS SYNTHESIS FOR DEVICE APPLICATIONS

Let us see how these principles actually work in synthesizing materials for device purposes. We will start with the chalcogens and end with tetrahedral materials. As was pointed out earlier, amorphous devices fall into two categories [17]. The first are unistable materials, whose bond strengths and steric hindrances act to prevent crystallization; this class is illustrated in Fig. 5 by an Ovonic Threshold Switch. The crosslinks are numerous and the bonding is strong, and therefore structural changes such as crystallization do not occur within the device operating range. The second are bistable materials, in which there are fewer crosslinks and the bond strengths are weaker so that the barrier to crystallization can be overcome. An example is the Ovonic Memory Switch, shown in Fig. 6. Note how flexible and elastic the chalcogenide bistable memory material is compared to the unistable threshold switch. The average coordination for each is significantly different, $C = 2.3$ in the memory material while $C = 2.9$ in the threshold material. Figure 2 shows how the unique structural changes in amorphous materials, ranging from subtle relaxations to changes of phase including crystallization, become the basis of a whole new field of information and encoding devices, including new types of photography. Especially interesting is the fact that structural reversibility characterizes the more flexible materials so that one can cycle from, e.g., the crystalline state back to the amorphous. These changes of phase can be driven reversibly for more than hundreds of billions of cycles without degradation.

In all materials, we know the origin of the normal structural bonds. We already pointed out that the primary origin of the
DECs in chalcogenides are the <u>lone-pair electrons</u>, either nonbonded or forced by the internal chemical and topological environment to assume a spectrum of bonding states, including one- and
three-electron states [39]. In an important paper, Kastner, Adler,
and Fritzsche [62] further developed this theme to explain the nature of the charged defects that result from these lone-pair interactions. They called the low energy one- and three-fold coordinated defect states <u>valence alternation pairs (VAPs)</u>. It is interesting that these VAPs have the property suggested by the original CFO
model [13], large and equal concentration of positively  and
negatively  charged centers which can act as efficient traps for
excess electrons and holes. The elucidation of the lone-pair nature of the chalcogenides by Kastner [63] allowed us [35,39,64] to
explain why there is no ESR signal in most chalcogenides despite
the typical presence of a high density of states in the gap [65].
The fact that lone pairs are spin-compensated in all their variety of
free or bonded conditions explains the above as well as how one
can have a negative correlation energy. The difference between my

Fig. 6   Model of an Ovonic Memory Switch showing fewer and
         weaker crosslinks and inherent flexibility which permit
         the reversible bistability. The light balls are Te atoms.
         The dark balls are Ge atoms. The darkest balls are Sb
         and S atoms. (Ref. 39.)

explanation [35-37] and those of Street and Mott [66] and of
Anderson [67] is that theirs are based upon disorder as sufficient
for the negative correlation energy and allow unduly for dangling
bonds, and therefore fail to distinguish between the chalcogenide
and the tetrahedrally-bonded materials. My position was, and is,
that it is the inherent flexibility of the divalent state which permits
the lone pairs to have the strong electron-phonon interactions that
are the basis of the induced spin pairing.

Since switching and memory are such basic functions in our
information-oriented society, it is important to note the special
characteristics of the chalcogenide-based Ovonic devices and correlate them with the explanations given above. In the Ovonic
Threshold Switch (see Fig. 7), we see a unique reversible transition between a high impedance and a low impedance state in less
than 120 picoseconds at room temperature. (I have never understood the interest in Josephson Effect switches for computers,
since they require liquid helium temperatures to achieve comparable
switching times.) Such a device is completely independent of
polarity and is made preferably in thin-film form, from less than
0.5 μm to many μm in thickness, depending upon the threshold
voltage required.

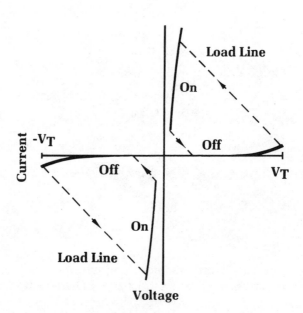

Fig. 7   Current-voltage characteristics of an Ovonic Threshold
         Switch.   (Refs. 17 and 68.)

When I first invented these devices, I called them Quantrols [68,69], since I believed that the switching mechanism was electronic in nature and that such speeds could be observed only if there were a quantum basis for the electronic change of state. From 1960 on, I described the electrical characteristics of these devices [70,71]. In my 1968 paper [17], I emphasized the electronic nature of the switching process, and explained the basis of the mechanisms of both threshold and memory switching phenomena. The application of a high electric field to specifically-designed chalcogenide glasses induces a rapid switching process to a nonequilibrium conducting state followed by injection. The electronic basis of the process has been proven [17-33], and it has many implications both to solid-state theory and to device potential. In the early 1960's, I performed a simple experiment to prove the electronic nature of the phenomenon by adding some selenium to the threshold materials (preserving the high-impedance state even in the liquid phase) and demonstrating switching above the melting point. Obviously, switching therefore was not based on a solid-to-liquid transition.

Now to discuss chemical-topological correlations. As pointed out, the Ovonic Threshold Switch is a heavily crosslinked material with strong bonds, and is therefore unistable; i.e., the electronic excitation does not change the basic structure. The Ovonic Memory Switch is deliberately made with fewer crosslinks and weaker bonds. Referring to Figs. 4-6, it can be seen that, e.g., if germanium is substituted for silicon, or antimony for arsenic, and the nonequilibrium threshold electronic switching effect is used to make the material reactive to electronic excitation, thus weakening or breaking the bonds, the subsequent thermal action which permits cooperative movement of atoms helps induce the memory state. Memory changes can be generated as well as accelerated by diffusion processes [3]. I have called materials which are based upon changes of local order bistable or phase-change materials. A typical current-voltage characteristic of an Ovonic Memory Switch is shown in Fig. 8.

Nowadays, the topic of artificial intelligence is of great interest. I feel that we are completing the grand circle that my wife and collaborator, Iris Ovshinsky, and I originally started in 1955 when we set out to understand the physical basis of intelligence, i.e., information, how it is encoded, switched and transmitted, and the energy transformations connected with it [3,72]. I proposed that this little-understood area of neurophysiology could

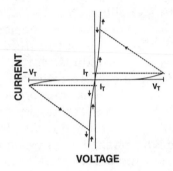

Fig. 8   Current-voltage characteristics of an Ovonic Memory
Switch.   (Refs. 17, 68, and 72.)

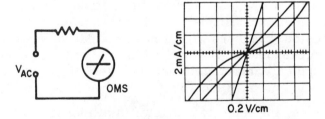

Fig. 9   Ovonic adaptive memory.   (Ref. 72.)

be illuminated by considering that "disorder," i.e., local order, could play a crucial role. I felt that the energy transformations, excitations, and structural changes associated with amorphous materials would be valid models for nerve-cell action, and I built my first nerve-cell switching model and memory to prove the analogy [72]. I was particularly interested in the adaptive memory aspect of my model [73] and have continued with the adaptive memory concept as a learning "machine" in micron thicknesses ever since. To illustrate my work in this area since 1960, consider Fig. 9. We are also pursuing the three-dimensional circuit potential of amorphous materials.

We predicted and observed memory effects in amorphous semiconductors in response to electrical or optical pulses, and associated them with either irreversible changes if the materials were strongly bonded and did not have inherent flexibility, e.g., in amorphous silicon, or reversible changes in cases where more flexible structures were generated by utilizing, e.g., more weakly-bonded divalent alloys such as the chalcogenides. The adaptive memory, therefore, reflects the ratio of the amount of order to the energy input. Before 1960, I showed switching, memory, and adaptive memory action in transition metal oxides [72].

I find it stimulating and fascinating to connect the new cosmological theories with the work being described here since they deal with the same types of problems, i.e., phase changes, supercooling, freezing-in of defects, nucleation, broken symmetry, etc., except that the time scale is a bit different when we are dealing with the origin of our universe (or universes)! Guth [74,75] assumes a liquid-to-crystal analogy whereas I would suggest that asymmetries of the amorphous state and the changes that can occur in it as described herein are more to the point; i.e., in the early transitional phase of the evolution of the universe, the theories of the amorphous phases discussed in this paper are more relevant than those of a crystalline phase. In fact, I believe that my multi-equilibria concept may have some connection and applicability to Guth's general theory. The unity of science is a marvel indeed!

As discussed previously, chalcogenide glasses have low values of the average coordination number. The network is not overconstrained and intermediate-range order is often observed. For nearly pure chalcogens, such as glasses in the Te-Se system, chemical crosslinking is very low. However, "mechanical" entanglements, especially as longer chains and rings are formed, serve as an energy barrier to crystallization, albeit a low one.

Recall our rule that as one generates stronger chemical bonds, the band gap goes up as well; e.g.; sulfur and oxygen both have stronger bonds to tellurium than does selenium, and thus the gap increases progressively as one replaces selenium by either sulfur or oxygen. If one combines several elements, then the bond strengths are averaged and the gap changes accordingly. Depending on the design of the material, especially the use of crosslinks involving particular bond strengths, crystallization can proceed at relatively rapid rates, especially in the presence of some activation such as increased temperature, incident light, or applied electric field. Similar energy input can be used in conjunction with rapid quench rates to return the material to the amorphous phase. Since the two phases are very distinct electrically and optically, and both phases are essentially completely stable at ambient conditions, such materials can be used as the basis for reversible nonvolatile memory systems. Either the crystalline or the amorphous phase can be used as the "zero" memory state. If the amorphous phase is considered the zero, writing can be accomplished by, e.g., applying a voltage pulse to crystallize a filament between two electrodes. As noted previously, chalcogenides ordinarily possess equal concentrations of positively and negatively charged defect centers which act as effective traps for injected free carriers. When an electric field is applied so that double injection takes place, the traps fill. Under these nonequilibrium conditions in an amorphous memory material, the large concentration of carriers weakens the structure, bond reconstructions take place, many covalent bonds are broken, and the rate of crystallization is enhanced (electrocrystallization). Typically, filaments of the order of $1\mu$m can be grown in times of less than 1 ms. The same results can be induced by optical excitation. It is important to point out that in the Ovonic Threshold Switch this filament is composed of carriers originating from nonbonding configurations, and therefore there is no structural change, i.e., no crystallization; in the Ovonic Memory Switch, the electronic threshold switching effect leads to desired structural changes.

A wide array of materials can be utilized for Ovonic memories. These include, but are not restricted to, chalcogenide alloys. Having worked on a particularly attractive chalcogenide system, i.e., tellurium-based materials, since 1960, we have been reporting on it in the scientific literature for many years. While multi-component alloys are ordinarily used, the memory mechanism can be understood by considering a simple example.

For an alloy such as $Te_{83}Ge_{17}$, the eutectic composition for the Te-Ge system, application of a voltage pulse leads to a phase separation into Te-rich and GeTe-rich regions. Since tellurium crystallizes even at room temperatures, the Te-rich regions quickly form crystallites. Both Te and GeTe under somewhat nonstoichiometric conditions are semi-metals with conductivities over 10 billion times larger than the Te-Ge glass at room temperature. We have found that Te crystallites always grow when sufficient energy is coupled to virtually any Te-based memory glass, which can include O, As, Sb, Pb, etc. The differences in physical properties which serve as the memory mechanism are due to the properties of the Te crystallites on the one hand and the amorphous matrix on the other. In the case of electronic memories, the written filament is highly conductive whereas the unwritten glass is highly resistive. One can also write, and it is often preferable, by amorphizing an originally crystalline film. This has been accomplished in the nanosecond range [76,77]. The memory thus can be easily read by applying a small voltage across the contacts. In the Ovonic memory, we utilize other parameters such as large changes in reflectivity [4,78]. The amorphous-crystalline transition is a completely reversible one, a very important attribute.

To electrically erase, application of a sharp current pulse with a rapid trailing edge is all that is necessary. The electronic effects plus the consequent Joule heating are localized within the conducting filament, while the surrounding medium remains at room temperature, thus quenching the active material and reforming the nonconducting glass. I have found it helpful to consider both the precipitation of the crystallites from the matrix and their dissolving back in from a chemical point of view [45,49]. As I pointed out in 1973 [2], "This is not only a 'melt' condition, but one in which the chemical affinities of the crosslinking atoms aid in the establishment of the amorphous state."

A system designed on this principle acts as a nonvolatile electrically erasable programmable read-only memory (EEPROM), an important link between volatile random access memories (RAMs) and unalterable read-only memories (ROMs). Ours were the first EEPROMs made, and were commercially available in the 1960's and 1970's. Their characteristics have been continually improved since then.

Because both the crystalline and amorphous phases of the material are completely stable at operating temperatures, it is evident that Ovonic memory switching can be used as the basis for

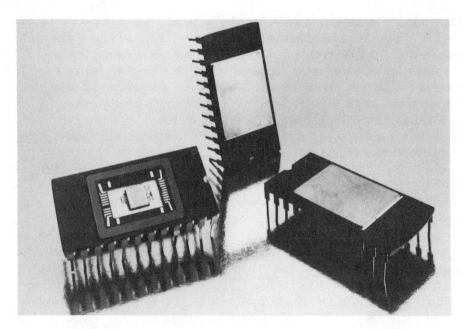

Fig. 10 Ovonic high-speed, high-density Programmable Read-Only Memory (PROM) manufactured by Raytheon.

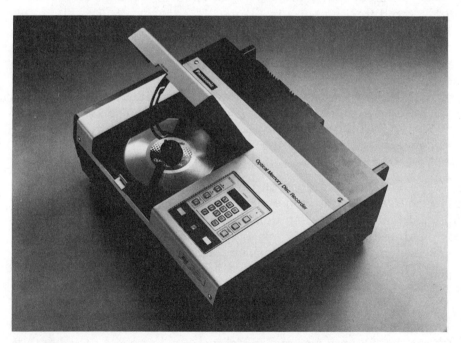

Fig. 11 Panasonic Optical Memory Disc Recorder.

ordinary ROMs and for archival applications by using a write-once mode. If one wants to assure irreversibility, it takes a change of chemistry, following the rules we have outlined, by utilizing stronger bonds than those in the reversible material. Figure 10 shows an Ovonic amorphous silicon-based, high-speed, high-density electrical PROM manufactured by Raytheon. Figure 11 is a commercially available (Panasonic) optical Ovonic memory based upon the chalcogenide crystalline-to-amorphous transition utilized as an optical PROM.

The use of light to induce structural and phase changes has been very rewarding for us. Not wanting to hear again the dreary litany of thermal versus electronic models as the mechanism for changes in amorphous materials, I decided to utilize light to produce new types of optical recording and photographic imaging with unique properties. In addition, I showed that one could use lasers, electron beams, and ordinary light to create information encoding systems, both series and parallel. We were the first to accomplish the laser crystallization of amorphous materials, i.e., to cause crystallization to occur by utilizing a laser interacting with the materials [4,79,80]. I took the position that the simple explanation of melting and recrystallization was not adequate to describe fast laser crystallization. Melting in amorphous materials is the first refuge of ignorance. I proposed that in a material that is unstable to a large amount of excited carriers (initiated by light or electric field), changes of conformation occur and can result in exceedingly fast configurational changes such as crystallization [3,80]. In elemental tetrahedral materials, such relaxations are minor since crystallization requires very little more than eliminating the distortions, which are primarily bond angle changes. In the more flexible chalcogenide alloy materials, bond switching can also take place. This view was supported by much experimental evidence and in a 1971 paper in Applied Physics Letters [81], we stated that "We have observed a high-speed crystallization of amorphous semiconductor films and the reversal of this crystallization back to the amorphous state using short pulses of laser light and evidenced by a sharp change in optical transmission and reflection. This optical switching behavior is analogous to the memory-type electrical switching effect in these materials which has received wide attention since the observation by S.R. Ovshinsky of both threshold and memory switching in amorphous semiconductors. ...we propose a model which closely relates the optical and electrical switching behavior, and shows that the phase change from amorphous to crystalline state is not only a thermal phenomenon but is directly influenced by the creation of excess electron-hole carriers by either the light, or, for

the electrical device, by the electric field. The reversibility of
the phenomenon in this model is obtained through the large difference in crystallization rates with the light on or off."

The idea of optical mass memory systems, for example,
with the entire contents of a library stored on several disks, is a
very appealing one, but was not seriously considered prior to our
work on reversible phase changes, such as amorphous-to-
crystalline or crystalline-to-amorphous transitions. That work
opened up the possibility of optically writing, erasing, and reading
via, e.g., the use of a laser which could be focused down to a
1-$\mu$m spot size. Resolution of this order of magnitude could provide information storage capacities of $10^8$ bits/cm$^2$ and dramatically higher densities with the use of electron beams. Present-day
technology allows the storage of about 250,000 pages of information on a single video disk, an even greater bit density. We
showed that a number of multi-component alloys, such as glasses
in the Te-Ge system, had very different reflectivities from the
same material in its crystallized form, irrespective of the particular components of the alloy. Similar properties can be attained
with a multitude of other amorphous alloys as well. One can optimize specific properties by varying the composition. With the use
of antireflection coatings tuned to either the crystalline or amorphous phase, the phase transition then provides many orders-of-
magnitude changes in transmission upon writing and erasing.

My collaborators and I showed [81] that an ordinary laser
pulse could both crystallize and amorphize a spot less than 1 $\mu$m
in diameter in under 1 $\mu$s, and that the same or another laser could
be used to read in either a reflection or a transmission mode. We
were operating optical disk systems based on these ideas in the
1960's and early 1970's. Exposure of the glass to a laser beam
with characteristic frequency greater than the energy gap excites
large concentrations of electron-hole pairs. These can have
several effects including recombination and trapping by the charged
defect centers, resulting in large densities of bond switching and
broken bonds. Under these conditions, crystallization proceeds
at an extremely enhanced rate (photocrystallization). This process
is similar to the electric field-induced crystallization discussed
previously. In the crystallized form, the material is more light ab-
sorbant than the glass. Exposure to the same laser beam thus can
transfer an increased amount of energy to the written spot, returning
it to the disordered state. Since the surrounding matrix is unaffected
by the focused laser beam, it serves to provide the proper thermal

as well as chemical environment, quenching in the anticrystalline configuration and thus reforming the glass. Consequently, either writing or erasing can be accomplished by the same laser pulse.

We have continually improved the parameters of reversible optical data storage disks, obtaining sharp increases in resolution, contrast, and lifetime. In addition, optical memory techniques other than amorphous-to-crystalline transitions have been developed. One such technique [81] uses the self-focusing property of many glasses to rapidly nucleate a vapor bubble at the interface between the chalcogenide glass and an inert transparent layer. Self-focusing occurs whenever the energy gap of the glass decreases with increasing temperature, a common phenomenon. If the laser has a characteristic frequency very near the energy gap at room temperature, a small amount of laser-induced heating just below the interface will cause ever-increasing absorption in the same region, rapidly nucleating a small bubble. The bubble scatters light effectively, enabling the spot to be read easily. The entire memory can not only be laser-erased but can be block-erased by gentle heating with an infrared lamp. The advantages of such a system are smaller spot sizes (and thus higher resolution), faster write times, and lower energy cost per bit.

Other ideas conceived by us for optical memory applications include photostructural changes such as photodispersion (utilized in our MicrOvonic File [60]), photodoping, photodarkening, and holographic storage [82]. There is now no question that the much-needed mass memories of the near future will be optically written, erased, and read. Finally, we note the present-day importance of laser crystallization of amorphous materials as an example of how new areas of technology can spring from basic scientific investigations. This use of amorphous solids as a vital step in preparing improved crystallites was not a subject of general scientific investigation until we demonstrated such phase changes. Laser printout is now accepted as a matter of course. We were the first to utilize lasers for such applications [2,4,46,51,53]. Figure 12 shows a printout of an early laser copying and printing demonstration. As the old Chinese proverb teaches, one should always leave a golden bridge of retreat so as not to humiliate one's opposition. In this vein, it is relevant to emphasize how the use of amorphous materials has proved to be crucial in the understanding, control, and operation of crystalline MOS devices. More and more crystalline scientists and technologists are appreciating the value of amorphous materials. To me, it has been a needless controversy since the understanding of disorder illuminates the inherent deviations from order in crystalline materials.

Looking farther in the future, still higher capacity memories will be essential. For such purposes, only x-rays, electron beams, or ion beams can yield the necessary resolution. The most promising technique at present involves the use of electron beams. Recent advances in electron optics suggest that 1000Å beams will soon be available, and even 100Å beams are a possibility. In the early 1960's, we showed that electron beams can be used to either crystallize or amorphize alloys. Furthermore, the crystalline and amorphous phases are quite distinct with regard to secondary-electron emission, so that the memory can be easily read by the electron beam. If 100Å resolution can be achieved, about $10^{15}$ bits of information can be stored on a 30-cm disk, more information than is contained in the books in all the libraries of, e.g., a highly literate country such as Japan.

The Ovonic memory concept forms the basis for preparation of many types of instant, dry, stable, photographic films with unique amplification, high resolution, and gradation of tones. This is accomplished by varying the fraction of the glass which has been crystallized and the grain size of the crystallites [45,46,55]. ECD has produced an array of films with either ultra-high contrast or exceptional continuous tones for imaging applications. Additional flexibility arises from the fact that the image can be obtained either directly after exposure, as discussed previously, or in latent form, to be developed subsequently when desired. One mechanism for

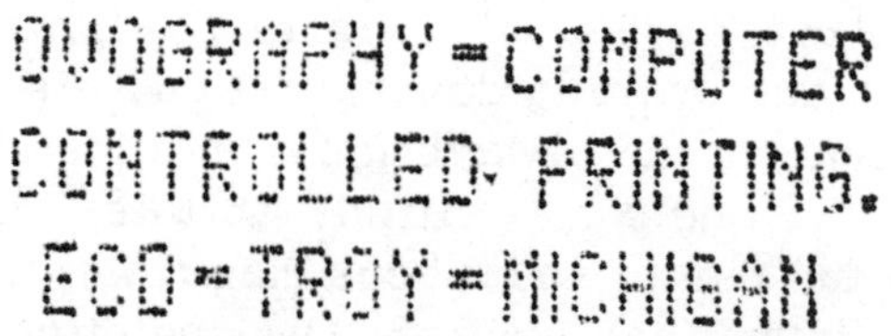

OVOGRAPHY = COMPUTER
CONTROLLED PRINTING.
ECD = TROY = MICHIGAN

Fig. 12  Electrostatic printout obtained with ECD photostructural film printer. Each of the typewriter-size characters were generated on a 5x7 matrix by computer tape. (Ref. 4.)

the latter approach is to use our proprietary organo-tellurides as the film material [83]. In this case, exposure to light induces nucleation centers which form the latent image. Subsequent annealing above the glass transition temperature then induces crystallization of the latent region which produces the desired image (see Fig. 13). Excitation also permits the diffusion of tellurium. Using these procedures, we have been able to attain significant amplification factors.

In addition to using the crystalline-to-amorphous/ amorphous-to-crystalline transitions, we have developed materials in which local structural changes can be induced and detected optically. These have proven useful in updating or correcting images well after exposure. While the materials described here are of the instant dry development type, an exciting feature by itself, we have also designed materials which have excellent etching properties. These have been used for high resolution masks (see Fig. 14) and other photographic applications [2].

## VIII.   CHEMICAL MODIFICATION

It was taken for granted that in amorphous materials certain important parameters were in lock step with each other, e.g., if one had a large band gap material, low electrical conductivity would necessarily result. I decided to challenge this dogma by showing that amorphous materials could be chemically modified, and that by controlling the states in the gap one could for the first time independently control the conductivity changes over many orders of magnitude (see Fig. 15). What was so exciting about these results was that we could obtain large conductivity changes in elemental materials and in alloys containing elements from Group III through Group VI, including materials with drastically different band gaps [8,9,42,43,84]. (In the Periodic Chart of the Elements, Fig. 4, various atoms are darkened to show most of those used in the modification process.) In many cases, a small amount of modifier could <u>increase</u> electrical resistance, while larger amounts <u>decrease</u> it.

I had previously shown that lithium could achieve the same effect in chalcogenide glasses [38]. This was during the same period of time that, following the work of Chittick et al. [85], Spear and LeComber [86] were demonstrating the possibility of substitutional doping in "amorphous silicon." (There still is a question about the effectiveness of p-doping in amorphous Si-H alloys [8].)

Fig. 13    Ovonic nonsilver photo-duplication film.    (Ref. 61.)

Fig. 14    Ovonic Continuous Tone Imaging Film exposed through a
high resolution test mask, exhibiting a resolution in excess of 1200 line pairs per millimeter.

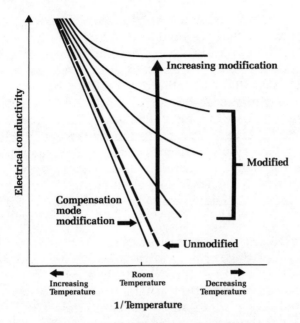

Fig. 15  Effect of chemical modification on the
electrical properties of amorphous
films.  (Refs. 8 and 43.)

| Host Material | Active Modifier |
|---|---|
| Ge Te Se As | Ni, Fe, Mo |
| As | Ni, W |
| $B_4 C$ | W |
| Si C | W |
| Si | Ni, B, C |
| $Si_3 N_4$ | W |
| BN | W |
| Te $O_2$ | Ni |
| Ge | Ni |
| Si $O_2$ | W |
| Ge Se$_2$ | Ni |
| Se$_{95}$ As$_5$ | Ni |

Fig. 16  Chemical modification of amorphous
semiconductors.  (Refs. 8 and 43.)

The fact that we could alter the conductivity of such a large variety of materials showed that we could outwit equilibrium and design a whole new family of materials with characteristics heretofore considered impossible. To put this in historical perspective, note the paper of Hamakawa [87], which states, "the electrical properties of chalcogenide glasses could not be controlled so widely before the sensational appearance of 'chemical modifications' proposed by Ovshinsky." While I appreciate the statement, I wish to re-iterate that my paper on modification covered elements and alloys from Columns III-VI in the Periodic Chart and was not limited to chalcogenides [8,84]. I was very pleased that Davis and Mytilineou corroborated chemical modification in amorphous arsenic with nickel as the modifier [88].

Figure 16 shows typical materials that were modified, and it can be seen that the various active modifiers are either d-orbital or multi-orbital elements. The d-orbitals act as "pin cushions" when co-deposited so that they interact with the primary elements being deposited in a manner so as to create new TIEs. These TIEs would not exist if the modifying elements were deposited conventionally [8,42]. I believe the achievement of modification proves my point about multi-equilibria, since the normal structural bonds need not be affected at all by the modifying element (although they can be, if desired), i.e., the optical gap remains the same while the electrical conductivity can increase by over 10 orders of magnitude. It should be quite clear that the three-dimensional freedom of the amorphous state permits unusual and stable orbital interactions of a highly nonequilibrium nature. As can be seen from Fig. 16, various multi-orbital elements can be used and new nonequilibrium TIEs can be generated even without cosputtering, since the very fact that they are multi-orbital permits several different configurations. We have also utilized excitation as a means of having an atom or molecule enter into and interact with the matrix in such a manner as to effect modification. It is of interest that we have accomplished modification through dual nozzle melt spinning as well [89]. It should be kept in mind that the quenching process itself is a method of achieving nonequilibrium configurations.

In our technique, substantial concentrations of an appropriately chosen modifier are introduced into the amorphous network in a nonequilibrium manner so that it need not enter in its "optimal" chemical configuration. The modifier in small amounts can decrease electrical conductivity, but in larger amounts ordinarily increases it. When the concentration of the modifier exceeds that of the intrinsic defect centers, the Fermi level begins to move. In other words, in small concentrations, the modifiers can compensate and convert positively charged DECs to negatively charged ones, or vice versa; however, in larger concentrations, the modifiers yield many more DECs than would have been present in an equilibrium material. Therefore, the chemical modifier alters the localized states in the gap that control transport, while alloying alters the optical gap without changing the transport properties significantly. We therefore can independently separate the electrical activation energy from the optical gap and control them individually [90]. In a sense, an alloying element is also a modifier since it modifies the overall band gap, but I have used the term chemical modifier to describe situations in which transport or active chemical sites are the properties of interest, for in such cases the purpose of modification is to alter the localized states within rather than the positions of the mobility edges.

I utilized surface chemical modification during the 1950's when I was mostly working with oxides, particularly those of the transition metals; I used amphoteric atoms and ions to change the conductivity by over 14 orders of magnitude [70], utilizing such interactions to design switches and memories, both digital and adaptive [3,72].

During the early 1960's, I investigated many amorphous and disordered phases, combining primary atoms with many types of alloying elements, and was the first to make amorphous gallium arsenide films. Our laboratory also made the first amorphous silicon carbide films [91].

Another method of modification in amorphous materials is doping. Following Chittick et al. [85], Spear and LeComber [86] reported doping experiments on what they considered to be amorphous silicon [92]. As we have pointed out, elemental amorphous silicon is not useful as an electronic material because free-energy considerations lead to an immense density of defects, including dangling bonds and voids. How is one then to utilize the silicon atom in amorphous materials for worthwhile electronic purposes? A means must be found to allow silicon atoms to be connected so that a completed tetrahedral structure ensues. The atoms that achieve such connections must play two roles. First, they must saturate the dangling bonds, i.e., they must be chemical compensators. Equally as important, they must also fulfill the role of structural links which act to provide <u>flexibility</u> to the matrix, relieving the stresses and strains of the pure silicon matrix and compensating it <u>structurally</u> so that the local order retains the electronic properties of the completed silicon configuration. I was therefore dubious about the usefulness of "amorphous silicon" since I thought that in its elemental form it held little electronic interest. When Fritzsche and his colleagues [93] showed that the dopable "amorphous silicon" really contained a large percentage of hydrogen and was therefore an <u>alloy</u>, I was pleased since it meant that my point of view and understanding were justified and correct, and, since alloys were where our talents lay, that we could make superior alloys based upon our chemical and structural concepts. This led me to suggest fluorine as a more suitable element since elements such as hydrogen and fluorine both terminate dangling bonds and at the same time enter into the structural network. I postulated that fluorine, due to its superhalogen qualities, i.e., its extreme electronegativity, small size, specificity and reactivity, not only terminates dangling bonds and can become a bridge, but also <u>induces new local</u> order and affects the TIE by several means, including controlling the way hydrogen bonds in the material since it can bond with silicon in several different ways, some of which produce defects [60]. Fluorinated materials are intrinsically different and fluorine is responsible for new TIEs. Lee, deNeufville and I [94] showed that fluorine does induce a new configuration when combined with silicon and hydrogen. Therefore, in amorphous silicon alloys, the addition of fluorine [41,56,95-97] minimizes defects, including dangling bonds, by generating new beneficial short-range order and TIEs.

I have utilized this concept for other materials such as germanium [97] and was able to solve the problem of "anomalous" density of states of amorphous germanium alloys, which most physicists consider to be tetrahedral materials. They, indeed, are tetrahedral in terms of their NSB; however, they are <u>not</u> in their DECs. There are various divalent and other configurations due to the "inert" lone pairs found in crystalline germanium compounds as well as in those of other elements in Group IV such as tin and lead [98]. Applying our chemical approach, I was able to show that this tendency away from tetrahedralness is even more prevalent in amorphous materials. The lack of tetrahedralness leads to increased DECs and unless compensated for can make germanium-containing alloys inferior as low density-of-states electronic materials. I consider that a very important attribute of fluorine is its

tendency to expand the valence of many atoms by making use of
the orbitals that are within its strong chemical attraction, and it
is therefore particularly valuable where defects are involved with
undercoordination.

The above is particularly relevant since in previous work I
had considered that germanium could be two-fold coordinated [39],
and that silicon under certain conditions in the amorphous state
could also have more than one orbital available that could result in
additional defects, and therefore that fluorine could terminate and
compensate as well the defect states that were not available to
hydrogen. Adler [99], using thermodynamic considerations, has
proposed that two-fold coordination plays a role in the defect centers of amorphous silicon-hydrogen alloys. I felt that in germanium-
containing materials, fluorine would interact with the "Sedgwick"
lone pairs to force germanium into a more tetrahedral structure,
thereby making an intrinsic material with an inherently low number
of DECs. Fluorine also introduces an ionic character to the bonding, helpful in relieving strains. It decreases the fluctuations in
potential on an atomic scale caused by the disorder of the amorphous state. The results are that we now make silicon-based
alloys with a concentration of localized states in the low $10^{15} cm^{-3}$
range, and achieving this quality with our germanium-based materials. By lowering the DEC noise level, we can more effectively
substitutionally dope these materials, and through the use of Raman
spectroscopy we have been able to show that they also have more
intermediate range order. The use of fluorine assures far more
stable amorphous materials and is crucial in making these and
similar materials into superior microcrystalline films [100]. Free
radical chemistry, the leitmotif of my work since the very beginning
in the 1950's, is involved in these processes. It is a subject that
cannot be covered comprehensively here. Suffice it to say that it
plays a very important role in the plasma decomposition processes
which lead to many of the condensed materials that are discussed
here [56]. We have performed experiments which clearly show the
important role that free radicals play in producing better tetrahedral
materials [101].

While amorphous silicon-hydrogen alloys can be substitutionally doped n-type, just as crystalline materials, and boron
doping yields p-type material, the boron doping is not very efficient. Following the chemical arguments of this paper, one can
see why. Instead of being constrained to enter the matrix in a
tetrahedral position, as in crystalline silicon, boron can form

three-center bonds especially with bridging hydrogen atoms, and therefore can generate nontetrahedral structures [8,9]. Having discovered the unusual properties of boron many years ago upon reading Lipscomb's brilliant and profound work [102], I was prepared to apply my understanding of it to amorphous materials, not only to explain boron's difficulty to adequately become a substitutional dopant in materials made from silane but also to utilize its various configurations as one of the elements to achieve chemical modification. Using this approach, we can see why it can be not only a chemical modifier even without co-deposition, because of the many configurations it can assume (this is one of the reasons it is a good glass former), but can also generate unneeded DECs, particularly in an alloy containing hydrogen. In fluorinated materials, due to the increase of intermediate-range order, both n and p substitutional doping is greatly improved. For tetrahedral materials, boron's empty orbital can also be used for coordinate bonding, making use of ordinarily nonbonded lone pairs with very interesting results. As I have pointed out, coordinate bonding can play an important role in amorphous materials [2,8,39]. Boron's natural glass-forming tendencies can be used to good avail since in proper amounts it is an excellent structural element and acts in its own way to affect coordination in a manner that can be as important as hydrogen and fluorine. These attributes give a structural stability to, for example, boron-containing tetrahedral amorphous materials. Boron and fluorine, therefore, are important elements in generating a superior photovoltaic material.

IX.    CONCLUSION

We might well ask, where are we now? We have come a long way in 30 years. Photovoltaic devices based upon the superb electronic properties of amorphous silicon alloys are now in production. Figure 17 shows a commercial tandem cell which has the highest energy to weight ratio of any amorphous photovoltaic device. New ultra-light devices have been developed [103]. Our new generation multi-cell devices are very stable and have solar conversion efficiencies similar to those of our single band gap cells [104] which are over 10% [105], and because of the principles which I have outlined above, we are now making small band gap materials which are approaching the low defect concentrations of our amorphous silicon alloys. This means that we can expect efficiencies as high as 30% when we optimize the different alloys in a three-layer cell.

The information industry, the semiconductor industry, and increasingly the telecommunication industry, all are presently tied to the crystalline structure of silicon. The battle of the electronic giants is taking place on wafers that are now close to their maximum size of about six inches. More and more investment is being made to achieve higher chip densities by photolithographic means. Gordon Moore of Intel, whose expertise in crystalline materials is well known, facetiously proposed that what is needed for the circuits of the mid-1980's and 1990's is an impossible chip the size of the cardboard "wafer" shown in Fig. 18. Since wafers larger than about six inches are not to be expected from the melt-pulling techniques used to grow crystalline silicon, it seemed obvious that this could never be realized, and the industry would be limited to the approximate size of the hand-held chip in the inset. However, the discussion of amorphous silicon alloys in this paper shows that we can make materials that are not only <u>analogous</u> in their circuit functions to crystalline materials but have unique attributes as well. The 1000-foot long, 16-inch wide tandem cell that is shown on the right is a complete photovoltaic cell (electrical contacts are

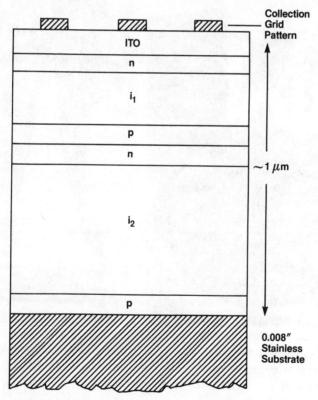

Fig. 17 Schematic cross section of 1.8eV/1.8eV ECD-Sharp Ovonic Tandem Solar Cell. (Refs. 106 and 107.)

later printed on). We are making an amorphous analog of an in-
finitely long "crystal" of any desired width. (The width is a matter only of machine design.) We are, therefore, taking up the
challenge and are designing large-area, totally integrated thin-
film circuits which we believe will soon transform the information
and telecommunication industries.

There are many other areas where I have utilized our synthetic materials approach. Our laboratory has worked successfully for years in fields as diverse as coatings, batteries, catalysts,
electrochemistry, hydrogen storage, thermoelectrics and superconductors. We are already in production in a number of these
areas. These fields are very important not only commercially and
technically but have significant theoretical implications. Obviously, these cannot all be covered here and will be the subject of
another paper; some of these areas have been described elsewhere [108].

Fig. 18  Cardboard model of an imaginary "crystal wafer" on the
left; a typical real wafer is shown in the inset; Ovonic
thin-film semiconducting material, an "infinite crystal,"
on the right.

My intention here has been to discuss the fundamental concepts involved in amorphous materials and to show how our basic understanding can be directly related to the devices that have been and are being developed. Ours is a synthetic materials approach. When one frees oneself from the restrictions of crystalline symmetry, then not only excellent crystalline analogs such as transistors can be made [109], but many new nonequilibrium phenomena and materials can be developed. In fact, the whole dogma of bulk homogeneity can be re-evaluated. I described exceedingly thin multi-layer and compositionally modulated devices in our patent literature years ago. Based on these concepts, we have been successfully developing devices, some of which are being utilized commercially [110]. We have reported in the scientific literature on the unusual effects seen in such materials [48,111-114]. This subject has now been rediscovered [115], and most likely will grow into a new field.

As can be seen, even though we have arbitrarily divided this paper into electronic, chemical, and mechanical sections, one cannot really speak of one parameter divorced from another. The electronic density of states can be the equivalent of active chemical sites, etc. Amorphous materials science is a synthesis of many different disciplines, and therefore has through this process been transmuted into a discipline of its own. Amorphous materials indeed are characterized by the total interactive environment.

The voyage into the amorphous field has been one of discovery and delight, and I take great pleasure in being represented in this institute and in this volume.

## ACKNOWLEDGEMENTS

Since it has been a personal odyssey and began at a time when there were few in the field, I have written this paper reflecting my own travels and travails. However, this work could not have been accomplished without my collaborators and colleagues of many years. As always, this work could not have been done without the partnership of my wife, Iris.

It is impossible to give adequate thanks to all the people with whom I have worked in the past 30 years. In the neurophysiological area where I first started, I wish to express my appreciation for their encouragement to both the late Ernest Gardner, Dean of

Wayne State University Medical School and the late Fernando Morin,
Chairman of the Department of Anatomy there. I wish to also express my thanks to I.I. Rabi, Nevill Mott, and Kenichi Fukui for
their encouragement through the years; also to Boris Kolomiets, an
early and major figure in Russian amorphous chalcogenide work,
for his statement at the IV Symposium on Vitreous Chalcogenide
Semiconductors in Leningrad in 1967, when I first scientifically
discussed switching. He said that this work would transform the
amorphous field whose progress to that date he denoted as a very
slowly-rising, essentially horizontal line to an almost vertical
one, a prediction that fortunately came true. It was at that same
meeting that I met Radu Grigorovici, a friend and important contributor to our field.

From the 1960's on, I benefitted immensely from the col-
leagual collaboration of Hellmut Fritzsche, Morrel Cohen, and
David Adler. I owe David a special debt of gratitude. Our close
association has been of great help to me. I acknowledge with
appreciation the collaboration of Arthur Bienenstock, John deNeufville, Heinz Henisch, Marc Kastner, and Krishna Sapru among
others. The work described in this paper was made possible
through the years by my colleagues, especially Wally Czubatyj,
Steve Hudgens, Masat Izu, and the rest of the superb group of
people that make up ECD.

REFERENCES

1.     D. Adler, Ed., Disordered Materials: Science and Tech-
       nology, Selected Papers by S.R. Ovshinsky, Bloomfield
       Hills, Michigan, Amorphous Institute Press, 1-296 (1982).
2.     S.R. Ovshinsky and H. Fritzsche, "Amorphous Semiconduc-
       tors for Switching, Memory, and Imaging Applications,"
       IEEE Trans. Electron Devices, ED-20, 91 (1973).
3.     S.R. Ovshinsky and I.M. Ovshinsky, "Analog Models for
       Information Storage and Transmission in Physiological Sys-
       tems," Mat. Res. Bull. 5, 681 (1970).
4.     S.R. Ovshinsky, "The Ovshinsky Switch," Proc. 5th Annual
       National Conference on Industrial Research, Chicago,
       86-90 (1969). The Ovographic work was done in collabora-
       tion with P. Klose.
5.     S.R. Ovshinsky, "An Introduction to Ovonic Research,"
       J. Noncryst. Solids 2, 99-106 (1970).

6.   F. Yonezawa and T. Ninomiya, Eds., "Topological Disorder
     in Condensed Matter," Proc. 5th Taniguchi Int. Symp.,
     Japan, 2 (1982).

7.   For a discussion of short-range order, see: A.F. Ioffe and
     A.R. Regel, "Non-Crystalline, Amorphous, and Liquid
     Electronic Semiconductors" in Progress in Semiconductors,
     vol. 4, John Wiley, New York, 237-291 (1960).

8.   S.R. Ovshinsky, "Chemical Modification of Amorphous
     Chalcogenides," Proc. of the 7th Int. Conf. on Amorph.
     and Liq. Semiconductors, Edinburgh, Scotland, 519-523
     (1977).

9.   S.R. Ovshinsky and D. Adler, "Local Structure, Bonding,
     and Electronic Properties of Covalent Amorphous Semicon-
     ductors," Contemp. Phys. 19, 109 (1978).

10.  S.R. Ovshinsky, "The Chemical Basis of Amorphicity:
     Structure and Function," Revue Roumaine de Physique 26,
     893-903 (1981). (Grigorovici Festschrift.)

11.  J.M. Ziman, Principles of the Theory of Solids, Cambridge
     University Press, 1 (1965).

12.  J.M. Ziman, "How Is It Possible To Have An Amorphous
     Semiconductor?" J. Noncryst. Solids 4, 426-427 (1970).

13.  M.H. Cohen, H. Fritzsche and S.R. Ovshinsky, "Simple
     Band Model for Amorphous Semiconducting Alloys," Phys.
     Rev. Lett. 22, 1065 (1969). See also N.F. Mott, Adv.
     Phys. 16, 49 (1967).

14.  J.M. Ziman, Models of Disorder, Cambridge University
     Press, ix (1979).

15.  I thank Dennis Weaire for bringing these contradictory re-
     marks to my attention.

16.  Thomas S. Kuhn, The Structure of Scientific Revolutions,
     University of Chicago Press, 6 (1962).

17.  S.R. Ovshinsky, "Reversible Electrical Switching Phenomena
     in Disordered Structures," Phys. Rev. Lett. 21, 1450-1453
     (1968).

18.  R.W. Pryor and H.K. Henisch, "Mechanism of Threshold
     Switching," Appl. Phys. Lett. 18, 324 (1971).

19.  H.K. Henisch and R.W. Pryor, "Mechanism of Ovonic
     Threshold Switching," Solid State Elec. 14, 765 (1971).

20.  R.W. Pryor and H.K. Henisch, "First Double Pulse Tran-
     sient Study of the On-State (TONC)," J. Noncryst. Solids 7,
     181 (1972).

21.    H.K. Henisch, R.W. Pryor and G.J. Vendura, "Characteristics and Mechanisms of Threshold Switching," J. Noncryst. Solids 8-10, 415 (1972).

22.    R.W. Pryor and H.K. Henisch, "Nature of the On-State in Chalcogenide Glass Threshold Switches," J. Noncryst. Solids 7, 181 (1972).

23.    W. Smith and H.K. Henisch, "Threshold Switching in the Presence of Photo-Excited Charge Carriers," Phys. Stat. Sol. A 17, K81 (1973).

24.    M.P. Shaw, S.H. Holmberg and S.A. Kostylev, "Reversible Switching in Thin Amorphous Chalcogenide Films - Electronic Effects," Phys. Rev. Lett. 31, 542 (1973).

25.    H.K. Henisch, W.R. Smith and M. Wihl, "Field-Dependent Photo-Response of Threshold Switching Systems," Proc. of the 5th Intl. Conf. on Amorph. and Liq. Semiconductors, Garmisch-Partenkirchen, Germany, September 1973, J. Stuke and W. Brenig, Eds., Taylor and Francis, London, 567 (1974).

26.    W.D. Buckley and S.H. Holmberg, "Nanosecond Pulse Study of Memory Material of Different Thicknesses," Sol. State Elec. 18, 127 (1975).

27.    K.E. Petersen and D. Adler, "Probe of the Properties of the On-State Filament," J. Appl. Phys. 47, 256 (1976).

28.    K.E. Petersen, D. Adler and M.P. Shaw, "Amorphous-Crystalline Heterojunction Transistor," IEEE Trans. 23, 471 (1976).

29.    D.K. Reinhard, "Response of the OTS to Pulse Burst Waveforms (Critical Power Density)." Appl. Phys. Lett. 31, 527 (1977).

30.    D. Adler, H.K. Henisch and N. Mott, "The Mechanism of Threshold Switching in Amorphous Alloys," Rev. Mod. Phys. 50, 209 (1978).

31.    D. Adler, M.S. Shur, M. Silver and S.R. Ovshinsky, "Threshold Switching in Chalcogenide-glass Thin Films," J. Appl. Phys. 51, 3289 (1980).

32.    M.P. Shaw and N. Yildirim, "Thermal and Electrothermal Instabilities in Semiconductors," Adv. In Elec. and Electron Phys. 60, 307-385 (1983).

33.    J. Kotz and M.P. Shaw, "Thermophonic Investigation of Switching and Memory Phenomena in Thick Amorphous Chalcogenide Films," Appl. Phys. Lett. 42, 199 (1983).

34.    S.R. Ovshinsky and D. Adler, to be published.

35. S.R. Ovshinsky, "Localized States in the Gap of Amorphous Semiconductors," Phys. Rev. Lett. 36, 1469 (1976).

36. S.R. Ovshinsky, "Amorphous Materials As Interactive Systems," Proc. 6th Intl. Conf. on Amorph. and Liq. Semiconductors, Leningrad, 1975: Structure and Properties of Non-Crystalline Semiconductors, B.T. Kolomiets, Ed., Nauka, Leningrad, 426-436 (1976).

37. H. Fritzsche, "Summary Remarks," Proc. 6th Intl. Conf. on Amorph. and Liq. Semiconductors, Leningrad, 1975: Electronic Phenomena in Non-Crystalline Semiconductors, B.T. Kolomiets, Ed., Nauka, Leningrad, 65 (1976).

38. S.R. Ovshinsky, "Lone-Pair Relationships and the Origin of Excited States in Amorphous Chalcogenides," AIP Conf. Proc. 31, 31 (1976).

39. S.R. Ovshinsky and K. Sapru, "Three-Dimensional Model of Structure and Electronic Properties of Chalcogenide Glasses," Proc. 5th Intl. Conf. on Amorph. and Liq. Semiconductors, Garmisch-Partenkirchen, Germany 1973; J. Stuke and W. Brenig, Eds., Taylor and Francis, London, 447-452 (1974).

40. A. Bienenstock, F. Betts and S.R. Ovshinsky, "Structural Studies of Amorphous Semiconductors," J. Noncryst. Solids 2, 347 (1970).

41. S.R. Ovshinsky and A. Madan, "A New Amorphous Silicon-Based Alloy for Electronic Applications," Nature 276, 482-484 (1978).

42. R.A. Flasck, M. Izu, K. Sapru, T. Anderson, S.R. Ovshinsky and H. Fritzsche, "Optical and Electronic Properties of Modified Amorphous Materials," in Proc. 7th. Intl. Conf. on Amorph. and Liq. Semiconductors, Edinburgh, Scotland, 524-528 (1977).

43. S.R. Ovshinsky, "The Chemistry of Glassy Materials and Their Relevance to Energy Conversion," Proc. Intl. Conf. on Frontiers of Glass Science, Los Angeles, California; J. Noncryst. Solids 42, 335 (1980).

44. For references to his work see Revue Roumaine de Physique 26, No. 809 (1981). (Grigorovici Festschrift.)

45. S.R. Ovshinsky, "Electronic and Structural Changes in Amorphous Materials as a Means of Information Storage and Imaging," Proc. 4th Intl. Congress for Reprography and Information, Hanover, Germany, 109-114 (1975).

46. S.R. Ovshinsky, "Amorphous Materials as Optical Information Media," J. Appl. Photo. Eng. **3**, 35 (1977).

47. S.R. Ovshinsky, unpublished data, 1975; S.R. Ovshinsky and K. Sapru, 1977.

48. The Francis Bitter National Magnet Lab. Annual Report for July 1982 to June 1983, 118.

49. S.R. Ovshinsky and H. Fritzsche, "Reversible Structural Transformations in Amorphous Semiconductors for Memory and Logic," Met. Trans. **2**, 641 (1971).

50. S.R. Ovshinsky, "Optical Information Encoding in Amorphous Semiconductors," Topical Meeting on Optical Storage of Digital Data, Aspen, Colorado, MB5-1-MB5-4,1973.

51. S.R. Ovshinsky and P.H. Klose, "Reversible High-Speed High-Resolution Imaging in Amorphous Semiconductors," Proc. SID **13**, 188 (1972).

52. The reason for the darkening of some of the elements in this figure will be given when we discuss chemical modification.

53. S.R. Ovshinsky and P.H. Klose, "Imaging in Amorphous Materials by Structural Alteration," J. Noncryst. Solids **8-10**, 892 (1972).

54. D.L. Staebler and C.R. Wronski, "Reversible Conductivity Changes in Discharge Produced Amorphous Si," Appl. Phys. Lett. **31**, 292 (1977).

55. S.R. Ovshinsky and P.H. Klose, "Imaging by Photostructural Changes," Proc. Symp. on Nonsilver Photographic Processes, New College, Oxford, 1973; R.J. Cox, Ed., Academic Press, London, 61-70 (1975).

56. S.R. Ovshinsky, "The Role of Free Radicals in the Formation of Amorphous Thin Films," Proc. Intl. Ion Engineering Congress (ISIAT '83 & IPAT '83), Kyoto, Japan, 817-828 (1983).

57. D. Adler, "Origin of the Photo-Induced Changes in Hydrogenated Amorphous Silicon," Solar Cells **9**, 133 (1983).

58. S. Guha, J. Yang, W. Czubatyj, S.J. Hudgens and M. Hack, "On the Mechanism of Light-Induced Effects in Hydrogenated Amorphous Silicon Alloys," Appl. Phys. Lett. **42**, 588 (1983).

59. H. Fritzsche, "Optical and Electrical Energy Gaps in Amorphous Semiconductors," J. Noncryst. Solids **6**, 49 (1971).

60. S.R. Ovshinsky, "The Shape of Disorder," J. Noncryst. Solids **32**, 17 (1979). (Mott Festschrift.)

61. S.R. Ovshinsky, "Principles and Applications of Amorphicity, Structural Change, and Optical Information Encoding," Proc. 9th Intl. Conf. on Amorph. and Liq. Semiconductors, Grenoble, France, 1981: J. de Physique, colloque C4, supplement au no. 10, 42, C4-1095-1104 (1981).

62. M. Kastner, D. Adler and H. Fritzsche, "Valence-Alternation Model for Localized Gap States in Lone-Pair Semiconductors," Phys. Rev. Lett. 37, 1504 (1976).

63. M. Kastner, "Bonding Bands, Lone-Pair Bands, and Impurity States in Chalcogenide Semiconductors," Phys. Rev. Lett. 28, 355 (1972).

64. S.R. Ovshinsky, "Electronic-Structural Transformations in Amorphous Materials - A Conceptual Model," July 13, 1972, unpublished.

65. S.C. Agarwal, "Nature of Localized States in Amorphous Semiconductors - A Study by Electron Spin Resonance," Phys. Rev. B 7, 685 (1973).

66. R.A. Street and N.F. Mott, "States in the Gap in Glassy Semiconductors," Phys. Rev. Lett. 35, 1293 (1975).

67. P.W. Anderson, "Model for the Electronic Structure of Amorphous Semiconductors," Phys. Rev. Lett. 34, 953 (1973).

68. M.P. Southworth, "The Threshold Switch: New Component for Ac Control," Control Engineering 11, 69 (1964).

69. J.R. Bosnell, "Amorphous Semiconducting Films," in Active and Passive Thin Film Devices, T.J. Coutts, Ed., Academic Press, 288 (1978).

70. J.D. Cooney, "A Remarkable New Switching Form," Control Engineering 6, 121 (1959).

71. "How Liquid State Switch Controls A-C," Electronics 32, 76 (1959).

72. S.R. Ovshinsky, "The Physical Base of Intelligence-Model Studies," presented at the Detroit Physiological Society, 1959.

73. E.J. Evans, J.H. Helbers and S.R. Ovshinsky, "Reversible Conductivity Transformations in Chalcogenide Alloy Films," J. Noncryst. Solids 2, 339 (1970).

74. A.H. Guth, "Inflationary Universe: A Possible Solution to the Horizon and Flatness Problems," Phys. Rev. D 23, 347-356 (1981).

75. A.H. Guth and P.J. Steinhardt, "The Inflationary Universe," Scientific American, 116, May (1984).

76.  R.J. von Gutfeld and P. Chaudhari, "Laser Writing and Erasing on Chalcogenide Films," J. Appl. Phys. 43, 4688-4693 (1972).

77.  A.W. Smith, "Injection Laser Writing on Chalcogenide Films," Appl. Optics 13, 795 (1974).

78.  J. Feinleib and S.R. Ovshinsky, "Reflectivity Studies of the Te(GeAs)-Based Amorphous Semiconductor in the Conducting and Insulating States," J. Noncryst. Solids 4, 564 (1970).

79.  S.R. Ovshinsky, presented at the Gordon Conf. on Chemistry and Metallurgy of Semiconductors, Andover, N.H., 1969.

80.  S.R. Ovshinsky, "Method and Apparatus for Storing and Retrieving Information," U.S. Patent No. 3,530.441.

81.  J. Feinleib, J.P. deNeufville, S.C. Moss and S.R. Ovshinsky, "Rapid Reversible Light-Induced Crystallization of Amorphous Semiconductors," Appl. Phys. Lett. 18, 254 (1971). Earlier, Laurence Pellier and Peter Klose worked with me in this area.

82.  J.P. deNeufville, "Optical Information Storage," Proc. 5th Intl. Conf. on Amorph. and Liq. Semiconductors, Garmisch-Partenkirchen, Germany 1973; J. Stuke and W. Brenig, Eds., Taylor and Francis, London, 1351-1360 (1974).

83.  Y.C. Chang and S.R. Ovshinsky, "Organo-Tellurium Imaging Materials," U.S. Patent No. 4,142,896.

84.  "Amorphous Materials Modified to Form Photovoltaics," New Scientist 76, 491 (1977).

85.  R.C. Chittick, J.H. Alexander and H.F. Sterling, "Preparation and Properties of Amorphous Silicon," J. Electrochem. Soc. 116, 77-81 (1969).

86.  W.E. Spear and P.G. LeComber, "Electronic Properties of Substitutionally Doped Amorphous Si and Ge," Phil. Mag. 33, 935 (1976).

87.  H. Okamoto and Y. Hamakawa, "Statistical Considerations on Electronic Behavior of the Gap States in Amorphous Semiconductors," J. Noncryst. Solids 33, 230 (1979).

88.  E.A. Davis and E. Mytilineou, "Chemical Modification of Amorphous Arsenic," Solar Energy Materials 8, 341-348 (1982).

89.  S.R. Ovshinsky and R.A. Flasck, "Method and Apparatus for Making a Modiﬁed Amorphous Glass Material," U.S. Patent No. 4,339,255.

90.   We know that there is controversy as to what constitutes
      an optical gap in amorphous materials, but we feel that our
      chemical examples are quite clear.

91.   E.A. Fagen, "Optical Properties of Amorphous Silicon
      Carbide Films," Silicon Carbide-1973, Proc. 3rd Intl. Conf.
      on Silicon Carbide, Miami Beach, Florida, R.C. Marshall,
      J.W. Faust, Jr. and C.E. Ryan, Eds., University of
      Southern Carolina Press, 542-549 (1973).

92.   For the work of others in the amorphous area, see, for
      example, Science and Technology of Noncrystalline Semi-
      conductors, H. Fritzsche and D. Adler, Eds., Solar Energy
      Materials 8, Nos. 1-3, 1-348 (1982).

93.   H. Fritzsche, M. Tanielian, C.C. Tsai and P.J. Gaczi,
      "Hydrogen Content and Density of Plasma-deposited Amor-
      phous Hydrogen," J. Appl. Phys. 50, 3366 (1979).

94.   H.U. Lee, J.P. deNeufville and S.R. Ovshinsky, "Laser-
      Induced Fluorescence Detection of Reactive Intermediates
      in Diffusion Flames and in Glow-Discharge Deposition
      Reactors, J. Noncryst. Solids 59-60, 671 (1983).

95.   S.R. Ovshinsky and A. Madan, "Properties of Amorphous
      Si:F:H Alloys," Proc. 1978 Meeting of the American Section
      of the Intl. Solar Energy Soc., K.W. Boer and A.F. Jenkins,
      Eds., AS of ISES, University of Delaware, 69 (1978).

96.   A. Madan and S.R. Ovshinsky, "Properties of Amorphous
      Si:F:H Alloys," Proc. 8th Intl. Conf. on Amorph. and Liq.
      Semiconductors, Cambridge, Massachusetts 1979; J. Non-
      cryst. Solids 35-36, 171-181 (1980).

97.   S.R. Ovshinsky and M. Izu, "Amorphous Semiconductors
      Equivalent to Crystalline Semiconductors," U.S. Patent
      No. 4,217,374; S.R. Ovshinsky and A. Madan, "Amor-
      phous Semiconductors Equivalent to Crystalline Semicon-
      ductors Produced by A Glow Discharge Process," U.S.
      Patent No. 4,226,898; S.R. Ovshinsky and M. Izu,
      "Method for Optimizing Photoresponsive Amorphous Alloys
      and Devices," U.S. Patent No. 4,342,044; S.R. Ovshinsky
      and A. Madan, "Amorphous Semiconductors Equivalent to
      Crystalline Semiconductors," U.S. Patent No. 4,409,605;
      S.R. Ovshinsky and M. Izu, "Amorphous Semiconductors
      Equivalent to Crystalline Semiconductors," U.S. Patent
      No. 4,485,389.

98.   E. Cartmell and G.W.A. Fowles, Valency and Molecular
      Structure, Van Nostrand Reinhold, New York, 1970.

99.   D. Adler, "Density of States in the Gap of Tetrahedrally Bonded Amorphous Semiconductors," Phys. Rev. Lett. <u>41</u>, 1755 (1978).

100.  R. Tsu, S.S. Chao, M. Izu, S.R. Ovshinsky, G.J. Jan, and F.H. Pollak, "The Nature of Intermediate Range Order in Si:F:H:(P) Alloy Systems," <u>Proc. 9th Intl. Conf. on Amorph. and Liq. Semiconductors</u>, Grenoble, France, 1981; J. de Physique, Colloque C4, supplement au no. 010, <u>42</u>, C4-269 (1981).

101.  R. Tsu, D. Martin, J. Gonzales-Hernandez and S.R. Ovshinsky, to be published.

102.  W.N. Lipscomb, <u>Boron Hydrides</u>, W.A. Benjamin, New York, 1963. We are honored to be working with him on some of these important problems today.

103.  J. Hanak, to be published.

104.  J. Yang, R. Mohr, R. Ross, "High Efficiency Amorphous Silicon and Amorphous Silicon-Germanium Tandem Solar Cells," to be presented at the First International Photovoltaic Science and Engineering Conference, Kobe, Japan, November 13-16, 1984. Our devices are the only ones that have both high efficiency and great stability.

105.  W. Czubatyj, M. Hack and M.S. Shur, to be published.

106.  S.R. Ovshinsky, "Commercial Development of Ovonic Thin Film Solar Cells," <u>Proc. of SPIE Symposium on Photovoltaics for Solar Energy Applications II</u>, Arlington, Virginia, vol. 407, p. 5-8 (1983).

107.  M. Izu and S.R. Ovshinsky, "Production of Tandem Amorphous Silicon Alloy Solar Cells in a Continuous Roll-to-Roll Process," <u>Proc. of SPIE Symposium on Photovoltaics for Solar Energy Applications II</u>, Arlington, Virginia, vol. 407, p. 42-46 (1983).

108.  S.R. Ovshinsky, <u>Problems and Prospects for 2004, Symp. Glass Science and Technology</u>, Vienna, 1984. To be published in J. Noncryst. Solids. (Kreidl Festschrift.)

109.  Z. Yaniv, G. Hansell, M. Vijan and V. Cannella, "A Novel One-Micrometer Channel Length a-Si TFT," 1984 Materials Research Society Symposium, Albuquerque, New Mexico, 1984.

110.  Ovonyx$^{TM}$ multi-layer x-ray dispersive mirrors.

111.    S.R. Ovshinsky, unpublished.

112.    L. Contardi, S.S. Chao, J. Keem and J. Tyler, "Detection of Nitrogen with a Layered Structure Analyzer in a Wavelength Dispersive X-ray Microanalyzer," Scann. Electron Microscopy II, 577 (1984).

113.    J. Kakalios, H. Fritzsche, N. Ibaraki and S.R. Ovshinsky, "Properties of Amorphous Semiconducting Multi-layer Films," Proc. Intl. Topical Conf. on Transport and Defects in Amorphous Semiconductors, Institute for Amorphous Studies, Bloomfield Hills, Michigan; J. Noncryst. Solids 66, 339-344, H. Fritzsche and M.A. Kastner, Eds., 1984.

114.    S.R. Ovshinsky and M. Izu, "Method for Optimizing Photoresponsive Amorphous Alloys and Devices," U.S. Patent No. 4,342,044.

115.    Proc. Intl. Topical Conf. on Transport and Defects in Amorphous Semiconductors, Institute for Amorphous Studies, Bloomfield Hills, Michigan, 1984; J. Noncryst. Solids 66, 1-392, H. Fritzsche and M.A. Kastner, Eds., 1984.

# THE CONSTRAINT OF DISCORD

D. Weaire*

Physics Department, University College

Dublin 4, Ireland

## I.    INTRODUCTION

Our enigmatic title, <u>The Constraint of Discord,</u> is taken
from a paper by J. Joly [1,2] which was published in 1886 (see
Fig. 1). Joly was interested in using accurate calorimetry to identify minerals by measuring their specific heats. This led him to
consider the difference between crystalline and amorphous forms of
the same mineral. He observed that "it may be stated as true
generally that in the crystal the specific heat is lower than that in
the amorphous state." With a boldness characteristic of Dublin's
scientific community during that period, Joly went on to suggest
fundamental reasons for such a finding. Picturing a solid as consisting of molecules, in accord with the general opinion of that
time, he considered the constraints acting on the molecular vibrations in an ordered and disordered environment. These arguments
predate the development of a proper understanding of structure (by
crystallography) and specific heat (by quantum theory), so they are
hopelessly misguided. But this is, so far as I am aware, the earliest example of the framing of the question: How is the local atomic
arrangement in an amorphous solid related to its properties?

There must be some older examples. After all, essentially
correct ideas regarding the microscopic structure of crystalline

---

*Address from October 1, 1984: Department of Pure and Applied
 Physics, Trinity College, Dublin, Ireland.

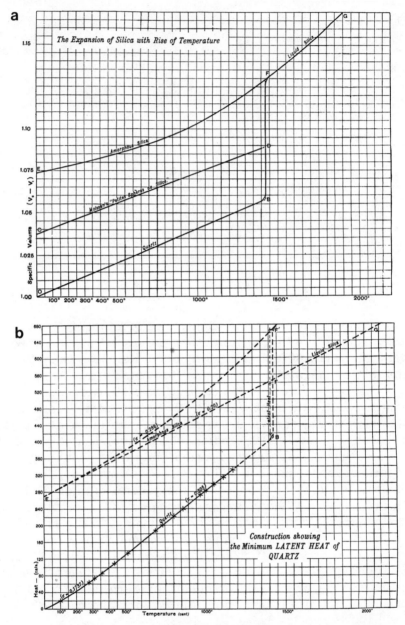

Fig 1     Joly's work on $SiO_2$ sparked off some controversy con-
          cerning the amorphous/crystalline nature of fused silica.
          These data are taken from a paper [2] by Cunningham in
          the Proceedings of the Royal Dublin Society, criticising
          some of Joly's ideas.

solids go right back to the Renaissance [3], and the comparison between crystals and glasses has often been discussed in the intervening centuries.  Cyril Stanley Smith tells a fascinating story of Porcelain and Plutonism [4], concerning the interaction between ideas in geology, glassmaking and fundamental science around 1800.  Much of this involved attempts to identify and distinguish crystals and glasses, but the astute comments of Keir [5] and others were based on observations of morphology, without much speculation on the internal structure underlying it.  Perhaps Newton's admonition that imaginative speculation about such matters was Begging the Question was generally taken to heart.

Joly's work is cited by his equally distinguished Dublin contemporary Thomas Preston in his Theory of Heat [6], but he shortly betrays the difficulty of properly classifying solids, by referring to "amorphous substances, such as glass and iron." Fine-grained polycrystalline metals were often considered amorphous at the time.  MacLean [7] describes how this idea lingered on in odd corners of metallurgy, long after it might have been discarded.  It is ironic that it was finally exorcised around 1960, just before amorphous metallic alloys were actually discovered!

Early in this century, x-ray crystallography swept away much of the prevailing confusion by solving the problem of crystal structure determination.  Nevertheless, the similarity of diffraction patterns from polycrystalline and amorphous solids caused persistent difficulties.  These were well discussed as early as 1934 in a book by J.T. Randall [8], and it has taken a long struggle to improve upon the methods used then.  Some of the hard-won conclusions regarding the crystalline or amorphous nature of particular solids are still the object of skepticism. Moreover, even if the prevailing continuous random models of amorphous solids are accepted, only their broad features (average coordination number, etc.) are directly determined by diffraction [9].  Thus, as far as structure is concerned, we are not really very far ahead of Joly.

Such progress as has been made owes much to model-building.  So, let us look at this next, transferring our attention from 1890 to 1960.

II.    A NEW WAY OF LOOKING

Perhaps we may be excused for choosing another Irishman

as examplar of his period. In the late 1950's, J.D. Bernal set
out to model the structure of liquids by building disordered models
with hard spheres and ball-and-stick constructions. He said that
"All I hope is that I have made a beginning at a new way of looking at liquids, and what becomes of it will lie on whether this beginning is something that can be built on or remains a scientific
curiosity" [10].

There may be isolated earlier examples of this sort of
thing. What is undeniable is that, at the time when Bernal's
idea reached its conclusion in the work of Finney [11], random
model-building was becoming an acceptable way for theorists
to do business. Thus, Bell and Dean [12], Polk [13] and others
built classic models at that time. With hindsight, it seems
strange that Zachariasen [14], who is generally given credit for
proposing the idea of a random network model back in 1932, did
not actually build one!

However, let us return briefly to Bernal's early work,
which was not very well received at the time. It was stimulated
by his fascination with the interplay of order and disorder in
random structures rather than by the fashion of the times, which
inclined towards more formal (but actually very limited) theoretical descriptions of structural correlations. Moreover, as a
model for typical liquids, Bernal's hard sphere packings had a
glaring shortcoming, which gave immediate grounds for dismissal,
namely the large difference (14%) between the density of the
random close-packed structure and the crystalline close-packed
structures, to be compared with experimental observations of
volume changes upon melting which are a few percent at most.

Thus, while much nonsense was talked about crystalline (!) arrangements in liquids, Bernal's very reasonable model
found little favor, for the time being. The resolution of the
problem concerning the density emerged many years later [15],
when the Bernal model was relaxed with reasonable interatomic
interaction potentials. Most of the discrepancy went away, so
that it was seen to be an artifact of hard spheres.

In the meantime, molecular dynamics had provided an
even better method for simulating liquid structure. The ultimate
success of the Bernal model lay elsewhere, in an application not
anticipated by him. The meeting of two graduate students at a
conference led to the realization that the model accords very well

with the structure of many amorphous metals [16]. So, Bernal became (unintentionally) the first builder of random structural models for amorphous solids.

Today model building retains much of the fascination which drew Bernal to it, and the computer offers fresh opportunities for demonstrating and analyzing models (Fig. 2). However, our mathematical framework for understanding them--for quantifying the "constraint of discord"--remains fragmentary. We can bring them into contact with experiment only in clumsy ways. As an example, consider the case of the quasi-linear specific heat of glasses and the supposed underlying "two-level systems" [17]. The phenomenology of this subject is impressive, but there is hardly any understanding of the actual nature of the two-level systems, that is, the local arrangements of atoms which have alternative metastable states, accounting for the anomalous specific heat. How can we identify these? How can we count them?

Many such questions await further inspiration. Two current directions of search for this may be identified. Some have

Fig. 2     A computer-generated picture of part of a random network
           model for hydrogenated amorphous silicon. (Courtesy of
           F. Wooten and Lawrence Livermore National Laboratory.)

chosen imaginative ways of using curved spaces to tame the complexities of random geometry in three dimensions [18]. Others, including the author, have retreated to the study of analogous problems in two dimensions, where life is much easier and there is more hope of progress. After all, even Zachariasen felt forced to sketch a two-dimensional structure in the first place in order to achieve reasonable clarity of thought and presentation.

The problem of two-dimensional random structures, once confronted, has some surprises in store for us.

III.      ORDER AND DISORDER IN TWO DIMENSIONS

If we turn to two-dimensional structures with the motivation stated above, two obvious candidates for study are those shown in Fig. 3. Both are based on soap films. Random arrangements of soap bubbles on a water surface [19] offer an analogue to Bernal packings, while random soap cells are analogous to covalent amorphous solids, since the coordination number of the vertices is fixed. We shall concentrate on the latter case.

Present interest in the 2d soap cell system can be traced back to Cyril Stanley Smith [20]. He, in turn, acknowledged a debt to D'arcy Wentworth Thompson [21], whose classic book On Growth and Form should not go unmentioned here either. Smith's particular interest was in the grain structure of metals; he saw that the equilibrium and evolution of this structure were governed by principles (essentially the minimization of surface energy) analogous to those of the soap cell system. However, he also saw wider implications, viewing the soap cell system as the prototype of a broad class of structures, including cellular structures in biology (Fig. 4). N. Rivier and the present author [22] have tried to review the field glimpsed by Smith, which spans physics, biology, geography, ecology, and materials science (at least). It is both wide and deep, so that what was at first envisaged as a relatively easy problem has acquired a complex life of its own.

What then is the problem with the soap cell system?

IV.      SOAP CELLS

We shall immediately set aside a crucial question: Just what is the precise relation between soap cells and metallurgical

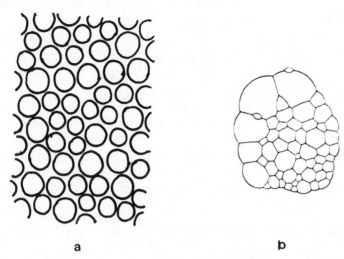

a                                        b

Fig. 3    Soap films can be used to make two kinds of random
          structures which are useful in thinking about amorphous
          solids.  (a)  Soap bubbles on a water surface [19].
          (b)  Soap cells formed by squeezing a froth between two
          plates [20].

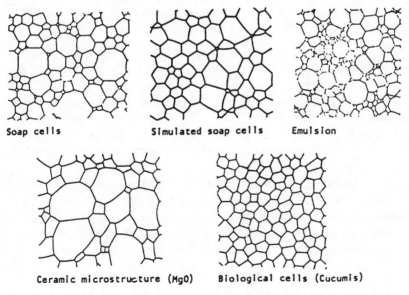

Soap cells          Simulated soap cells      Emulsion

Ceramic microstructure (MgO)    Biological cells (Cucumis)

Fig. 4    Random cellular structures.

grains? This raises too many issues for consideration here and
is unresolved. Let us therefore consider the soap cell system in
its own right.

It is (at least according to Smith's description) simple
enough in principle. The total energy is just the total length of
the cell sides, which is minimized at equilibrium under the constraint of constant cell volumes. However, there is in practice
a slow diffusion of gas between cells, driven by the pressure
differences associated with surface tension. This causes small
cells to shrink, while large ones grow. Cells continually shrink
to zero area and disappear, being replaced by a vertex. This
causes the overall scale of the structure to increase since the
density of cells is decreasing. The effect is analogous to grain
coarsening in metallurgy.

Curiously, the rate of change of area of an individual
cell at any instant is given (on the basis of reasonable physical
assumptions) by a simple equation which at first escaped attention. This is

$$\frac{dA}{dt} = k(n-6) \qquad (1)$$

where A is the area of a given n-sided cell. The constant k is the
same for all cells. This was pointed out by no less a person than
Von Neumann [23], but it is nevertheless quite elementary.

That is not to say that the evolution of the soap froth, which
is driven by the above growth law, is at all trivial. On the contrary, its asymptotic fate is quite unresolved. Smith's original
work [20] suggested that, after an initial period of adjustment,
the structure at all later times was (in a statistical sense) identical, with only a change of scale due to the continued loss of
cells. This was contradicted by the analysis of Smith's own later
data by Aboav [24]. Also, while Smith suggested that the average
diameter d varied (asymptotically) as the square root of time t,
Aboav found d ~ t. There is no theoretical justification for either
conclusion. When we add to this the corresponding questions for
the metallurgical grain growth problem (also unresolved), we are
confronted with an intriguing nest of related puzzles [22].

In the absence of a theory of this and other aspects of
soap froth structure, Weaire and Kermode [25,26] set out to simu-

late the system by means of a computer program. This work has been successful, but has not yet been adapted to large enough mainframe computers to probe the asymptotic behavior definitively (see Fig. 5).

The essence of the simulation program is simple enough-- the equilibrium configuration is found by an iterative technique, and the cell areas are then changed according to (1). The new equilibrium configuration is found, and so on. One snag lies in the discontinuous nature of the rearrangements involved in both equilibrium and growth. When a cell side vanishes, the four adjacent cells rearrange as in Fig. 6.

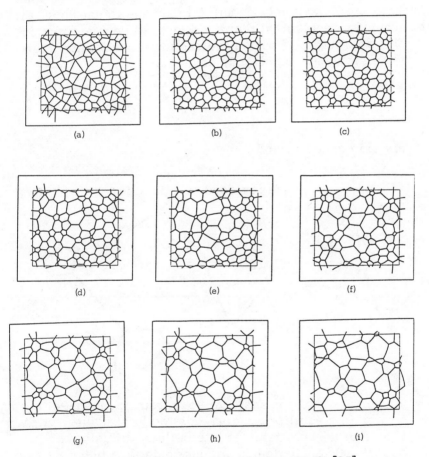

Fig. 5    Evolution of the simulated 2d soap froth [25].
(a) Voronoi network used as a starting configuration.
(b) Equilibrated structure based on (a).
(b) – (i) Evolving structure (at equal time intervals).

This work has been extended [26] to the calculation of stress-strain relations and the associated structural changes in the soap froth (which have been used as an analogue for those in superplastic deformation of metals).

The process shown in Fig. 6 (together with the vanishing of cells) is the key topological event in the evolution of the soap froth and its response to stress. This observation also led Wooten and Weaire [27,28] to develop a method for generating 3d random networks from diamond cubic by similar rearrangements (Figs. 7 and 8). At the level of mere model-building in an ad hoc spirit, this has been quite interesting, and it provides a (conceptually and computationally) simple modus operandi. It does not seem to have previously occurred to anyone to use this, the most elementary topological rearrangement, to generate random tetrahedral structures. Perhaps the useful feedback into three-dimensional problems has already begun....

This side vanishes and

is replaced by this side

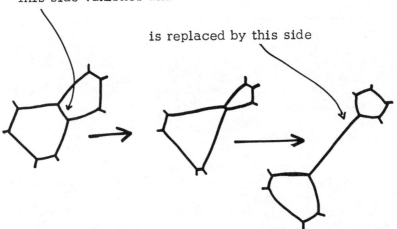

Fig. 6  Whenever a two-dimensional soap froth changes--either in coming into equilibrium from a distorted structure, or in adjusting to the effects of intercellular diffusion-- occasional topological rearrangements take place as shown. These are provoked by the vanishing of a cell side, creating a fourfold vertex which immediately splits into two three-fold ones, different from the previous vertices.

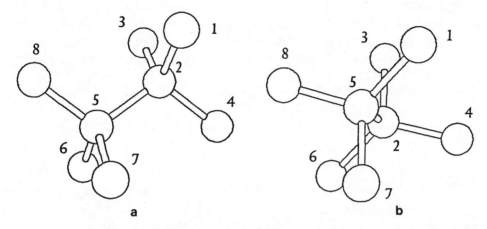

Fig. 7    The bonds of a 3d random network can be rearranged in a
          manner analogous to that illustrated in Fig. 6 for 2d.
          The local arrangement of atoms around a bond in the dia-
          mond cubic structure is as in (a). The rearranged struc-
          ture is shown in (b).

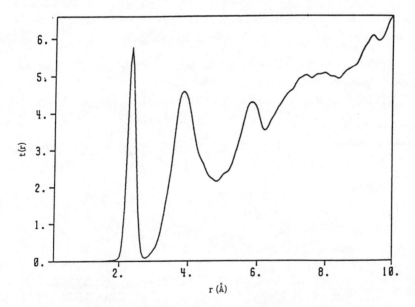

Fig. 8    Radial distribution function (divided by r) of a random net-
          work [27,28] generated from diamond cubic by the process
          shown in Fig. 7.

## ACKNOWLEDGEMENTS

This paper is based on a lecture given at the Institute for Amorphous Studies. I would like to thank Stan and Iris Ovshinsky for their encouragement and interest in the academic side of the study of amorphous solids. Research support by NBST (Ireland) is acknowledged.

## REFERENCES

1.    J. Joly, Proc. Roy. Soc. 41, 250(1886).
2.    J.A. Cunningham, Scientific Proc. Roy. Dublin Soc. 9, 383(1901).
3.    C.J. Schneer, Am. Sci. 71, 254(1983).
4.    C.S. Smith in Toward a History of Geology, edited by C.J. Schneer, MIT Press, Cambridge, MA, p. 317.
5.    J. Keir, Phil. Trans. Roy. Soc. 66, 530(1776).
6.    T. Preston, Theory of Heat, MacMillan, London (1894).
7.    D. MacLean, Grain Boundaries in Metals, Oxford University Press (1957).
8.    J.T. Randall, The Diffraction of X-rays and Electrons by Amorphous Solids, Liquids and Gases, Chapman and Hall, London (1934).
9.    A.C. Wright in Coherence and Energy Transfer in Glasses, edited by Golding and P. Fleury, to be published (1984).
10.   Quoted in M. Goldsmith, "Sage, a life of J.D. Bernal" Anchor Press, London (1980).
11.   J.L. Finney, Proc. Roy. Soc., London, A319, 479(1970).
12.   R.J. Bell and P. Dean, Phil. Mag. 25, 1381(1972).
13.   D.E. Polk, J. Noncryst. Solids 5, 365(1971).
14.   W.H. Zachariasen, J. Am. Chem. Soc. 54, 3841(1932).
15.   D. Weaire, M.F. Ashby, J. Logan and M. Weins, Acta. Met. 19, 779(1971).
16.   G.S. Cargill, J. Appl. Phys. 41, 2248(1970).
17.   W.A. Phillips, J. Low Temp. Phys. 7. 351(1972).
18.   D. Nelson in Topological Disorder in Condensed Matter, edited by T. Ninomiya and F. Yonezawa, Springer-Verlag (1983), p. 60.
19.   A.S. Argon and L.T. Shi, Phil. Mag. A46, 275(1982).
20.   C.S. Smith in Metal Interfaces, Cleveland ASM (1952) p. 65.

21.     D'A.W. Thompson, <u>On Growth and Form</u>, Second Edition,
          Cambridge University Press (1942).
22.     D. Weaire and N. Rivier, Contemp. Phys. <u>25</u>, 59(1984).
23.     J. Von Neumann in <u>Metal Interfaces</u>, Cleveland ASM (1952),
          p. 108.
24.     D.A. Aboav, Metallography <u>13</u>, 43(1980).
25.     D. Weaire and J.P. Kermode, Phil. Mag. <u>48</u>, 245(1983).
26.     D. Weaire and J.P. Kermode, Phil. Mag., to be published,
          (1984).
27.     F. Wooten and D. Weaire, J. Noncryst. Solids, to be pub-
          lished, (1984).
28.     F. Wooten and D. Weaire, to be published, (1984).

# STRUCTURAL STUDIES OF AMORPHOUS MATERIALS

Arthur Bienenstock

Stanford Synchrotron Radiation Laboratory
Stanford University
Stanford, California 94305

## I. INTRODUCTION

This paper is concerned with the progression from the initial studies of atomic arrangements in amorphous materials to those which have become possible recently as a result of the availability of synchrotron radiation. I will tend to concentrate on the work that I know well, and pay less attention to those aspects of structure with which I am less familiar or to which my group has contributed less.

The ambiguities associated with determining the structure of an amorphous material are almost made clear by the dictionary definition of amorphous as "having no determinate form." Indeed, the contrast with crystalline materials is quite striking.

### A. Description of Crystalline Systems

The description of the three-dimensionally periodic structure of a crystalline material is relatively simple, since it is built on a lattice. Hence, the description begins with the lattice itself, using the repeat vectors $\underline{a}_1$, $\underline{a}_2$, and $\underline{a}_3$, which describe the periodicity. These repeat vectors also describe a unit cell which is repeated indefinitely in three dimensions to form the crystal.

To get a more complete description of the atomic arrangement, we must next determine the atomic positions within a single unit cell. The number of atoms within the unit cell, $N_c$, ranges from just a few for simple elements or compounds to the thousands for proteins. Even in the latter case, however, $N_c$ is small compared to the approximately $10^{24}$ atoms which make up a typical crystal or amorphous material. At any rate, the determination of the crystal structure just involves the determination of the $3N_c$ positional coordinates of the atoms within the unit cell. Once the repeat vectors and the positions of the atoms within the unit cell are known, the positions of all the atoms in the crystal can be described simply because of the periodicity in three dimensions.

To understand the structure more deeply, we must examine the relative positions of and the bonding between the atoms. For example, an examination of the crystal structure of selenium would reveal linear chains of atoms in the $\underline{a}_3$ direction. That is, each atom is closely bonded to two atoms, one on each side, to form the chain. In this sort of analysis, one would determine the distances to the near neighbors, the atomic species of the near neighbors, and the number of neighbors. Then, one would examine further neighbors along the chain as well as interchain coordination.

That class of description would be the final part of the description except, perhaps, if one was concerned with defects. In that case, one would ask what are the probable defects. Here, again, periodicity simplifies because one can ask the questions: "What are the most common vacancies?" and "What are the most likely interstitial sites?" That is, one asks the simple question: "Is there an atom missing from its site (which is well defined), or do we have an atom in a site where it should not be?"

B.      Description of Amorphous Systems

All this simplicity is lost in the transition to amorphous materials. The lattice is lost and with it, the ability to describe the position of every atom. To achieve such a description of an amorphous material, we would need three coordinates for every atom in the material. Clearly, that is impossible with $10^{24}$ atoms. Even if one had those coordinates, very little order could be made of them. As a result, one asks simpler questions. In particular, one asks about the average coordination of each atomic species in the material. What is the average distance to the near neighbors? What are the types of near neighbors? How many neighbors are there? These same questions are asked about the second neighbors, and the third neighbors, and so on. Thus, a type of statistical description of the atomic arrangement is sought. This is, in fact, the sort of information one gets from radial distribution functions (RDFs) obtained from x-ray, electron or neutron diffraction.

## II. RADIAL DISTRIBUTION FUNCTIONS

These RDFs may be written as:

$$D(r) = 4\pi r^2 \rho(r)$$

$$= N^{-1} \Sigma_i \Sigma'_j K_i K_j \delta(r - r_{ij}). \qquad (1)$$

Here, each summation runs over all the atoms in the sample with the exclusion, indicated by the prime in the sum over j, of terms with i = j. $r_{ij}$ is the separation between atoms i and j. N is the total number of atoms, while the $K_i$ are weighting factors. For x-rays, $K_i$ is approximately equal to the number of electrons in the atom or ion i. For neutrons, it is the coherent scattering cross section.

D(r) can be visualized in the following manner. Each i-j pair of atoms in the sample contributes two identical terms to the function. Each term is a delta function at r equal to $r_{ij}$, multiplied by the product of the weighting functions, $K_i K_j$. The delta functions add together to form peaks in D(r) at those values of r which are common interatomic distances. Consequently, the positions of these peaks yield the common interatomic separations. Since the number of delta functions contributing to each peak is twice the number of atomic pairs contributing to the peak, the areas of the peaks yield information about coordination numbers.

### A. Amorphous Silicon and Germanium

RDFs are obtained from diffraction data by Fourier transformation of the intensities, after effects like incoherent and multiple scattering have been removed and geometrical phenomena like the polarization effect have been taken into account. The book by Warren [1] has a very good discussion of this process. It will not be treated in detail here. The interpretation of RDFs does, however, call for some analysis.

RDFs are most readily interpreted for elemental materials. One obtained by Kortright [2] for amorphous Ge is shown in Fig. 1. The function is zero for small r because atoms do not approach less than a certain distance. It then rises sharply with increasing r to peak at the nearest-neighbor distance of 2.44A, drops down to zero, and then rises again for distances corresponding to second neighbors, and so on.

The area of the first neighbor peak indicates that the average coordination number is close to 4, its crystalline value. Similarly, the peak position and area for the second neighbors are consistent with each Ge atom being at the center of a regular tetrahedron of Ge atoms. Thus, the RDF shows that the structural picture obtained from crystalline Ge is a good starting point for understanding the structure of amorphous Ge.

Moss and Graczyk [3] have shown clearly, however, that there are serious limitations to the application of crystalline models to amorphous Si (these limitations hold as well for amorphous Ge). They compared the RDF of an amorphous Si film with

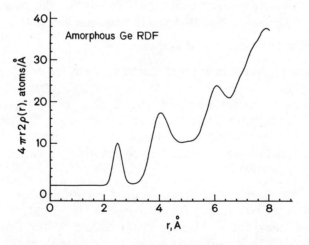

Fig. 1      Radial distribution function for amorphous Ge.  (From ref. 2.)

one obtained from the film after it had been crystallized thermally. They found that the first neighbor peaks in the two materials had both the same position and width. The second neighbor peak in the amorphous material had the same position, but was slightly broader than the corresponding peak in the crystalline material. But the very strong third neighbor peak of the crystalline material, due to 12 third neighbors, was almost completely missing from the amorphous material's RDF.

This was an extremely important paper because it showed that amorphous silicon and germanium are not composed of small crystallites arranged at random. If such crystallites were present, a strong third neighbor peak would have been apparent in the RDF.

It tended, instead, to give credence to a model proposed by Polk [4] for these amorphous elements. Polk constructed a model of these systems physically by linking together the little plastic pieces that are used to define tetrahedral coordination in "stick" models of atomic arrangements. To achieve an amorphous arrangement, Polk introduced odd-membered rings (five and seven), rather than just the six-membered rings that one finds in crystalline germanium, into the structure. The result was a large amorphous array of tetrahedrally coordinated "atoms."

Polk then constructed an RDF by measuring the coordinates of all the "atoms" in the model and then determining the pair distribution function. He found that he obtained the first and second neighbor peaks which are characteristic of the crystal, but that the third neighbor peak was missing from the radial distribution function defined by his model, as it was from the Moss and Graczyk RDF. Polk said that this feature was a consequence of having those odd-membered rings. Thus, this "random network model" was the most successful that we had for interpreting the RDF, and describing the atomic arrangements in amorphous Si and Ge. The success of the model led most of us to believe that such a random network is the most appropriate description of amorphous silicon or germanium.

B.      Amorphous Compounds with Tetrahedral Coordination

Those odd-membered rings created a problem, however. While they may be appropriate for germanium and silicon, they are troublesome for gallium arsenide or gallium phosphide, which can also be made amorphous by vapor deposition. Odd-membered rings in binary, tetrahedrally-coordinated systems imply that there will be like-atom near neighbors. Because the number of members are odd, you cannot alternate plus or minus charges around the ring and not come out with a situation where you have like neighbors. To this day, there is a question as to whether those odd-membered rings and like neighbors appear in films of compounds with tetrahedral coordination like gallium phosphide, gallium arsenide, and indium phosphide. One cannot tell from RDF analysis as the chemical pairs contributing to the first neighbor peak cannot be identified. Since Polk did his work, however, people have made random network models of tetrahedrally coordinated materials which consist entirely of even-membered rings and which also do not show that third peak in the RDF. This example shows us immediately that new questions arise in the interpretation of radial distribution functions and amorphous materials when more than one atomic species is present in the material. That topic will form a major portion of this paper. But, before we leave the elemental case, let us take a look at what we can understand easily of the short-range structures of the elemental amorphous semiconductors. The discussion will be a brief version of what is described elegantly by Rawson [5].

## C.     Elemental Amorphous Semiconductors

The elemental amorphous semiconductors are all on the right hand side of the Periodic Table. Silicon and germanium are in Column IVA, phosphorus and arsenic are in Column VA, and sulphur, selenium and tellurium are in Column VIA. They are all characterized by the fact that they have at least a half-filled outer electron shell and their bonding can be described easily by the 8-N rule, where N is the number of the column of the Periodic Table in which the element appears. In all these elements, each atom completes an outer octet of electrons by covalently bonding and sharing a pair of electrons with each of its neighbors. As a result, the number of atoms to which it is coordinated is 8-N. For example, selenium starts out with six outer electrons. By bonding to two selenium atoms, its outer shell is filled with eight electrons. Similarly, arsenic starts out with five outer electrons and bonds to three other arsenic atoms. Finally, of course, silicon and germanium have four outer electrons and bond to four other atoms to complete their outer shell. This change in bonding going from group VIIA to group IVA has an immense effect on the structure. The VIIA elements form chains or rings. VA elements form two-dimensional layers while the fourfold coordinated atoms in group IVA form structures which are three-dimensionally connected.

## D. Simple Metallic Glasses

There is another group of elemental materials that is relatively easy to understand. Our first picture of it came from Bernal [6] who literally took a spherical container and dropped ballbearings into it until it was filled up. You know that if you drop ballbearings into a flat box, a hexagonal network is formed on the bottom which leads to a "crystal" as the box is more completely filled. A spherical container breaks up that hexagonal network, however, so that the resulting ballbearing arrangement is "amorphous." He then filled the container with hot wax so that the system would hold together, broke the glass apart and measured the positions of each ball using a transit. Later, this approach was simulated on a computer but the first measurements were obtained in this manner.

Bernal found that there were many configurations of atoms which are essentially identical to those found in simple metals or crystals of the rarer gases. There were, in addition, local arrangements where the symmetry was fivefold, or sevenfold, or some other noncrystallographic symmetry.

No one has succeeded in maintaining a metallic or rare gas elemental material amorphous at room temperature. There are, however, many metal-metalloid systems which can be made amorphous through rapid quenching of a melt or vapor deposition. Virtually all of these are characterized by having a phase diagram which shows a deep eutectic at approximately 20 atomic percent metalloid. When the eutectic composition melt is cooled into the deep eutectic region, the system becomes relatively viscous so that it is possible to quench it into the glassy state. Since the metal typically has a significantly larger atomic number than the metalloid, x-rays typically see only the metal arrangement and do not see the metalloids. As a result, a RDF obtained by x-rays shows only the metal-metal distribution function. RDFs obtained in this manner are in extremely good agreement with the RDF function obtained from the hard sphere model discussed above. This is illustrated in Fig. 2, due to Cargill [7] in which a reduced RDF for this dense random packing (DRP) of hard spheres model is compared with an observed reduced RDF for $Ni_{76}P_{24}$. (In the reduced RDF, the radial distribution function is taken as proportional to $r$, rather than $r^2$, and, in addition, the average value of that function has been subtracted out.

It is striking that this beautiful agreement is obtained between a model radial distribution for an elemental system and a physical system which is a binary. It is also striking that almost every metal that can be made by quenching of the melt has about 20 percent of something else in it, and yet we have no pure elemental glasses. Polk [8] addressed that problem too, and came up with a very simple picture. Bernal had noted that there were large numbers of simple polyhedra within his model glass. The largest of these polyhedra have very large spaces at the center and could incorporate smaller atoms within them. That smaller atom is typically the metalloid. If all of these polyhedra holes were filled by metalloids, they would constitute about 20 percent of the glass. What Polk noted was that if the system were pure elemental metal, it would be easy for it to rearrange and form a crystalline material. The presence of the smaller atoms at a composition of about 20 percent tends to lock up the structure and keep it in the glassy state after it has been quenched into that state.

## III.    POLYATOMIC SYSTEMS

The description of a polyatomic system is somewhat more complex than that of an elemental system. Since there is more than one atomic species, it is not sufficient to designate how many atoms there are on the average, at a distance r from a given atom. The atomic species involved in the pair must also be identified. As a result, the radial distribution function is described in terms of partial distribution functions which contain two additional labels, $\alpha$ for the central atom whose coordination is being examined, and $\beta$ for the atomic species surrounding the atom of interest. If, for example, the system being studied was molten sodium chloride, we would be interested in the sodium-sodium, sodium-chlorine, chlorine-sodium, and chlorine-chlorine distribution functions. This is as complete a description of the system as can be obtained from x-ray diffraction. In analogy to Eq. (1), one defines partial radial distribution functions by the equation:

$$4\pi r^2 \rho_{\alpha\beta}(r) = N_\alpha^{-1} \sum_{i=1}^{N_\alpha} \sum_{j=1}^{N_\beta} \delta(r - r_{ij}). \tag{2}$$

One major problem with x-ray diffraction is that these individual partial distribution functions cannot be measured directly. Instead, one measures a normal RDF which is a weighted sum of these partial distribution functions. It is:

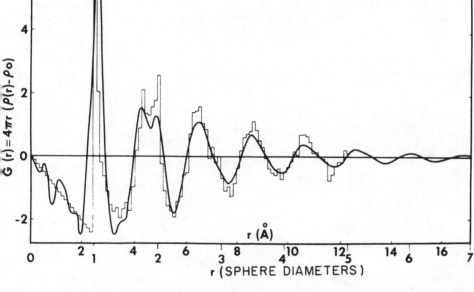

Fig. 2    Comparison of reduced radial distribution functions G(r)
for Finney's DRPHS structure and for amorphous $Ni_{76}P_{24}$.
(From ref. 7.)

$$4\pi r^2 \rho(r) = 4\pi r^2 \sum_{\alpha} \sum_{\beta} x_\alpha Z_\alpha Z_\beta \rho_{\alpha\beta}(r). \tag{3}$$

Here, $x_\alpha$ is the concentration of species $\alpha$. Each pair is weighted by the product of the atomic numbers, $Z_\alpha$, in the pair. That is because the scattering factor of each atom is roughly proportional to the product of the number of electrons. Since this weighted sum is measured, very little information about the individual partial distribution functions is obtained.

Nevertheless, scientists have gone a long way in determining arrangements in polyatomic systems. Some of the most beautiful work was performed by Warren and coworkers [9] many years ago on silicon dioxide, a more recent version of which is shown in Fig. 3. Warren was able to take this radial distribution

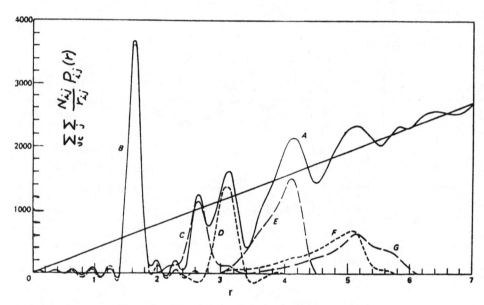

Fig. 3    Radial distribution function for amorphous $SiO_2$. A is the measured curve, B is the sum of the calculated curves for the first six contributions: Si-O, O-O, Si-Si, Si-2nd O, O-2nd O, and Si-2nd Si. C is the difference between A and B. [B.E. Warren and R.L. Mozzi, Acta Cryst. 21, 459(1966).]

function and show, first of all, that the shortest distance is a silicon-oxygen distance, the next distance is an oxygen-oxygen distance, and the third distance is a silicon-silicon distance. He could do that quite readily and easily for two reasons. First of all, the silicon-oxygen distance at about 1.4 Ångstroms is much shorter than any common silicon-silicon or oxygen-oxygen distance. The oxygen-oxygen distance is exactly what it is in many silicon dioxide crystals, and corresponds to a tetrahedron of oxygen atoms around each Si, as is found in crystalline $SiO_2$. Similarly, the silicon-silicon distance was closely related to what it is in a particular phase of $SiO_2$, cristobalite. He had two features working for him in this study. One was that the inter-atomic distances of interest were quite distinct. That is, the silicon-oxygen distance is quite a bit less than the oxygen-oxygen distance which is clearly less than the silicon-silicon distance. The other is that the structure of the amorphous material is very close to that of the structure of a common crystal of the same material. That allowed him to explore the structural chemistry of amorphous silicates extensively, and build up understanding considerably.

The short-range atomic arrangements in many other amorphous compounds have been determined in a similar way using RDFs. For example, that arrangement in amorphous $As_2Se_3$ seems almost identical to that in the crystalline compound. As a result of these studies, we have a fair picture of what is happening in many of these systems. In each of them, the 8-N rule is operative. For example, in $SiO_2$, the silicon with four outer electrons is surrounded by a tetrahedron of four oxygens, and the oxygen with six outer electrons is bonded to two silicons. Similarly, in $As_2Se_3$, the arsenic with five outer electrons is bonded to three seleniums, and the selenium with six outer electrons is surrounded by two arsenic atoms.

In the late 1960's, however, we started to have difficulty understanding the RDFs of some chalcogenide glasses made at ECD. One of the first systems that we examined was the Ge-Te system. It was interesting then because we were trying to understand why the conductivity of the material changed so much when it was crystallized. This simple system which Betts, Ovshinsky and I [10] studied was an "academic version" of some of ECD's memory materials. Figure 4 shows the RDF function of $Ge_{.11}Te_{.89}$ and then $Ge_{.5}Te_{.5}$ (GeTe). These radial distribution functions had a number of features that were unusually interesting and which they shared with Ge–S and Ge–Se amorphous alloys. First of all, the radial distribution functions changed very smoothly and simply with composition. They were, in addition, very well fit by a model that Stan Ovshinsky first put forth for them in which tellurium chains were cross-linked by germanium atoms. This is a simple manifestation of the 8–N rule. That is, the tellurium, having six outer electrons, is twofold coordinated, while the Ge, having four

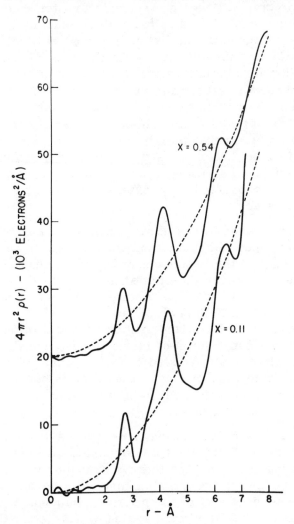

Fig. 4    Radial distribution functions for samples of $Ge_xTe_{1-x}$, with $x = 0.11$ and $0.54$. The curve for $x = 0.54$ has been displaced upwards by $20 \times 10^3$ electrons$^2$/Å. (From ref. 10.)

outer electrons, is fourfold coordinated. This picture, which Ovshinsky put forth for the low Ge concentration alloys, matched the RDFs beautifully at even higher Ge concentrations when we put numbers in models based on it.

A perplexing feature of this system was that the first neighbor peak in the GeTe RDF was about 0.2 Ångstroms shorter in distance than the corresponding crystalline peak. This difference between the crystalline amorphous first neighbor interatomic distances was too large to be attributed to any minor perturbation of the system. In addition, the first neighbor peak was significantly smaller than what would be predicted by the average of sixfold coordination in the crystalline GeTe. If we took an average coordination number for the GeTe near neighbor bond, it was something of the order of three.

As a result, we were in a double bind in our attempts to understand this amorphous system. The first was that the amorphous compound appeared to be quite different from the crystalline compound. The near neighbor distance was significantly shorter, and the coordination number seemed considerably lower than in the crystalline material. What's more, we could not even say that the first neighbor peak corresponded to a Ge-Te bond since that distance was also appropriate for Ge-Ge and Te-Te bonds. The peak could have been attributed to any of those bonds, and there seemed to be no way of getting a unique determination of even the near neighbor bonding.

Indeed, we very rapidly had two quite distinct models for amorphous GeTe. In one, the Ge was fourfold coordinated and the Te was twofold coordinated. In the other, both the Ge and the Te were threefold coordinated.

At first, the threefold coordinated model seemed a bit odd. We soon realized, however, that there were an average of five electrons per atom so that having coordination number of three is not all that unusual. We are accustomed to thinking that way about amorphous GaP or GaAs where the average number of electrons per atom is four and each atom is fourfold coordinated. Thus, we truly had two competing models for this structure.

Our structure analysis procedure, RDF analysis, was running into trouble because it yields a weighted sum of the partial distribution functions whereas we really needed to know the individual partial distribution functions in order to obtain a unique description of the system. We soon found that the same models and the same ambiguity applied to the Ge–S and Ge–Se systems. We could find no way to distinguish between the two models with x-ray diffraction.

Many people in the field seemed enchanted with the four-fold coordination model for a number of reasons. First of all, it gave a simple description of the amorphous system across an immense composition range from pure Ge to pure Te with everything changing continuously and slowly in keeping with the changes of the RDFs with composition. In addition, Turnbull had conjectured that if some systems were constrained to be chemically homogeneous, then the amorphous materials might be more stable than the crystalline. The Ge-Te might be an example of such a system. For example, the equilibrium system consisting of 40 atomic percent Ge and 60 atomic percent Te is a mixture of two phases: pure tellurium and GeTe. If they were constrained to be homogeneous, there is no obvious crystalline phase that could be formed, whereas it would be quite simple to form and amorphous phase with four-fold coordination of Ge described above.

The system was also attractive because the amorphous phase was so different from any crystalline phase. That seemed to rule out any microcrystalline model of the amorphous material. Finally, the model was consistent with a picture that Mott [11] had put forth for why many impurities have very little effect on the electrical conductivity of amorphous semiconductors. He conjectured that the impurity atoms would be coordinated such that their bonding requirements were fulfilled. As a result, virtually every electron is a bonding electron in the system and occupies a valence band state. That is, no donor or acceptor states are formed by the impurities. Such coordination is possible in the amorphous material because there are none of the constraints normally imposed by crystallinity. As a result, one does not obtain the impurity conductivities normally brought about by donors or acceptors in crystalline semiconductors.

After awhile, we gave up on obtaining a unique description of the structure of that system and went on to another system that was even harder to interpret, copper in $As_2Se_3$. Having satisfied ourselves that we seemed to understand why it was that some impurities do not influence the electrical conductivity of amorphous semiconductors very much, we wanted to understand why others do have a large effect. A good example of that was the system copper in arsenic triselenide. The very nice RDFs on the system obtained by K.S. Liang [12] in his Ph.D. thesis are shown in Fig. 5. Here, we see copper ranging from 0 to 30 atomic percent in the arsenic triselenide glasses. Again, precisely the same problems that occurred in the Ge-Te system occurred here. The first neighbor peak gets bigger and bigger as copper is added. There is no distinct peak associated with copper bonding, since appropriate copper-arsenic and copper-selenium distances are about the same as the arsenic-selenium distance. The one conclusion that we could draw is that the average coordination number of the system had to be going up. This is because copper has a smaller atomic number than either arsenic or selenium, while the peak area is increasing with increasing copper coordination. One thing that we

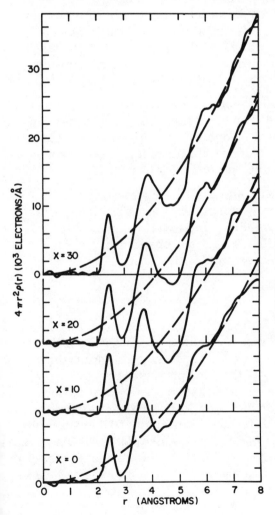

Fig. 5
Radial distribution functions for amorphous
$Cu_x(As_{.4}Se_{.6})_{100-x}$ alloys, normalized so that the total number of atoms in each sample is unity.
(From ref. 12.)

realized right away was that the odds were very high that the band gap was decreasing since the average coordination number was increasing, so that this might be the reason for the increase in conductivity. Liang came up with a number of models that could fit his RDFs. The one which he most favored had fourfold coordinated copper. Nevertheless, there were many other models which he could not rule out because he did not have the individual partial distribution functions.

Thus, we might summarize the situation for RDF analysis in 1972 when Liang finished his work as the following: First of all, RDFs are great for elemental amorphous materials. They give the distances and the coordination numbers out to third or fourth neighbors. These can be used to compare with detailed structural models. They are also extremely good for many oxide glasses and other systems where there are distinct distances that can be recognized, and where the structures of the amorphous materials are close to those of the crystalline. Under these circumstances, good models can be generated which fit the RDFs.

In many other polyatomic systems, however, the situation is far from good. The A-A, A-B, and B-B near neighbor distances are about the same, and all lead to one big broad first neighbor peak. As a result of that, one cannot obtain unique coordination numbers and unique structural interpretations of the RDFs.

I reacted to this situation and the bleak job market for students by becoming a university administrator. Very soon after that, the situation changed markedly. In particular, we obtained x-ray synchrotron radiation. Hence, let me spend a bit of time describing the properties of synchrotron radiation.

IV.    SYNCHROTRON RADIATION

In modern facilities, synchrotron radiation is produced by accelerating electrons to energies of approximately $3 \times 10^9$ eV or 3GeV, and then transferring them to a storage ring in which they circulate. The storage ring itself is a complex electron optical device. For our purposes, it may be considered as an array of straight lengths of ultrahigh vacuum pipe joined to one another by bends, so that the electron path is quasi-circular. When the electron path is bent by bending magnets, these charged particles give off an immense amount of radiation. Since the electrons at these energies are travelling virtually at the speed of light, the radiation

given off comes out in the very flat pattern shown in Fig. 6. As
the figure illustrates, the vertical divergence is extremely small,
some seconds of arc, while the horizontal divergence angle is
equal to the angle through which the electron path is bent. Because
the electrons travel in bunches, the light comes out in pulses. At
the Stanford Synchrotron Radiation Laboratory, the pulses are
typically 200 picoseconds long and come out about every micro-
second. In addition, the light is almost completely polarized in
the plane of the electron orbit.

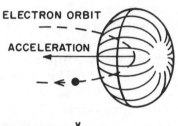

**ELECTRON ORBIT**

**ACCELERATION**

CASE I : $\frac{v}{c} \ll 1$

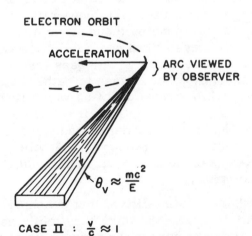

**ELECTRON ORBIT**

**ACCELERATION**

} ARC VIEWED
  BY OBSERVER

$\theta_v \approx \frac{mc^2}{E}$

CASE II : $\frac{v}{c} \approx 1$

Fig. 6    Pattern of synchrotron radiation emitted by electrons in
circular motion. Case I: nonrelativistic electrons,
$v/c \ll 1$. Case II: relativistic electrons, $v/c \sim 1$.
See, e.g., H. Winick and A. Bienenstock, Ann. Rev.
Nucl. Part. Sci. <u>28</u>, 33(1978).

Perhaps the most important feature of the radiation is its spectrum, which is shown in Fig. 7 for SSRL's storage ring, SPEAR, running at 3GeV. As the figure shows, the spectrum varies continuously from photon energies of the order of volts out to tens of kilovolts. That is, it runs from the infrared out to x-ray wavelengths of interest in x-ray diffraction. In viewing this, remember that the very common copper K-alpha radiation has a photon energy of about 8 kilovolts.

The radiation from such bending magnets formed the basis of virtually all x-ray synchrotron radiation experimentation until the late 1970's when SSRL introduced wiggler magnets. A wiggler is an array of magnetic poles which are inserted around one of those straight sections between two bending magnets of the storage ring. The array of magnets force the electrons to undergo a quasi-sinusoidal path with very sharp curves. The consequence of the sharp curves, due to large magnetic fields, is to shift the spectrum to higher photon energies, as shown in Fig. 7. In addition, each of the multiple poles acts as a bending magnet type source so that the spectrum is multiplied again by the number of poles. As a result, at high photon energies one can get one to two orders of magnitude increases in the intensity of the radiation compared to a bending magnet. This, in turn, leads to a factor of about a million increase in intensity over what is obtained from a copper K-alpha x-ray tube for many experiments. The availability of tunable radiation with an intensity of a hundred thousand to a million times that previously available always makes possible experiments which just could not be performed previously. One sees that readily when one realizes that it frequently takes a week to gather data at SSRL in some experiments. While almost all of us would find such a period of time appropriate for experimentation, it would be difficult to run an experiment that took a hundred thousand weeks. I should mention that our latest wiggler magnet has 54 poles, and is so intense that the x-rays could melt a piece of metal if they were incident upon it normally.

At any rate, the availability of that synchrotron radiation gives x-ray scientists a freedom which they really never had previously. That is, one can choose the wavelength at which one performs a scattering experiment, rather than being dependent upon those wavelengths which are available from x-ray tube anode materials. Alternatively, one can perform experiments as a function of photon energy in a fairly simple manner.

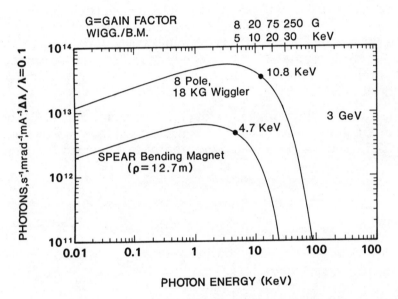

Fig. 7    SPEAR bending magnet and wiggler magnet spectrum at 3 GeV. Supplied by H. Winick.

A.        Extended X-Ray Absorption Fine Structure

          The first new structural tool that came out of synchrotron
radiation was based on the analysis of extended x-ray absorption
fine structure--EXAFS. This analysis had been proposed and utilized by Sayers, Lytle, and Stern [13], but had not been satisfac-
torily performed utilizing laboratory x-ray sources. With tunable
synchrotron radiation, it became an extremely valuable tool. The
basic experiment is illustrated in Fig. 8. The x-rays emerge from
the vacuum of the storage ring through a beryllium window, pass
through a slit and then are monochromatized by double Bragg reflection. The double Bragg reflection is used for two reasons.
The first is that the exit path is essentially parallel to the en-
trance path so that the remainder of the experiment need only be
translated as the wavelength is changed through rotation of the
monochromator crystals. The second reason is that harmonics of
the desired wavelength can be eliminated by slightly detuning the
two crystals.

          The x-rays then pass through a first ion chamber which
measures the intensity incident upon the sample, then through
the sample and are measured, and finally in a second ion chamber.

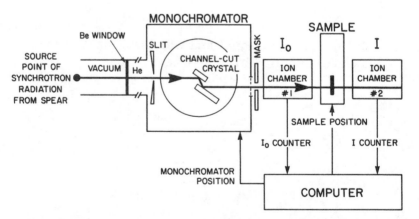

Fig. 8       Block diagram of EXAFS experimental apparatus.
             (From ref. 14.)

As a result, one obtains the x-ray absorption coefficient for the
wavelength at which the monochromator is set. By slowly rotating
the monochromator crystals and translating the ion chambers and
the sample, one can measure the absorption coefficient as a function of x-ray wavelength or photon energy.

Figure 9 shows the absorption coefficient as a function of
energy, between 8 and 10KeV, from a sample of crystalline
$CuAsSe_2$, as reported by S. Hunter [14]. The absorption coefficient decreases with increasing photon energy until 9KeV, which
is the Cu K-absorption edge. At that energy, the absorption coefficient increases markedly as the x-rays have energy to eject an
electron from the K-shell to the continuum. Within increasing
photon energy the absorption coefficient falls off almost monotonically. Superimposed on that fall-off is a fine structure which
extends for 500 to 1000eV. This fine structure is the Extended
X-ray Absorption Fine Structure known as EXAFS. Its origin can be
understood in the following manner.

Near the absorption edge, the absorption coefficient is
very simply related to the probability of exciting an electron from
this K-shell into the continuum. As shown in Eq. (4), this proba-

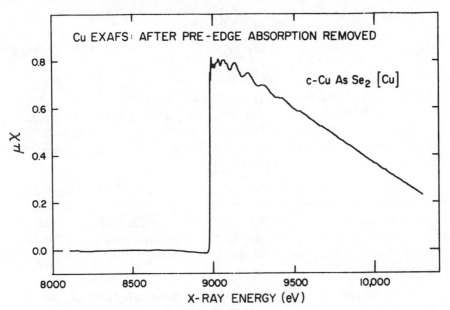

Fig. 9    Relative x-ray absorption coefficient of crystalline
          $CuAsSe_2$ as a function of x-ray energy near the Cu
          absorption. (From ref. 14.)

bility is proportional to the absolute value squared of a matrix element of the interaction, $H_{e.m.}$, between the light and the atom. It is taken between the initial state, which is a tightly-bound state close to the nucleus, and the final state, which is almost a free electron state:

$$\mu \propto |<f|H_{e.m.}|i>|^2. \tag{4}$$

Since the initial state is so tightly bound, it is essentially a delta function at the nucleus of the absorbing atom. $H_{e.m.}$ is smoothly varying and can be neglected in this discussion. Consequently, one is sampling the final state at the nucleus of the absorbing atom in this matrix element.

The final state may be described, as it is in Eq. (5), as primarily an outgoing spherical wave. This outgoing spherical wave, however, is reflected by the atoms surrounding the absorbing atom so that there are waves reflected back to the original absorbing nucleus:

$$\psi_f \propto (e^{ikr} + \text{scattered waves}). \tag{5}$$

$$|<f|H_{e.m.}|i>| = |\int \psi_f^*(\underline{r}) H_{e.m.} \psi_e(\underline{r}) d^3r|$$

$$\propto |\psi_f(\underline{r}_N)|, \tag{6}$$

where $\underline{r}_N$ is the coordinate of the absorbing atom. Thus, the backscattered waves interfere with the outgoing spherical wave at the original absorbing nucleus. As the x-ray energy gets larger and larger, the electron wavelength gets shorter and shorter so that there is constructive and destructive interference with changing photon energy.

Thus, the fine structure results from a diffraction effect in the photo-excited electron's wave function. Since it is such a diffraction effect, we know the type of information that it will yield. It is: (a) the distances to the surrounding atoms; (b) the number of neighbors at each distance, and (c) the identification of the species of the near neighbors.

One of the beautiful things about this approach is that it is species specific. That is, the absorption edge is associated with a particular atomic species in the material. Hence, analysis of the EXAFS yields information about the average atomic arrangement around a particular atomic species in the material.

In addition, the approach does not require periodicity, since the phenomenon involved is short range. Consequently, EXAFS is very valuable for the analysis of atomic arrangements in amorphous materials as well as biological materials like metallo-proteins and metalloenzymes.

My first student to explore this technique was Sally Hunter, who returned to the $Cu-As_2Se_3$ system that Liang et al. had pre-viously examined. She studied samples containing 5, 10, 20, and 30 atomic percent copper, and found that, in all cases, the coor-dination number looked like that of copper in crystalline $Cu-AsSe_2$. That is, the copper is fourfold coordinated [14].

Subsequently, Kastner [15], following up on a picture that Ovshinsky had suggested some time ago, explained the fourfold coordination. The basic idea is the following. Copper has one outer electron which can take part in covalent bonding. This leads to a simple bond with one Se atom. The Se atoms have, however, lone-pair electrons whose energy can be lowered if they also are shared, through a coordination bond, with a copper atom. To complete an outer electron octet, the copper forms coordination bonds of this sort with three selenium atoms, leading to a total of four near neighbors.

A couple of years later, electrical measurements were made [16] on $Cu-As_2Se_3$ alloys with considerably lower copper concentrations. It was found that the electrical properties do not change monotonically with decreasing copper concentrations. We began to wonder if it was possible that, at low copper concentra-tions, the copper coordination is different from the fourfold coor-dination at the concentrations we had studied. This sort of phenomenon had been seen in other systems. This meant going to work again with EXAFS, but with very much lower copper con-centrations. The possibility of doing that came about with a variation of EXAFS.

If we just analyze the x-ray absorption coefficient, and have only a one part in $10^4$ or $10^5$ copper in the system, then there would be an extremely small jump in the absorption coeffi-cient at the copper K edge, so that we would not be able to measure the fine structure at all. It would be too small to observe in the background of the absorption coefficient of the other atom in the system.

Instead, Laderman [17] used another approach that had been developed previously at SSRL. As indicated above, the fine structure in the absorption coefficient comes about because there are modulations in the probability of exciting a K electron up into the continuum. With the excitation, almost every K electron excited into the continuum emits K-alpha fluorescence radiation. Thus, the K-alpha fluorescence shows the same fine structure when measured as a function of excitation energy as does the absorption coefficient. The advantage of measuring the fluorescence (Cu K-alpha in this case) is that it is unique to the atomic species whose coordination is being determined. Thus, for dilute species, there is an immense gain in signal-to-noise since one can discriminate against the photons emitted by all the other atoms in the species in the detection process.

Using this approach, Laderman was able to study samples with Cu concentrations as low as 0.05 atomic percent and to show that the predominant bonding is fourfold throughout this concentration range. Indeed, he could not observe any deviation from fourfold coordination.

It is apparent that EXAFS has expanded markedly our ability to understand atomic arrangements in amorphous materials. In particular, it provides the means by which the coordination of individual atomic species in complex, polyatomic glasses may be determined. It does, however, have drawbacks. One of them, which Hunter noted and explained [14], is that EXAFS provides information only about near neighbors in amorphous materials. Indeed, as discussed by Kortright et al. [18], EXAFS may even be misleading for near neighbors when there is a broad distribution of those neighbors. Thus, for example, Hunter was never able to come up with a consistent interpretation of all the EXAFS data that she took on the Ge-Se system, so that we could not resolve the previously-discussed question as to whether the Ge is threefold or fourfold coordinated in amorphous GeSe using that technique. We were obliged to turn to a completely different approach, again using synchrotron radiation.

B.    Anomalous X-Ray Scattering

In most x-ray diffraction circumstances, the atomic scattering factor is simply the Fourier transform of the atom's electron density and is denoted $f_o$. When, however, the x-rays have a photon energy close to that of an absorption edge of the atom, the atomic scattering factor changes into a complex quantity shown in

Eq. (7):

$$f = f_0 + f' + if''. \tag{7}$$

Under these circumstances, the atom behaves almost like a classical harmonic oscillator with the K electrons being weakly bound to the atom with a spring whose spring constant gives the binding energy. The x-rays drive the electron at its resonant frequency so that there is a large change in its motion. As with a mechanical system, the driven system no longer stays in phase with the driving force close to the resonant frequency. The phase change yields the imaginary part of the scattering factor.

The behavior of $f'$ and $f''$ are shown in Fig. 10 as a function of a normalized photon energy. $f'$ is negative as the edge is approached from below and then rises up again to small positive

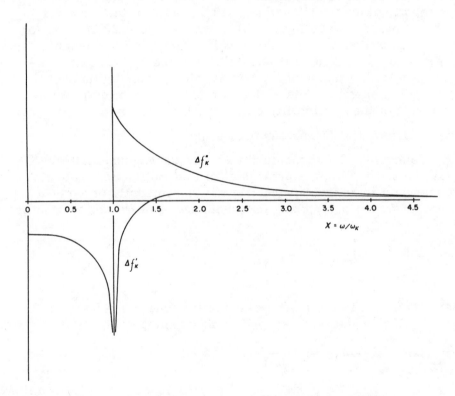

Fig. 10    Real, $f'$, and imaginary, $f''$, shifts of the atomic scattering factor as a function of angular frequency, $\omega_K$, near an absorption edge, K.

values above the edge. $f''$ is simply related to the x-ray absorption coefficient. As a result, it is extremely small below the absorption edge, rises sharply at the edge, and then falls off with increasing photon energy.

These changes of the scattering factors with photon energy offer immense possibilities for the simplification of structure determination using x-ray scattering. They were not utilized extensively prior to the availability of synchrotron radiation because it was not possible to tune the x-rays close to the absorption edge and get appreciable magnitudes of $f'$. Indeed, $f'$ was not well known close to the absorption edge for many elements because of the unavailability of tunable x-ray sources.

Hence, one of the first things we did at SSRL was to start to measure a number of these $f'$ and $f''$ values for the elements. Much of this work is summarized in the paper by Fuoss et al. [19]. They report values for $f'$ between -10 and -16 for the elements in the fourth row of the Periodic Table, and between -20 and -30 for elements in the sixth row. These values are quite appreciable compared to $f_0$, and make it possible to markedly change the effective scattering factor of a given atomic species by tuning the x-ray photon energy. This option, of course, becomes possible as a result of the availability of synchrotron radiation.

C.      Differential Anomalous Scattering

Shevchik [20] recognized that this anomalous scattering could be used to obtain a great deal of information about atomic arrangements in amorphous materials. The intensity of x-ray scattering from a polyatomic amorphous material may be written as:

$$I(k) = \sum_\alpha \sum_\beta N_\alpha f_\alpha^* f_\beta \int_0^\infty 4\pi r^2 \rho_{\alpha\beta}(r) \frac{\sin(kr)}{(kr)} dr, \tag{8}$$

where $4\pi r^2 \rho_{\alpha\beta}(r)$ is the partial distribution function defined in Eq. (2). This equation is simplified by defining:

$$S_{\alpha\beta}(k) = \int_0^\infty 4\pi r^2 \rho_{\alpha\beta}(r) \frac{\sin(kr)}{(kr)} dr. \tag{9}$$

$S_{\alpha\beta}(k)$ is the Fourier transform of the partial distribution function involving species $\alpha$ and $\beta$. Knowledge of it, therefore, yields the partial distribution function. With it, Eq. (8) can be written as:

$$I(k) = \sum_{\alpha} \sum_{\beta} N_\alpha f_\alpha^* f_\beta S_{\alpha\beta}(k). \tag{10}$$

Let us consider now an experiment performed with photon energies near to the absorption edge of element A in a polyatomic material. At those photon energies, in general, only the scattering factor of element A will change markedly if there is a slight change in photon energy. All the other atomic scattering factors are varying very slowly at that photon energy region. Hence, the difference between two sets of intensity measurements taken at photon energies very close to an absorption edge may be written as:

$$\Delta I(k) = N_A \Delta f_A^* \sum_{\beta} f_\beta S_{A\beta} + \Delta f_A \sum_{\beta} N_\beta f_\beta^* S_{\beta A}$$

$$= N_A \sum_{\beta} \left\{ \Delta f_A^* f_\beta + f_\beta^* \Delta f_A \right\} S_{A\beta} \tag{11}$$

where we have used the identity:

$$N_\alpha S_{\alpha\beta} = N_\beta S_{\beta\alpha}. \tag{12}$$

Equation (11) shows that this difference in intensity depends only on those partial distribution functions involving atom A. Consequently, the environment of atomic species A can be determined through appropriate Fourier transforms of the right-hand side of Eq. (11) to yield a differential distribution function (DDF) as first obtained by Fuoss et al. [21].

Fuoss et al. studied both amorphous GeSe and amorphous GeSe$_2$. The DDFs for GeSe$_2$ yielded exactly what was expected. The Ge is fourfold and the Se is twofold coordinated. One other interesting thing that they found is that the Se-Ge-Se second neighbor bond is slightly longer than the Ge-Se-Ge second neighbor bond. This is because the Se, if it were purely covalent, would have a 90° bond angle, whereas the Ge has a tetrahedral bond angle which is well over 100°. The same situation has been observed in crystalline GeSe$_2$. There would be no way of seeing this sort of difference in a normal radial distribution function, nor could it be observed with EXAFS. The observation shows the power of differential anomalous scattering.

In GeSe, the differential distribution function obtained from the Ge edge data is virtually identical to that obtained at the Se edge. Both indicate threefold coordination of the Ge and Se. Thus, at last we have a structural technique for distinguishing the bonding situations in the germanium monochalcogenides. This constitutes, I believe, a major advance in our ability to determine atomic arrangements in amorphous materials.

V.     FUTURE SYNCHROTRON RADIATION RESEARCH ON THE
       STRUCTURE OF AMORPHOUS MATERIALS

In principle, it should be possible to use anomalous scattering to get the individual partial distribution functions. Thus far, however, there has been little success in such attempts because the equations which yield the partial distribution functions from the x-ray intensity data are ill-conditioned, as discussed by Bienenstock [22]. Recent work of my students indicates, however, that reliable partials are likely to be obtained soon.

In addition, we may soon expect significant advances in our ability to study the dynamics of atomic arrangements. That is, we will soon be determining atomic arrangements and their changes in real time for amorphous materials.

For example, Stephenson [23] has watched the early stages of phase separation of a molten oxide after it was quenched into a glassy region in which there is metastable immiscibility. He was able to observe a small angle scattering pattern every 30 seconds through the quench and for approximately an hour after the quench. With the higher intensities obtained from wigglers, such studies will soon be performed in the fractions of a second time scale.

Similar time scales are being evolved for EXAFS studies. In these studies a dispersive technique is used with a position sensitive detector so that an entire EXAFS pattern is measured without any angular scanning [24].

There is every reason to believe that these time-resolved studies, which could be performed on viscous systems, will yield a great deal of understanding of the rearrangements which take place as glasses go through the glass transition and, in some cases, separate into two or more phases. It will be, I believe, a very exciting period for those of us interested in the structure of amorphous materials.

## ACKNOWLEDGEMENTS

It has been an immense pleasure to prepare this paper on our structural studies of amorphous materials. My interest in this area resulted primarily through my association with the research at ECD and Stan Ovshinsky. I have also benefitted through my association with David Turnbull.

## REFERENCES

1.  B.E. Warren, X-ray Diffraction, Addison Wesley, 1969, Chapter 10.
2.  J. Kortright, Ph.D. thesis, Stanford Universiy, 1984.
3.  S.C. Moss and J.F. Graczyk, Phys. Rev. Lett. 23, 1167 (1969); Proceedings of the Tenth International Conference on the Physics of Semiconductors, edited by S.P. Keller, J.C. Hensel and F. Stern, U.S. Atomic Energy Commission, Division of Technical Information, p. 658.
4.  D.E. Polk, J. Noncryst. Solids 5, 365(1971).
5.  H. Rawson, Inorganic Glass-Forming Systems, Academic Press, London, 1967, Chapter 16.
6.  J.D. Bernal, Nature 183, 141(1959); 185, 68(1960).
7.  G.S. Cargill, III, J. Appl. Phys. 41, 12(1970).
8.  D.E. Polk, Acta Met. 20, 485(1972).
9.  B.E. Warren, H. Krutter and O. Morningstar, J. Amer. Ceram. Soc. 19, 202(1936).
10. F. Betts, A. Bienenstock and S. R. Ovshinsky, J. Noncryst. Solids 4, 554(1970).
11. N.F. Mott, Adv. Phys. 16, 49(1967).
12. K.S. Liang, C.W. Bates, Jr. and A. Bienenstock, Phys. Rev. B 10, 1528(1974).
13. D.E. Sayers, F.W. Lytle and E.A. Stern, in Advances in X-Ray Analysis, edited by G.R. Mallett, M.J. Fay and W.M. Mueller, Plenum, New York, Volume 13, p. 248.
14. S. Hunter, Ph.D. thesis, Stanford University, 1977, published as Stanford Synchrotron Radiation Laboratory Report No. 77/04; S.H. Hunter, A. Bienenstock and T.M. Hayes, The Structure of Noncrystalline Materials, edited by P.H. Gaskelly, Taylor and Francis Ltd., London, 1977, p. 73.

15.    M. Kastner, Phil. Mag. B 37, 127(1978).

16.    G. Pfister, M. Morgan and K.S. Liang, Solid State Commun.
       30, 277(1979); M. Kitao, H. Akao, T. Ishikawa and
       S. Yamada, Phys. Stat. Sol. (a) 64, 493(1981).

17.    S. Laderman, A. Bienenstock and K.S. Liang, Solar Energy
       Mats. 8, 15(1982).

18.    J. Kortright, W. Warburton and A. Bienenstock, in EXAFS
       and Near Edge Structure, edited by A. Biancone, L.
       Incoccia and S. Stipich, Springer-Verlag, Berlin, 1983,
       p. 362.

19.    P. Fuoss and A. Bienenstock, in Inner-Shell and X-Ray
       Physics of Atoms and Solids, edited by D.J. Fabian,
       H. Kleinpoppen and L.M. Watson, Plenum, 1981, p. 875.

20.    N.J. Shevchik, Phil. Mag. 35, 1289(1977).

21.    P.H. Fuoss, P. Eisenberger, W.K. Warburton and
       A. Bienenstock, Phys. Rev. Lett. 46, 1537(1981).

22.    A. Bienenstock, in The Structure of Non-Crystalline Mate-
       rials, edited by P.H. Gaskell, Taylor and Francis Ltd.,
       London, 1977, p.1.

23.    G.B. Stephenson, Ph.D. thesis, Stanford University, 1982,
       published as Stanford Synchrotron Radiation Laboratory
       Report No. 82/05.

24     R.P. Phizackerey, Z.U. Rek, G.B. Stephenson, S.D.
       Conradson, K.O. Hodgson, T. Matsushita and H.
       Oyanagi, J. Appl. Crystallogr. 16, 220(1983).

# EXAFS OF DISORDERED SYSTEMS

E.A. Stern

Department of Physics, University of Washington

Seattle, Washington 98105

## I. INTRODUCTION

The basic theoretical foundation for condensed matter physics is well known. The Schrödinger equation of electrons and nuclei with the Coulomb interaction and relativistic corrections such as the spin-orbit interaction contains all of the basic physics. The complication in applying this basic knowledge is the large number of particles involved, of the order of $10^{23}$. To reduce this problem to a manageable size, simplifying assumptions or approximations are required. Usually, the initial assumption is the structure of the matter. For example, the knowledge that crystalline solids have a periodic array of atoms permits the exploitation of the periodicity to reduce the number of particles from $10^{23}$ down to the order of 1, namely, the ion cores and the valence electrons in a unit cell. In disordered systems the structure is more complicated since the long-range periodicity is absent, but knowledge of structure is still a basic requirement for any detailed understanding of their properties. For crystalline samples, diffraction, using sources with wavelengths of the order of 1Å, is the extremely productive technique by which to determine the long-range order. Sources typically used for diffraction are x-rays, neutrons, and electrons. When long-range order is absent, as is the case for disordered systems, diffraction loses its pre-eminence as the structural tool, and other techniques of the coherent diffuse scattering of x-rays and neutrons and the extended x-ray absorption fine structure technique (EXAFS) fill the gap.

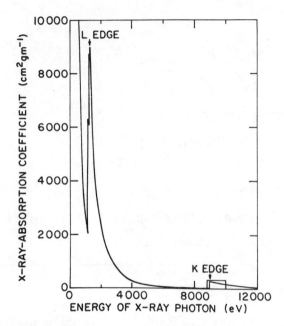

Fig. 1   The x-ray absorption coefficient of copper metal as a
function of x-ray photon energy.

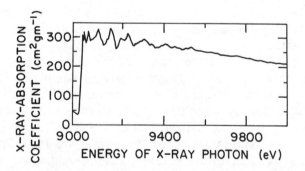

Fig. 2   The expanded view of the absorption coefficient around the
K-edge enclosed in the rectangle of Fig. 1.

In Section II, the EXAFS technique [1] will be described, and its applicability to disordered systems as compared with diffuse scattering will be discussed in Section III. An example of the application of EXAFS to amorphous Ge(a-Ge) will be presented in Section IV, focusing on the process of the initiation of crystallization from the amorphous state [2]. A summary and conclusion are given in Section V.

II.    EXAFS

In modern physics courses, the absorption coefficient of x-rays in matter is described as shown in Fig. 1 for copper metal. The absorption coefficient decreases monotonically as the x-ray photon energy increases except at edges where a sudden increase occurs. These absorption edges occur when the photon energy equals the binding energies of the various shells of electrons in the atom, opening up the new channels of absorption where the bound electrons are excited into escaping photoelectron states. The L-edge corresponds to exciting the n=2 or L-shell electrons, while the K-edge corresponds to exciting the most tightly bound n=1 K-shell electrons.

The absorption of x-rays actually has a much richer variation, as can be seen in Fig. 2, which shows an expansion of the rectangular region in Fig. 1 around the K-edge. The absorption past the edge is not monotonic but has a complicated fine structure. It is convenient, for reasons presented below, to divide the fine structure into a near-edge region within 20eV of the edge and an extended x-ray absorption fine structure (EXAFS) region extending from the near-edge limit to the order of 1000eV past the edge.

A clue to the origin of the fine structure can be obtained by the K-edge absorption coefficient of Kr vapor and of Kr vapor absorbed on graphite shown in Figs. 3 and 4, respectively. The EXAFS is absent for the vapor which consists of isolated atoms, but is present when the Kr is in the vicinity of the carbon atoms of the graphite. It is clear that the EXAFS is introduced when the Kr is in a condensed state with other atoms in its vicinity. It will be shown that it is possible to unfold the EXAFS to obtain structure information on the location of the atoms in the Kr vicinity.

It is of particular interest to understand how the presence of neighboring atoms modifies the absorption to produce the EXAFS. Fermi's golden rule governs the absorption of the x-rays. The x-ray

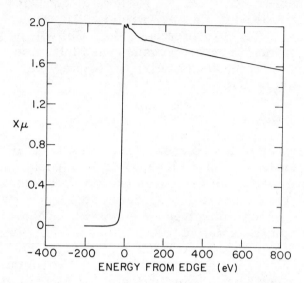

Fig. 3  The x-ray absorption (in arbitrary units) for krypton atoms
in the vapor form.

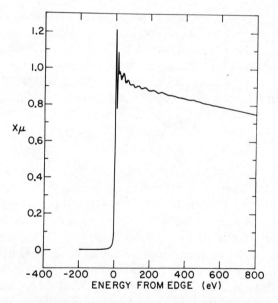

Fig. 4  The x-ray absorption coefficient (in arbitrary units) for
krypton atoms absorbed on graphite planes.

photon field produces perturbation $H'$ which causes the photo-electron to make a transition from the initial core state $|i >$ to the final photoelectron state $|f>$.

The number of photons absorbed per sec by an atom is

$$w = \frac{2\pi}{\hbar} |< i |H' f>|^2 \rho (E_f),\qquad\qquad (1)$$

where $\rho(E_f)$ is the density of final states of the photoelectron of energy $E_f$. The initial state $|i>$ is a deep core state independent of the environment and localized at the atom's center. The $\rho(E_f)$ can be approximated by the smoothly varying free electron value because at EXAFS energies the kinetic energy of the photoelectron is large compared with its binding in the material. For this same reason, the scattering of the photoelectron by surrounding atoms is small and need be treated to first order only. Thus, only variations in the final state $|f>$ with $E_f$ remain to explain the EXAFS.

Figure 5 illustrates the physical mechanism of the variation of $|f>$. The outgoing photoelectron wave function illustrated by the solid circles centered above the absorbing atom is scattered by the surrounding atoms as indicated by the dashed circles. The $|f>$ is the sum of the outgoing and scattered parts and that sum depends on their relative phases at the center of the absorbing atom, the only region where $|i>$, and thus the matrix element, is non-zero. The phase varies with the wavelength of the photoelec-tron as its kinetic energy changes with the photon energy. When the phase is destructive $|f>$, and thus the x-ray absorption is decreased from the isolated atom case, and vice versa when the phase is constructive. The fine structure simply reflects the sum of the outgoing and scattered parts of the photoelectron state from the various surrounding atoms. Depending on the various relative phases, this sum can be larger or less than that of the isolated atom as seen in Figs. 2 and 4.

Putting into a mathematical description the above simple physical picture, one obtains for the EXAFS

$$X(k) = \frac{\mu(k) - \mu_o(k)}{\mu_o(k)}$$

$$= \sum_j \frac{N_j\, t_j\, (2k)}{R_j^2} \exp(-2k^2\sigma_j^2) \exp(-2R_j/\lambda) \sin(2kR_j+\delta_j)\qquad (2)$$

Here $\mu(k)$ is the absorption coefficient per atom of interest in the material, $\mu_0(k)$ is its value for an isolated atom. Although the mathematical expression appears to complicate a simple picture, each term can be directly related to the physical picture just given.

The strength of the EXAFS is proportional to the number of atoms $N_j$ of the jth type contributing to the backscattering at a coordination shell of average distance $R_j$. The spherical outgoing wave decreases as $R_j^{-1}$ and the backscattered spherical wave also decreases as $R_j^{-1}$ combining to the $R_j^{-2}$ dependence.

The wave number $k = p/\hbar = 2\pi/\lambda$, where p is the photo-electron momentum, $\hbar$ is Planck's constant divided by $2\pi$, and $\lambda$ is the de Broglie wavelength of the photoelectron. The $2kR_j$ argument of the sine function gives the phase shift as the electron wave travels $R_j$ out and $R_j$ back in again. The phase shift $\delta_j$ accounts for the fact that the photoelectron is not really free but has to

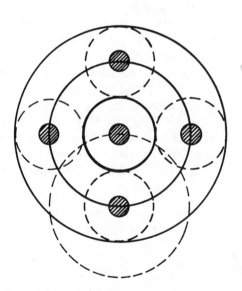

Fig. 5  A schematic representation of the outgoing photoelectron (solid circles) and its scattered contribution from the surrounding atoms (dashed circles).

travel through the potential of the center atom and suffers a phase shift in backscattering from the $\underline{j}$th atom. The backscattering from the $\underline{j}$th atom will have a characteristic dependence on k which varies with the atom type and is denoted by $t_j(2k)$. The $\exp(-2k^2\sigma_j^2)$ is a Debye-Waller factor, taking into account the fact that the $N_j$ atoms at average distance $R_j$ are not actually all at the same distance and their backscattered waves will not be adding exactly in phase. Thermal vibrations or structural disorder will cause a mean square variation $\sigma_j^2$ about $R_j$, and if the disorder is Gaussian, the result is the Debye-Waller factor. Finally, the $\exp(-2R_j/\lambda)$ factor accounts for the finite lifetime of the photoelectron. The backscattered wave will interfere with the outgoing part only if it remains coherent, which occurs if the photoelectron has not been scattered in another state. The lifetime can be translated into a mean free path $\lambda$ between collisions such that the probability of the photoelectrons not scattering varies as the factor $\exp(-2R_j/\lambda)$.

Note from (2) that the "frequency" in k (its coefficient in the sine function) locates the various coordination shells of atoms. This is modified by the k-dependence of $\delta_j$; e.g., if $\delta_j =$ a-bk, then the "frequency" is a $2(R_j-b/2)$. The effects of $\delta_j$ can be calibrated by comparison with standards of known structure, permitting the determination of $R_j$ directly. In addition, the amplitude of $\chi(k)$ is proportional to $N_j$.

The "frequency" spectrum of $\chi(k)$ can be obtained by Fourier transforming with respect to 2k. Figure 6 shows the magnitude of such a transform for metallic copper. As predicted, the peaks locate the coordination shells of the atoms surrounding the center atom. The first four shells are clearly visible. Note that the EXAFS probes about 5Å from the center atom. The mean free path limits this probing distance.

The features of EXAFS that make it useful for structural analysis are the following:

1.      EXAFS can determine the local atomic arrangement about each type of atom separately. Since the x-ray energy can be tuned to the absorption edge of the atom of interest, its local environment can be isolated.

2.      Long-range order is not required for EXAFS. It, thus, can be used on disordered systems such as amorphous solids.

3.      EXAFS analysis can determine the numbers, types, average distance, and variation about the average distance of the sur-

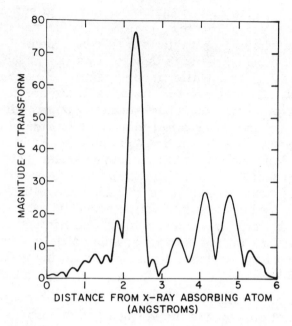

Fig. 6    The magnitude of a Fourier transform in k-space of the
          EXAFS for copper metal.

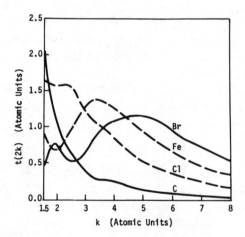

Fig. 7    The backscattering amplitude as a function of photoelectron
          wave number k for the indicated atoms with varying atomic
          number.  From B.K. Teo and P.A. Lee, J. Am. Chem. Soc.
          101, 2815(1979).

rounding atoms. The types of atoms can be distinguished by their characteristic backscattering variation $t(2k)$ as illustrated in Fig. 7. Note the important feature that low Z atoms are not masked by high Z atoms as in x-ray diffraction. The high Z atoms dominate in scattering at high $k$, but the low Z atoms dominate at low $k$.

4.      Although this is usually not pertinent to bulk disordered systems (though it is for absorbed monolayers, even in a disordered array), the angular dependence between atoms can be determined in some special cases: when the system has twofold or less rotational system or when three or more atoms are nearly aligned along a straight line.

5.      The analysis is relatively simple and straightforward, <u>not</u> requiring elaborate calculations such as solving the Schrödinger equation as in LEED. The information is more directly available as indicated in the Fourier transform of Fig. 6.

6.      The near-edge structure, which has been neglected until now, contains added electron binding information that complements EXAFS. To mine this information fully requires detailed solutions of the Schrödinger equation since the kinetic energy of the photoelectron is comparable to the binding energy. As our sophistication in calculating near-edge structure progresses, the information present in these data will become more readily available.

        EXAFS has some limitations that restrict its usefulness for structure determination:

1.      EXAFS contains no long-range order information and thus is not a replacement for the diffraction technique but complements it.

2.      Because the EXAFS regime starts 20eV above the edge, the lowest photoelectron wave number normally available is $k \simeq 2\text{Å}^{-1}$. As we discuss below, this restricts EXAFS in some cases to only being able to analyze systems with $\sigma < 0.15\text{Å}$. This could be a serious restriction for some highly disordered systems.

III.    EXAFS VERSUS DIFFUSE SCATTERING FOR DISORDERED
        SYSTEMS

        The other major structural determining technique for disordered systems besides EXAFS is diffuse scattering of x-rays and neutrons. To be specific we will focus on x-rays. Diffuse scattering in principle can determine a weighted average of

$\Sigma_{A,B} P_{AB}(r)$, where $P_{AB}(r)$ is the pair correlation function which gives the probability of a type B atom being a distance $r \to r + dr$ from a type A atom [3]. EXAFS in principle can determine $P_{AB}(r)$ alone because the A atom can be isolated by tuning to its absorption edge and the B atom can be distinguished by its characteristic backscattering. Obviously, when the material is composed of several types of atoms there is no unique way to unfold $\Sigma_{A,B} P_{AB}(r)$ to obtain $P_{AB}(r)$, and EXAFS has a decided advantage.

It is possible to modify the diffuse scattering technique to obtain $P_{AB}(r)$. Use of anomalous scattering of x-rays [4] or isotopic substitution [5] in the material for neutron scattering introduces variations in scattering about one type of atom and can, in principle, introduce enough information to obtain $P_{AB}(r)$ separately. However, the measurements become more tedious and lose accuracy, especially at high wave numbers, and the applicability of these techniques is limited.

The main advantage of diffuse scattering over EXAFS is its ability to obtain low wave number information which translates to large r information or slow variations in $P_{AB}(r)$ as a function of r. EXAFS may lose low k information because of the near-edge structure. Since the EXAFS regime starts at $\sim$ 20eV past the edge, the smallest k available to probe the structure is $k_{min} \simeq 2\text{Å}^{-1}$. The Debye-Waller factor corresponding to this smallest $k_{min}$ is $\exp(-2k^2_{min}\sigma^2)$. The EXAFS will be seriously attenuated by this factor when $2k^2_{min}\sigma^2 \gtrsim 0.5$ or when $\sigma \gtrsim 0.2\text{Å}$. Thus, the loss of low k information in EXAFS means that highly disordered systems will produce no EXAFS and cannot be probed by EXAFS.

There is an important special exception for disordered systems, one in which low k information in EXAFS can be resurrected and the above limitation can be overcome, namely, for covalently bounded amorphous materials. This case will be discussed in detail in the next section. Suffice it to say here that an accurate value at k = 0 can be obtained and, in addition, the EXAFS analysis can be extended into the nominally near-edge structure regime up to about 2eV from the edge, obtaining $P_{AB}(r)$ information on highly disordered shells of atoms.

Besides its capability to directly obtain the individual $P_{AB}(r)$, another advantage of EXAFS over diffuse scattering of x-rays is the higher k information obtainable. In making the comparison, account must be taken of the different manner in which wave number is defined in the two fields. For diffuse scattering the wave number q is defined as varying as qr, while in EXAFS the wave number k is defined as varying as 2kr. Thus, $q = 2k$, and in diffuse scattering where the maximum $q_{max} = 15\,Å^{-1}$, it corresponds in EXAFS to a $k = 7.5\,Å^{-1}$. Maximum k values in the EXAFS of Ge are $k_{max} \simeq 12\,Å^{-1}$, a substantially larger value.

The importance of large k can be illustrated as follows. Consider a heterogeneous material with regions of high order embedded in an amorphous matrix, a situation which will be encountered with Ge in the next section. The P(r) for this case is illustrated in Fig. 8(a). When high k information is missing, the measured P(r) will be as shown by the dashed curve in Fig. 8(b), and it will not be possible to resolve the heterogeneous nature of the distribution function. Another example is illustrated in Fig. 8(b), where a narrow distribution cannot be resolved with the high k missing.

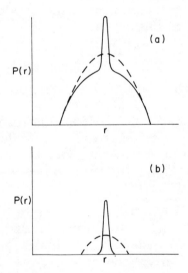

Fig. 8 Illustrating the effect of the absence of high k values in determining the pair distribution function P(r). The solid lines are the actual P(r) and the dotted lines are what would be obtained if large k were missing. In (a), P(r) consists of the sum of two phases, one with an ordered distribution and the other with a highly disordered one. In (b), P(r) consists of just an ordered distribution.

The lack of resolution when high k information is missing is familiar from optics. It is well known that it is not possible to resolve dimensions less than the wavelength of light being employed. High k corresponds to small wavelength, and when it is missing, the short wavelengths are not available to give high resolution.

To summarize, in general, diffuse scattering has the advantage over EXAFS for highly disordered systems because of low k information. EXAFS, however, obtains $P_{AB}(r)$ directly and has higher k information which gives more resolution in $P_{AB}(r)$. In the important special case of covalently bounded amorphous materials EXAFS also can obtain more low k information and still retains its other advantages over diffuse scattering. However, this feature is limited to only the first few coordination shells of atoms because the finite mean free path $\lambda$ of the photoelectron restricts EXAFS to a local probe within 5Å of the center atom. The next section illustrates these points in detail for amorphous Ge.

This example will point out another advantage of EXAFS over diffuse scattering, namely, that the determination of coordination number in EXAFS is absolute while diffuse scattering requires knowledge of the density, introducing the uncertainty of which density to use, the macroscopic one with voids or the microscopic one neglecting the voids [6].

## IV.    APPLICATION OF EXAFS TO AMORPHOUS GE

In the application of EXAFS to amorphous Ge there were two main points investigated [2]. One was to determine the structural relationship between amorphous and crystalline Ge. The other was to investigate the transition as amorphous Ge begins to crystallize.

Crystalline Ge has a diamond structure in which each atom is tetrahedrally bound to 4 first neighbors and has 12 neighbors in the second coordination shell. Experimentally, it has been found that the tetrahedral bonding angle of 109.5° is disordered in the amorphous state with about a 10° rms disorder. In addition, the adjacent distorted tetrahedrons become twisted with respect to one another. Since these are both average results, they can occur in two extreme ways. One, called the microcrystalline, or soliton, model, assumes that the disorder is not homogeneous but is small in the microcrystalline regions and large in the grain boundaries between . The microcrystallites have to be of the order of 10Å in order that they give diffuse diffraction peaks as required by experiment. The other, called the continuous random network (CRN), is a homogeneous disorder model in which the disorder is on the average the same about each atom. Although it is generally believed that the CRN model is the correct one, there is no direct experimental proof since diffuse scattering does not have enough resolution to discriminate between the two competing models. We will show that EXAFS does have the necessary resolution.

Germanium samples were sputtered in 5m Torr of argon on $25\mu$ thick Kapton substrates at a rate of $0.6\mu$/hour in a vacuum system of a base pressure of $5 \times 10^{-7}$ Torr. The samples were about $1\mu$ thick and about 10 layers were placed in series to obtain a K-absorption edge step of $\Delta\mu x \simeq 1$. Samples were made at substrate temperatures $T_S$ of 175°C, 265°C, 300°C, 335°C, and 370°C. Crystalline samples were also prepared from a fine Ge powder that passed through a $5\mu$ mesh. The above samples were homogeneous and thin enough to avoid the thickness effects that can distort EXAFS amplitude. Measurements were made at the Stanford Synchrotron Radiation Laboratory (SSRL), where precautions were taken to minimize higher harmonic content in the x-ray beam by detuning the two-crystal monochromator. These precautions eliminated systematic errors, permitting EXAFS amplitudes to be measured to an accuracy of about 1%.

Figure 9(a) shows the K-edge data of crystalline Ge, while Fig. 9(b) shows similar results for an amorphous Ge sample with $T_S = 175$°C. Note that the fine structure of the crystalline sample is sharper, corresponding to higher frequencies and indicating more contributions from shells farther out. A Fourier transform verifies this, as indicated in Fig. 10. Also shown in Fig. 10 are the Fourier transforms for the samples prepared at the other $T_S$ values.

The transforms indicate little change in the first-neighbor peak, which is the largest peak. However, dramatic changes occur in the farther-out peaks. The farther-out peaks are not present for the lowest $T_S$ sample and grow as $T_S$ is increased. A careful analysis of these data gives the result that the coordination number of the first shell in all samples is 4 to within 1% and the distance to the first shell changes by less than 0.004Å. The only discernible change is a small increase in the disorder of the first shell.

The large change in the second shell could be due to two extreme possibilities. One is that the sample is changing homogeneously, with the second shell becoming more ordered with increasing temperature. The other possibility is that the sample is changing heterogeneously and only part of the sample is crystallizing, leading to the increasing second-shell amplitude.

To analyze the second shell, the transform at the second-shell position is isolated from the rest of the data and back-transformed into k-space to find the second-shell contribution to the EXAFS. By this means, the second-shell term $\chi_2$ in the sum of Eq. (2) is isolated:

$$\chi_2 = \frac{N_2 t(2k)}{R_2^2} \exp(-2k^2 \sigma_2^2) \exp(-2R_2/\lambda) \sin[2kR_2 + \sigma(k)]. \qquad (3)$$

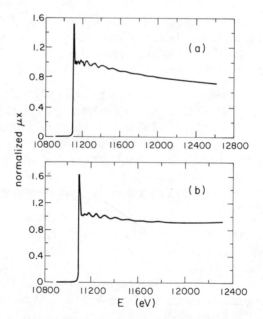

Fig. 9   The normalized x-ray absorption coefficients at the K-edge for (a) crystalline Ge and (b) amorphous Ge.

By an appropriate analysis $\chi_2$ can be separated into an amplitude $A_2(k)$ and a phase $\Phi_2(k)$:

$$A_2(k) = \frac{N_2 t(2k)}{R_2^2} \exp{-(2k^2\sigma_2^2)}\exp(-2R_2/\lambda) \tag{4}$$

$$\Phi_2(k) = 2kR_2 + \delta(k) \tag{5}$$

Isolating the second shell in the Fourier transform of the crystalline sample, one also can obtain an $A_{2c}(k)$ and $\Phi_{2c}(k)$. Taking the $\log_e$ of the ratio of these two amplitudes leads to:

$$\ln\frac{A_2(k)}{A_{2c}(k)} = \ln\frac{N_2 R_{2c}^2}{R_2^2 N_{2c}} + 2k^2(\sigma_{2c}^2 - \sigma_2^2). \tag{6}$$

From the transforms of Fig. 10 it is clear that $R_{2c} = R_2$. Also, from the fact that the first-shell coordination is the same in all

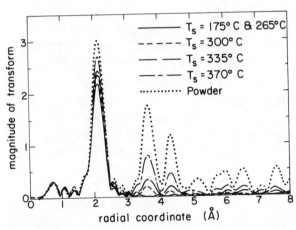

Fig. 10    The magnitude of the Fourier transform of $k^2$ times the EXAFS for various Ge films produced by sputtering at the substrate temperatures $T_s$ indicated. The dotted curve is a crystalline fine powder sample. The k-range of the transforms are from 0.5 to $15.1\text{Å}^{-1}$, and the EXAFS were all measured at 80K.

samples and the second-shell atoms are themselves first neighbors to the first-neighbor atoms of the center atom, it follows that the second-shell coordination numbers are the same, i.e., $N_2 = N_{2c}$, and Eq. (6) reduces to

$$\ln \frac{A_2(k)}{A_{2c}(k)} = 2k^2(\sigma_{2c}^2 - \sigma_2^2). \tag{7}$$

The right side of Eq. (7) is valid if the disorder is homogeneous and a Gaussian. Such is the case for the crystal, whose disorder can be accurately described by harmonic vibrations. However, the noncrystalline samples can have a more complicated disorder which will become apparent when their data are analyzed. If the log ratio is plotted as a function of $k^2$, its deviation from a straight line will reveal the more complicated behavior.

Such a plot is shown in Fig. 11 for the $T_5 = 300^{\circ}C$ sample, where the experimental points are supplemented by the point at the origin since $N_2 = N_{2c}$ and $R_2 = R_{2c}$. It is clear that the disorder in this sample is not simply a Gaussian. The behavior can be explained by the sample being heterogeneous, a portion having a small disorder and the rest having a large disorder. The solid line in Fig. 11 is such a fit. The results of this type of analysis on the samples are listed in Table 1. The disorder found for the highly disordered portions of the sample agrees with the value of the amorphous sample. For values of $T_s = 300^{\circ}C$ and above, the samples are a mixture of crystalline and amorphous parts, while below they are completely amorphous.

X-ray diffraction measurements on these samples show no crystalline peaks for the three samples with $T_s = 300^{\circ}C$ and lower. Optical absorption measurements indicate, in agreement with the x-ray measurements, that these three samples are amorphous. EXAFS measurements also indicate that the $T_s = 175^{\circ}C$ and $265^{\circ}C$ samples are amorphous but a discrepancy occurs for the $T_s = 300^{\circ}C$ sample. EXAFS indicates that about 20% of that sample is more ordered. The EXAFS results can be reconciled with the diffraction results if it is assumed that the crystallites are very small, less than 20Å in dimension. Then the size effect will greatly broaden the diffraction peaks, making them blend into the background from the amorphous matrix.

The EXAFS measurements also indicate that the crystallites are minute. The disorder of the crystalline component of the

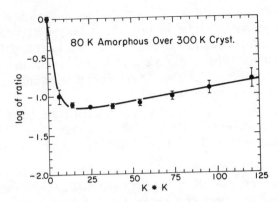

Fig. 11    Shown by points with error bars, the natural log of the
ratio of the second-shell amplitudes of a Ge film with
$T_S = 300°C$ to that of crystalline Ge. The curve is a fit
to the data. The film appeared amorphous by x-ray diffraction, but is actually two phase with a microcrystalline component. The $T_S = 300°C$ film was measured at
80K, while the Ge was measured at 300K.

Table 1.   Results of second-shell analysis at various substrate
temperatures $T_S$ on Kapton film. The rms disorder measured
at 80K is $\sigma^2$ measured in $Å^2$, and $\alpha$ is the fraction of the
sample that has crystallinity order.

| $T_S(°C)$ | $\alpha$ | $\sigma^2(x10^{-3}Å^2)$ |
|---|---|---|
| 175 | 0 | $100 \pm 20$[a] |
| 265 | 0 | $100 \pm 20$[a] |
| 300 | 0.20 | $9.5 \pm 1.0$[b] |
| 335 | 0.45 | $7.8 \pm 0.7$[b] |
| 370 | 0.86 | $6.6 \pm 0.5$[b] |
| Crystalline | 1.0 | $3.5 \pm 0.1$[b] |

[a]For amorphous portion of sample (same for all samples).
[b]For crystalline portion of sample.

samples is listed in Table 1. All sputtered samples show a larger disorder than the crystalline value. This larger disorder can occur only because of the strains introduced in the crystallites by the surrounding matrix. Since the strains in the surrounding amorphous matrix vary on an atomic scale, their decay into the crystallites also must be on an atomic scale. In order that the disorder in the crystallites of the $T_s = 300^{\circ}C$ sample be its measured value, the crystallites have to be of atomic dimensions. The estimate of the crystallites' size in the $T_s = 300^{\circ}C$ sample is about $10\text{Å}$.

The higher $T_s$ samples have diffraction peaks, indicating that they contain crystallites larger than $40\text{Å}$. However, the EXAFS results indicate that even the largest $T_s$ sample still has 14% amorphous content and a significant percentage of its crystallites are small, of the order of atomic dimensions. The latter result follows from the significant increase of disorder above the crystalline value.

The EXAFS results produce the most direct information to date about the structure of the amorphous state of Ge. The fact that EXAFS can distinguish between the amorphous state and $10\text{Å}$ microcrystallites clearly shows that the amorphous state of Ge is a continuous random network.

## V.    SUMMARY AND CONCLUSIONS

EXAFS is a new tool for structure determination of materials. It complements the standard diffraction techniques and, because of its characteristics, opens up new possibilities for study of systems not amenable to other methods. In an application to amorphous Ge, EXAFS showed conclusively that its structure is a continuous random network and not a microcrystalline state. However, in transforming to the crystalline state amorphous Ge was found to pass through a mixed amorphous-microcrystalline state where the microcrystallites are about $10\text{Å}$ in diameter. The crystallization of amorphous Ge is a continuous transition extending over a $T_s$ range of at least $70^{\circ}C$. Even at the highest $T_s$ of $370^{\circ}C$ the sample has not fully crystallized, still containing 14% amorphous and a significant amount of microcrystallites.

EXAFS was able to discern the distinction between the microcrystalline and amorphous states because of high resolution due to its capability to detect data to higher k-values than diffuse x-ray scattering. This high resolution also led to a more accurate characterization of the first coordination shell of amorphous Ge relative to the crystalline state.

The results of this study have introduced many questions, perhaps more than they have elucidated. What are the nucleation sites about which the microcrystallites and crystallites grow? How sensitive is the crystallization topology to the method of preparation, substrate, etc.? Why is the crystallization process so inhomogeneous and spread over such a large temperature region? Further studies to elucidate these questions are planned.

REFERENCES

1.      Some general reviews of EXAFS are: E.A. Stern, Contemp. Phys. 19, 289(1978); P.A. Lee, P.H. Citrin, P. Eisenberger and B.M. Kincaid, Rev. Mod. Phys. 53, 769(1981); E.A. Stern and S.M. Heald in Handbook on Synchrotron Radiation, edited by E.E. Koch, North Holland, N.Y. (1983), vol. 1, chapter 10.

2.      E.A. Stern, C.E. Bouldin, B. von Roedern and J. Azoulay, Phys. Rev. B 27, 6557(1983).

3.      See, for example, B.E. Warren, X-ray Diffraction, Addison-Wesley, Reading, MA (1969), chapter 10.

4.      P. Fuoss, P. Eisenberger, W.K. Warburton and A. Bienenstock, Phys. Rev. Lett. 46, 1537(1981).

5.      J.E. Enderby, D.M. North and P.A. Egelstaff, Phil. Mag. 14, 961(1966).

6.      R.J. Temkin, W. Paul and G.A.W. Connell, Adv. Phys. 22, 581(1973).

7.      C.E. Bouldin, E.A. Stern, B. von Roedern and J. Azoulay, Phys. Rev. B, October 15, 1984, to be published.

# MÖSSBAUER SPECTROSCOPY--A REWARDING PROBE OF
# MORPHOLOGICAL STRUCTURE OF SEMICONDUCTING GLASSES

Punit Boolchand

Physics Department, University of Cincinnati

Cincinnati, Ohio 45221-0011

## I.    INTRODUCTION

It has been fashionable to discuss the structure of stoichiometric melt-quenched network glasses in terms of chemically-ordered continuous random networks (CRN) since Zachariasen's pioneering work on the subject nearly 50 years ago [1]. Glasses of the type $AB_2$, such as $SiO_2$ and $GeSe_2$ in analogy to a-Ge, for example, have been described as random networks of geometrically well-defined $A(B_{1/2})_4$ tetrahedral units.

The availability of new spectroscopic results in the past four years has shown, however, that the structure of these network glasses is not all that random. Specifically, Raman [2] and Mössbauer [3] experiments show that some fraction of like-atom bonds A-A and B-B appears to be an intrinsic feature of the completely relaxed stoichiometric melt-quenched $GeSe_2$ and $GeS_2$ glasses. The presence of a finite and reproducible broken chemical order and particularly its y composition dependence in $A_{1-y}B_{2+y}$ glasses (A=Ge, B=S or Se) indicates [3] that the microscopic origin of these like-atom bonds cannot be due to isolated bonding defects in a completely polymerized and chemically ordered $A(B_{1/2})_4$ network. These homopolar bonds appear to be clustered indicating, as we will show, a phase separation of the network on a molecular scale.

This new conceptual approach has been stimulated by the general recognition that although near-neighbor covalent bonding

forces are the most important forces contributing to the cohesive
energy of a glass network, Van der Waals forces also play a significant role [4-6]. In chalcogen-based materials, Van der Waals
forces derive from the resonance of lone pair electrons and these
forces are known to promote chalcogen-chalcogen pairing as well
as a tendency to form chain-like or ring-like or layer-like structures. That these forces must play a very special role in $GeSe_2$
glass may be seen from measurements [7] of molar volumes in the
$Ge_xSe_{1-x}$ binary which display a striking local maximum at the
stoichiometric composition $x = 1/3$. These data underscore the
importance of network packing forces (intercluster interaction) over
covalent near-neighbor bonding forces (intracluster interaction)
which lead to a low atomic density network for $GeSe_2$. One may
understand why the lowest energy network configuration of a chalcogenide glass has a finite number of homopolar bonds in the
following terms. Apparently, the loss in cohesive energy of a network upon forming a few percent of homopolar bonds is more than
compensated by the increased Van der Waals contribution to this
energy term upon promoting some chalcogen-chalcogen pairing [6].

To quantitatively understand the observation of broken
chemical order, it has been suggested [6] that these $GeSe_2$ and
$GeS_2$ glasses consist of at least two types of morphologically
and stoichiometrically distinct large molecular clusters, analogous
to donor and acceptor molecules in molecular crystals. In this
molecular cluster network model (MCN) approach, cluster surfaces
are believed to play an integral role in determining the glass-
forming tendency. In this approach, the degree of broken chemical
order is derived from the surface to volume ratio, namely, the size
of the clusters. In this review, we will highlight some new experimental developments that shed light on the idea of molecular
clustering in network glasses. These new experimental developments have utilized Mössbauer spectroscopy [8] as a probe of
chemical order of both anion and cation sites in the present glasses.

Broadly speaking, several thousands of experiments on the
micro-molecular structure have utilized one of the four general
probes listed in Table I. The methods falling in categories I,
II, and III, namely, diffraction methods [10], vibrational spectroscopy [11], and photo-emission spectroscopy [12], have been re-
viewed in the literature from time to time. Mössbauer spectroscopy
as a probe of network structure of semiconducting glasses forms
the scope of the present review. Two previous reviews on the current subject, one by P.P. Seregin et al. [8] and the other by

Table I.  Experimental methods of studying the structure of inorganic glasses.

| Method | Experimental Observable | Connection With Structure |
|---|---|---|
| I.  Diffraction | | |
| (a) e,n, X-ray diffuse scattering | Structure Factor | Pair Correlation Function; |
| (b) EXAFS | Interference Function | Coordination number and |
| (c) XANES | | bond lengths of near- |
| (d) SANS | | neighbors |
| II.  Vibrational Spectroscopy | | |
| Infrared | Normal modes of | Identify building blocks of |
| Raman | characteristic building | a network glass |
| Mossbauer | blocks | |
| III.  Photoelectron Spectroscopy | | |
| X-ray Photoemission | Density of electron- | Ring statistics of network |
| UV Photoemission | valence or core states | |
| IV.  Hyperfine Interactions | | |
| (a) NQR (Nuclear Quadrupole Resonance) | EFG, Relaxation times | Local bonding chemistry |
| (b) NMR (Nuclear Magnetic Resonance) | Chemical shifts | |
| (c) $\mu$SR (Muon Spin Resonance) | Chemical shifts | |
| (d) ME (Mössbauer Effect) | EFG, $\lvert \psi(o) \rvert^2$ contact charge density | |
| (e) TDPAC (Time-Dependent Perturbed Angular Correlation) | EFG | |

W. Müller-Warmuth and H. Eckert [8], appeared nearly two years ago. For completeness, it may be also mentioned that Mössbauer spectroscopy has been widely utilized to characterize Fe sites in a variety of metallic as well as insulating glasses [9]. Some of the other site spectroscopies, particularly NMR [13] and NQR [14], as probes of structure in glasses are well documented.

Our presentation in this review is as follows: In Section II, we provide some background material on Mössbauer spectroscopy. The intention is to familiarize the reader with the method, thus making easier the interpretation of the spectra of chalcogenide network glasses. In Section III, we present experimental results on g-GeSe$_2$, first as revealed by diffraction measurements and vibrational spectroscopy, and then as extended by the present Mössbauer results. Instead of describing results on a variety of glass systems, we have thus chosen to focus on g-GeSe$_2$. This approach will permit not only illustration of the application of the present method but also a complete structure discussion of the prototypical glass. We conclude our illustrative review with some cautionary general remarks on the scope and limitations of the present method in Section V, where we also summarize our conclusions.

## II.    MÖSSBAUER SPECTROSCOPY

Neither Se nor Ge [15] offers the prospect of a suitable Mössbauer probe for glass work. The heavier isovalents of these elements, namely, Te and Sn, have, however, suitable Mössbauer resonances. The experimental approach, therefore, requires that ternary alloys of Ge, Se and Sn or Ge, Se and Te be investigated, and this is a point we will return to later. Upon alloying trace amounts of Sn or Te in GeSe$_2$ glass, one can expect these impurity atoms to mimic respectively the bonding chemistry of the cation and anion sites provided no phase separation occurs. In this section, we introduce the method focussing principally on $^{119}$Sn absorption (probe of cation sites) and $^{129}$I emission (probe of anion sites) spectroscopy. We have taken examples out of our glass work to illustrate the principles of the method. Excellent review articles on various aspects of the spectroscopy are available in the literature [16].

The principal components of a Mössbauer spectrometer are shown in Fig. 1, and consist of a velocity drive (transducer), an emitter attached to the drive, an absorber that is stationary, and a low energy x-ray detector. To display the lineshape of the nuclear resonance, one records the transmission of γ-rays through an absorber as a function of energy of the emitted γ-rays. The latter is accomplished using the Doppler effect. By imparting a precise velocity either positive (motion toward absorber) or negative (motion away from absorber), one increases or decreases the energy of the emitted γ-ray ($E_\gamma$) as seen in the frame of reference of the stationary absorber. Constant acceleration drives used in conjunction with multichannel analysers permit one to continuously scan a velocity range, i.e., a narrow energy window $\pm \frac{V}{c} E_\gamma$ centered around the emission line, and observe the resonance lineshape directly. Because of the recoil-free nature of the effect,

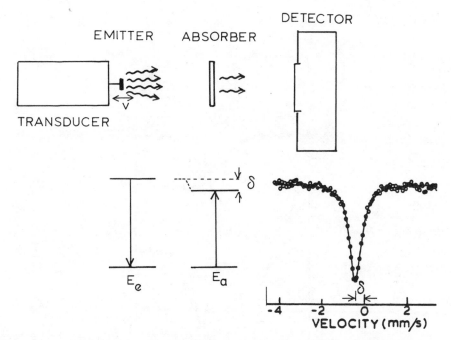

Fig. 1    Elements of a Mossbauer spectrometer: Transducer, emitter, absorber, and detector. Shown below the experimental arrangement is the nuclear level splitting. If the gamma ray energy in the emitter ($E_e$) exceeds the transition energy in the absorber ($E_a$) by an amount δ, the resonance centroid will be observed at a negative Doppler velocity.

the linewidth of the nuclear resonance is the natural width given by Heisenberg's uncertainty principle. The linewidth $(2\hbar/\tau)$ is determined by the lifetime $(\tau)$ of the nuclear excited state and this width is typically of the order of $10^{-7}$ eV. This linewidth translates into a Doppler velocity which is usually a fraction of a mm/s. Figure 2 shows a spectra of some Sn-bearing crystals taken with such a spectrometer.

There are basically three pieces of information that can be obtained from an analysis of the spectra shown in Fig. 2. This information is derived from:

(a)    Lineshift
(b)    Line-shape or splitting
(c)    Integrated intensity

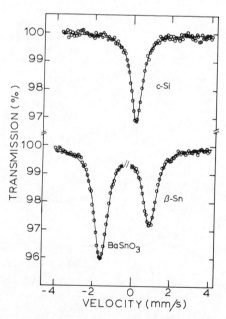

Fig. 2    $^{119}$Sn spectra of indicated absorbers taken with an emitter of $^{119}$Sn$^m$ in vanadium metal. Lineshifts of c–Si and β–Sn relative to BaSnO$_3$ are plotted in Fig. 3.

These experimental observables permit one to microscopical-
ly characterize the chemical environments of the probe atoms either
in the emitter or in the absorber matrix. The chief utility of this
spectroscopy derives from the fact that the strength of hyperfine
interactions (i.e., lineshift and linesplitting) usually is larger
than the natural linewidth of the nuclear resonance, enough to
permit separation of signals from different sites.

A.        Probe of Ge Sites in Network Glasses

In favorable cases, one may hope to isolate the chemical
bonding configuration of Ge sites in a network glass by alloying
traces of Sn and pursuing $^{119}$Sn absorption measurements. In such
measurements, the spectrum of a Sn-bearing glass or crystal, used
as an absorber, is taken with a mono-energetic emitter of the
23.8keV γ-ray (3/2 → 1/2 transition). The absorption spectra of
indicated hosts shown in Fig. 2 were taken with a source of 250
days $^{119}$Sn$^m$ diffused in vanadium metal.

The location of the absorption line on the velocity axis
measured in relation to some standard host is known as the isomershift or lineshift. The standard host taken for this purpose include
either $BaSnO_3$ or $CaSnO_3$. Both these materials are examples of
$Sn^{4+}$, and show a narrow line whose center of gravity is the same.
As can be seen from Fig. 2, the shifts of β-Sn and c-Si are both
positive and approximately 2.5mm/s and 1.7mm/s, respectively.
These shifts in energy units directly provide the difference in
transition energy of the 23.8keV γ-ray in various hosts. Specifically, the spectra of Fig. 2 show that the transition energy in
$BaSnO_3$ to be the smallest while that in β-Sn is the largest.

The isomer shift δ is generally written as

$$\delta = \frac{2\pi}{5} Ze^2 \Delta <r^2> [|\psi(o)|^2 - |\psi(o)|^2_{ref}] . \tag{1}$$

The term in square brackets represents the electron charge density
at the nuclear site in the host of interest measured relative to the
reference host. In Eq. (1), $\Delta <r^2>$ represents the change in the
nuclear charge radius between the ground and excited state, and
this nuclear moment has been established [16] to be $+3.3 \times 10^{-3} fm^2$
for $^{119}$Sn.

Figure 3 gives a $^{119}$Sn isomer shift scale in which we have
projected shifts of selected Sn bearing crystals as well as glasses.
On this plot, we find that shifts characteristic of tetrahedral Sn

reside in the region of +1.3mm/s to +2.0mm/s, a region which lies in-between the shifts of $Sn^{4+}$ and $Sn^{2+}$. The large positive shifts of $Sn^{2+}$ can be understood in terms of the two 5s-like valence electrons that are localized on Sn. These contribute overwhelmingly to the contact charge density $|\psi(o)|^2$. Just the reverse is true for $Sn^{4+}$, which has the lowest $|\psi(o)|^2$ because of the absence of these 5s electrons. Sn present in a local tetrahedral symmetry, as in c-Si, is described in terms of $sp^3$-like covalent bonds. This configuration has a shift that resides in-between $Sn^{4+}$ and $Sn^{2+}$, primarily because only one 5s-like electron contributes to $\cdot|\psi(o)|^2$.

$\alpha$-Sn and Sn as an impurity in the group IV elemental semiconductors [17] represent some of the few examples of tetrahedrally coordinated species found in crystalline hosts. Because of its more metallic character, Sn tends to choose octahedral over tetrahedral coordination in most Sn-bearing crystals. In covalent glasses, which are less dense than their crystalline analogues, just the reverse is true, and we find, for example, that Sn is predominantly tetrahedral in $GeX_2$ glasses (X=S, Se, and Te). As discussed later, spectra of these glasses display an intense Sn

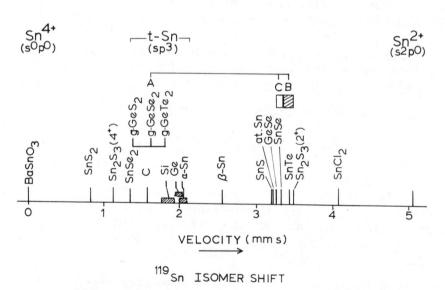

<sup>119</sup>Sn ISOMER SHIFT

Fig. 3    $^{119}$Sn isomer shifts of selected crystals and chalcogenide glasses.

site (usually labelled as A site) characterized by a narrow line.
We have found [18] that if one plots the lineshifts of these A sites
and the shifts of several Sn tetrahalides where the cation is known
to be fourfold coordinated in a tetrahedral symmetry, against the
Pauling electronegativity difference $\Delta X_P = X_X - X_{Sn}$, then a universal correlation results (Fig. 4). We have shown elsewhere [18]
that this correlation can be quantitatively understood in terms of
covalently bonded interactions which are modified by charge transfer effects. This correlation serves as conclusive evidence that
the A sites seen in the $GeX_2$ glasses represent geometrically and
chemically tetrahedrally coordinated Sn species. This is a point
that has largely gone unappreciated by Soviet workers in the field
(see P.P. Seregin et al in ref. 8) who have in our view erroneously
ascribed this site A to $Sn^{4+}$, an octahedrally coordinated species,
as found in corresponding crystals.

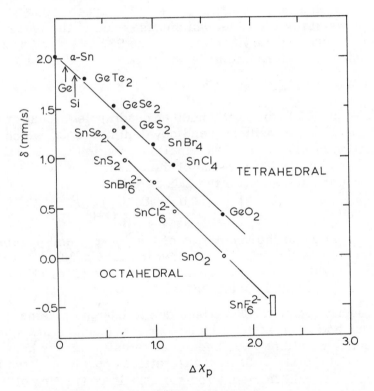

Fig. 4    $^{119}Sn$ isomer shifts of $SnX_4$ tetrahedral species plotted
as a function of Pauling electronegativity difference
$\Delta X_P = X_X - X_{Sn}$. Figure is reproduced from ref. 18.

The isomer shift of $Sn^{2+}$ species in covalent networks usually tends to be in the vicinity of 2.9mm/s to 3.4mm/s and furthermore is correlated with the quadrupole splitting that invariably accompanies such species. This correlation has been discussed by a number of previous workers in the field and the reader is referred to ref. [19] for a more complete discussion.

(i)      Line Splitting

It is usual not to observe single lines but a multiplet structure when probe atoms reside in a host that is magnetic and/or noncubic. The origin of this multiplet structure can be traced to either an electric quadrupolar and/or magnetic dipolar interaction which may be static or dynamic in origin. Specifically, when probe atoms occupy sites of noncubic local symmetry, the presence of an Electric Field Gradient (EFG) tensor leads to a splitting of the resonance line into a multiplet. In $^{119}Sn$ spectroscopy, this multiplet consists of a doublet which arises due to the presence of an electric quadrupole interaction in the $3/2^+$ 23.8keV state (Fig. 5), given by the relation

$$\Delta = \frac{e^2 Q V_{zz}}{2} (1 + \eta^2/3)^{1/2} \tag{2}$$

where $eQ = -0.065$ barns is the nuclear quadrupole amount of the $3/2^+$ states [16]. In insulating and covalent networks, which may be crystalline or glassy, the principal contribution to the EFG $(eV_{zz})$ arises due to an imbalance of the atomic-like 5p charge cloud of Sn and one can write the EFG as follows:

$$eV_{zz} = -\frac{4e}{5<r^3>} \left[ U_z - \frac{U_x + U_y}{2} \right] (1+R) \tag{3}$$

where $U_{x,y,z}$ denote the population of the $p_x$, $p_y$, and $p_z$ orbitals, while R represents the Sterheimer shielding factor [20]. The EFG physically provides a measure of the asphericity of the charge distribution about probe nuclear sites.

A second source of EFG arises due to charges external to probe atoms and in crystals of high symmetry (such as hexagonal) lattice sum calculations have been used to estimate this host contribution of the EFG. Several theoretical approaches have been used to estimate EFG in disordered solids. In conducting glasses which are described in terms of closed packed structures, distribution of EFG parameters has been estimated for the case of dense random packing [21]. In semiconducting glasses, an atom-

istic approach [22] (extended Huckel procedure) and also a band
approach [22] has been used to calculate EFGs. It is beyond the
scope of this review to discuss these approaches in any detail.

(ii)    Integrated Intensity

        Information pertaining to atomic dynamics of probe atoms is
contained in the area under the resonance line. Specifically,
temperature variation of the integrated intensity provides a means
to obtain the mean square displacement of the probe atoms. In
glasses, this can be related to specific low frequency vibrational
modes in the one phonon density of states. Mossbauer spectroscopy can, thus, serve as a chemically specific vibrational spectroscopy.

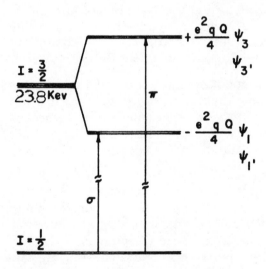

Fig. 5       Doublet spectra in $^{119}$Sn spectroscopy usually result
             due to an electric quadrupole interaction in the $3/2^+$
             nuclear state. When the EFG is axially symmetric ($\eta=0$),
             the $\pi$ ($\pm 1/2 \to \pm 3/2$) and $\sigma$ ($\pm 1/2 \to \pm 1/2$) are pure trans-
             itions.

(iii)    Multiple Line (Site) Intensity Ratios

In instances when a spectrum yields multiple sites that are chemically inequivalent, Mossbauer spectroscopy provides a means to establish directly the relative population of these sites by comparing the ratio of the areas under respective multiplet structures. In glasses, site intensity ratios have proved to be a powerful probe of network morphology [3,23], as we shall demonstrate by some examples in the present work.

B.    Probe of Chalcogen Sites in Network Glasses

It is possible to probe the chemistry of chalcogen sites in crystals or glasses by using the 35.5keV ($3/2^+ \rightarrow 1/2^+$) $\gamma$-resonance in $^{125}$Te in absorption spectroscopy, or the 27.8keV ($5/2^+ \rightarrow 7/2^+$) in $^{129}$I, using emission spectroscopy [24]. In the latter approach, the glass of interest forms the emitter matrix and its spectrum is taken with a monoenergetic absorber. There are several advantages to using the latter approach.

1.    The $^{129}$I resonance offers an order of magnitude greater resolution than the $^{125}$Te resonance because of its narrower natural linewidth (0.69mm/s versus 5.2mm/s).

2.    Second, because the quadrupole moments of the nuclear states in $^{129}$I are a factor of approximately three larger [16] than in $^{125}$Te, one has greater sensitivity. One can probe, for example, all other things being equal, smaller EFGs.

3.    In emission spectroscopy, a change in chemistry occurs on going from a Te parent atom to an I daughter atom which considerably simplifies the microscopic interpretation of the NQI parameters. Ths simplification can be traced to the tendency of I to be onefold coordinated and is a point we shall return to later.

The chief disadvantage of this method is that the noncrystalline host of interest has to be labelled by $^{129}$Te$^m$ atoms (radioactive). While this is not a problem in working with bulk glasses, this does pose difficulties in examining sputtered or evaporated samples unless provision for ion implantation exist [25], in which case, the probe atoms can be implanted after the fact of sample preparation.

In practice, one can combine both absorption and emission spectroscopies to gain insights in the chemistry of the chalcogen site in covalent network glasses. We have used both approaches to investigate in detail melt-quenched $GeSe_{2-x}Te_x$ ternary glasses. In Section III, we shall describe results obtained from $^{129}I$ emission spectroscopy on this ternary. The reader is referred to ref. [26] for discussion of the $^{125}Te$ absorption spectroscopy results on this system.

The central idea of $^{129}I$ emission spectroscopy is that one infers the bonding chemistry of parent $^{129}Te^m$ atoms (usually alloyed or implanted) in the chalcogen-based host by measuring the nuclear hyperfine structure of the daughter $^{129}I$ atoms that are formed by nuclear transmutation ($\beta$-decay). In this approach, the matrix of interest is the emitter matrix and one records its spectrum with a monoenergetic $^{129}I$ absorber, which is usually taken to be cubic NaI. This is convenient because the latter host which contains I in a closed shell configuration ($I^-$) is also used as a reference [27] for isomer-shifts measurements.

In instances when $^{129}I$ probe atoms are present in a locally noncubic environment, one observes a 12 line multiplet structure arising out of a NQI in both the $5/2^+$ ground and $7/2^+$ excited state as shown in Fig. 6. The multiplet structure can be theoretically

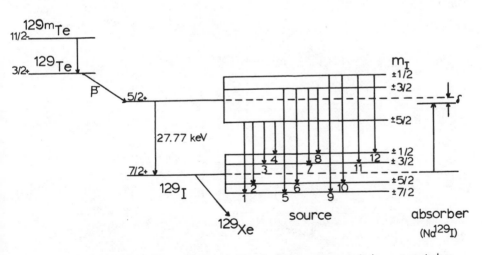

Fig. 6    Electric quadrupole interaction in the $7/2^+$ and $5/2^+$ state of $^{129}I$ showing the 12 transitions that result on account of the mixed (MI + E2) nature of the 27.77keV $\gamma$-transition.

analyzed in terms of 3 parameters: a centroid or isomer shift ($\delta$), ($e^2 Q_g V_{zz}$), nuclear quadrupole coupling in the ground state and $\eta = |V_{xx} - V_{yy})/V_{zz}|$, the asymmetry parameter of the EFG tensor. The nuclear quadrupole moment ratio R ($= Q_e/Q_g$) of the excited ($Q_e$) to ground state ($Q_g$) has been established [28] to be 1.238(1). In Fig. 7, we show the $\eta$ dependence of the multiplet structure for a fixed $e^2 Q_g V_{zz}$ value, and note that generally the hyperfine structure yields a pattern that lacks inversion symmetry about its centroid. This has the important consequence that the sign of the quadrupole coupling constant, and therefore the EFG, can be uniquely established from such a spectrum. Only for the case

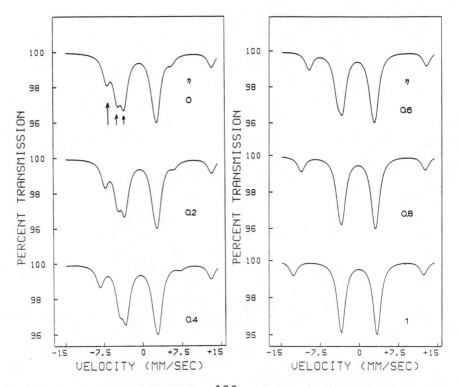

Fig. 7      $\eta$ dependence of the $^{129}$I electric quadrupole multiplet structure for a fixed value of $e^2 Q_g V_{zz}$. Note that when $\eta = 0$, the spectra reveal a characteristic triplet structure (shown by arrows) on one of the main lines.

when $\eta = 1$ can this not be done because the pattern becomes completely symmetric and the sign of EFG remains no longer defined.
In particular, when $\eta = 0$, one observes a characteristic triplet
structure (shown by arrows) on one of the principal lines seen in
the spectrum. There are primarily two types of spectra we shall
encounter in our glass work, one where $\eta = 0$ and the other where
$\eta = 0.8$ or so.

Chemical bonding information in this spectroscopy is best
obtained by plotting the isomer shift as a function of quadrupole
coupling as shown in Fig. 8. Detailed discussion of this plot
appears elsewhere in the literature [28-30] and will not be given
here. Different parts of this plot pertain to I present in different
chemical states. Specifically, a onefold coordinate I site, as in
$I_2$ dimer for example, resides along the line drawn due north-east
on this plot. Note that the sign of the quadrupole coupling of such
a onefold coordinated species is always negative, and this is a
point we will return to later.

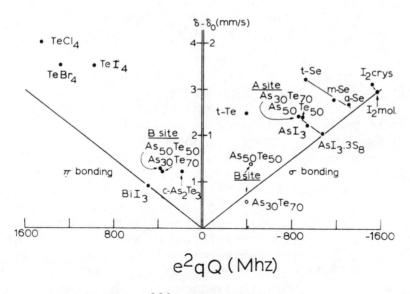

Fig. 8     Correlation of $^{129}I$ isomer shift with quadrupole coupling
taken from ref. 29. Note that when I is onefold coordinated, the sign of the coupling is always negative.

Within the context of NQI as a probe of chalcogen chemistry, two central ideas need to be introduced at this stage. It is well known that the highly directional nature of bonds that chalcogens form arise due to the open p valence shell ($s^2p^4$). Specifically, the twofold coordination of these elements arise because two of the four p valence electrons enter in covalent bonds with two near-neighbors while the remaining two p electrons form the so-called nonbonding lone-pair electrons. Of course, the two s valence electrons can also participate in bonding by hybridization and their principal effect is to increase the chalcogen bond angle from $90^\circ$ to about $102^\circ$ as is found in the elemental chalcogens for example.

Because of the lack of cubic symmetry around a twofold coordinated Te site, one expects in general a large EFG. The magnitude and sign of this EFG can be inferred using an atomistic approach. Because of the local $C_{2v}$ symmetry, the principal z axis of the EFG tensor is expected to lie along the lone-pair electron lobes, while the principal x and y axis to lie along the bonding orbitals. This distribution of the four p valence electrons requires the net EFG to be negative and of magnitude $\frac{4e}{5\langle r^3\rangle}$. For the case of crystalline Te, the sign of the EFG has been explicitly established [31], and it is indeed found to be negative as expected. We expect the sign of the EFG in general to be negative so long as Te is twofold coordinated as in the elemental chalcogens (p-S, a-Se).

Experiments on a wide variety of crystals have indicated the following general pattern: whenever the parent $^{129}Te^m$ atoms are twofold coordinated, as in the elemental chalcogens, the daughter $^{129}I$ is nominally onefold coordinated. This is inferred from the sign of the EFG which undergoes a change from being negative at a Te probe to positive at an I probe. We have suggested [24] that this pattern is the consequence of a bond rearrangement on account of a change in chemical valence. Specifically, the principal axis of the EFG rotates by $90^\circ$ in going from a Te probe to an I probe (as shown in Fig. 9) as one of the Te $\pi$-bonds breaks and the other $\pi$-bond becomes an I-$\sigma$ bond. Because all nuclear quadrupole moments are negative, the quadrupole couplings ($e^2QV_{zz}$) change sign from being positive at Te to negative at I. The positive sign of the EFG for onefold coordinated I site is best understood to arise, as in the case of an $I_2$ dimer, on account of $\sigma$ bonding of $p_z$-like hole in the closed shell $5\,s^2p^6$ configuration.

Experiments also indicate that in instances when the parent Te atom has a coordination that is higher than twofold, such as threefold or distorted octahedral, the daughter I continues to remain $\pi$-bonded, i.e., the sign of the EFG remains unchanged and continues to remain negative [29]. Apparently, in such a case it is energetically unfavorable for more than one of the $\pi$-bonds to break to realize a onefold coordinated $\sigma$-bonded species.

There are two additional features of NQI parameters that serve as microscopic signatures of Te coordination. First, the $^{129}$I quadrupole coupling of a $\pi$-bonded species is nearly half of a $\sigma$-bonded species for the same isomer shift. This is related to the fact that although each of the $p_x$, $p_y$, and $p_z$ I-orbitals shields the 5s contact charge density by the same amount, the magnitude of EFG produced by a $p_z$ orbital is twice that produced by a $p_x$ or $p_y$ orbital. Second, experiments reveal that in covalently-bonded networks, the ratio R of the $^{125}$Te quadrupole coupling to the $^{129}$I quadrupole coupling equals nearly 1/2, and is a point discussed elsewhere in detail [24].

To summarize, this spectroscopy makes accessible through the $^{129}$I NQI parameters several tests of the Te parent coordination which derive from (a) sign of the $^{129}$I $e^2Q_gV_{zz}$, (b) magnitude of $^{129}$I isomer shift and $e^2Q_gV_{zz}$, and (c) ratio of $^{125}$Te/$^{129}$I Quadrupole couplings, as discussed above.

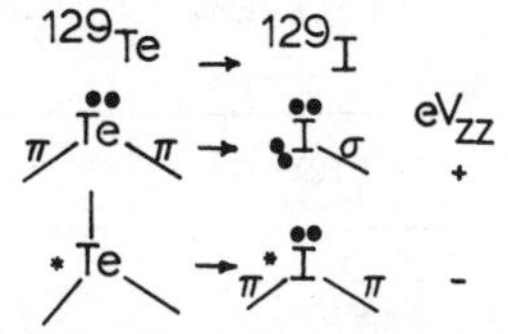

Fig. 9    Bonding configuration of $^{129}$I daughter atoms formed from (a) twofold Te atoms, and (b) threefold Te atoms. The filled circle and asterisk designate lone-pair and anti-bonding electron states. Our usage of the terms $\pi$ and $\sigma$ bonds here differs from the usual chemical language in that these are defined in the principal axes of the EFG tensor of Te or I. See ref. 29 for additional details.

III.    EXPERIMENTAL RESULTS ON $GeSe_2$ GLASS AND DISCUSSION

To probe the cation site in the glass of interest, we have examined ternary $(Ge_{0.99}Sn_{0.01})_x Se_{1-x}$ glasses prepared by melt quenching in water in the usual way [3]. Our approach is to examine the compositional evolution of the spectra as x approaches 1/3. For probing the chalcogen sites, we have chosen to look at ternary $GeSe_{2-x}Te_x$ glasses in a similar fashion, focussing on the nature of sites prevailing as x approaches 0. In what follows, we shall summarize results of these experiments which have been published elsewhere [23,30].

A.    Ge Chemical Order

$^{119}Sn$ Mossbauer spectra at selected compositions near x = 1/3 in the ternary $(Ge_{0.99}Sn_{0.01})_x Se_{1-x}$ are displayed in Fig. 10. The most striking result to emerge from these spectra is the presence of two types of Sn sites: a symmetric site A which shows a single line and an asymmetric site B which exhibits a quadrupole doublet (Table II). The site intensity ratio, $I_B/(I_A + I_B)$, increases with x in a manner that is sketched in Fig. 11. This figure also shows the $T_g$ of the corresponding glasses for comparison which serves as a check of sample stoichiometry. The observed line-widths in the glasses $\Gamma$ (full width at half maximum) = 0.93(3)mm/s

Table II    $^{119}Sn$ isomer shift ($\delta$) and quadrupole splitting ($\Delta$) in indicated glass (g) and crystalline (c) samples. The shifts are quoted relative to $BaSnO_3$.

| Sample | | $\delta$(mm/s) | $\Delta$(mm/s) |
|---|---|---|---|
| g–$GeSe_2$ | A | 1.55 (3) | ... |
| | B | 3.20 (3) | 2.13 (3) |
| c–$SnSe_2$ | | 1.36 (2) | ... |
| c–$SnSe$ | | 3.31 (2) | 0.74 (2) |

are just as narrow as the ones seen in the $SnSe_2$ and SnSe crystals. This indicates that the sites (A,B) seen in the glasses are chemically well defined.

We identify site A in our spectra with Sn atoms that replace Ge in a symmetric tetrahedral $Ge(Se_{1/2})_4$ unit. The evidence to support this identification includes (i) the single-line nature of this site which is consistent with a vanishing EFG in a local tetrahedral coordination, (ii) the isomer shift of the single line (see Fig. 3 and Table II) which is characteristic of a Sn site that is geometrically and chemically tetrahedrally coordinated to four Se near-neighbors.

The dominant nature of site A at $x = 1/3$ (shown by $I_B/I = 0.16$ in Fig. 11) is in accord with results of Raman vibrational spectroscopy which reveal that $Ge(Se_{1/2})_4$ units comprise the principal

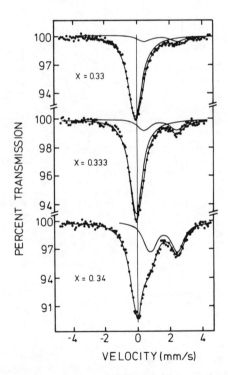

Fig. 10    $^{119}Sn$ spectra of indicated $(Ge_{0.99}Sn_{0.01})_xSe_{1-x}$ glasses showing the presence of two Sn sites: Site A is the intense single line near v = 0mm/s, while Site B is the quadrupole doublet centered at about +1.5mm/s. Figure is taken from ref. 26.

building block of a $GeSe_2$ glass. In our spectra, we identify site B with Sn atoms that replace one of the Ge sites in an ethane-like $Ge_2(Se_{1/2})_6$ unit. The evidence in support of this identification includes the doublet nature of this site which we believe results due to the locally asymmetric Sn coordination in such a cluster, and

$$Ge-Sn{\nearrow}^{X}_{\searrow X}{}^{X}$$

in which the broken tetrahedral symmetry causes a finite EFG and therefore a quadrupole splitting. Secondly, the $I_B/I$ data of Fig. 11 shows that site B dominates as x approaches 2/5. This constitutes strong evidence in favor of the proposed identification because it has been shown by optical [32] and by chemical [33]

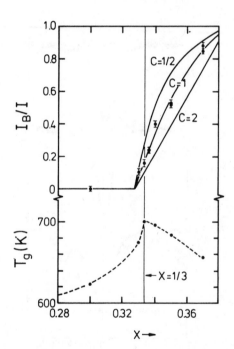

Fig. 11    Observed x-dependence of the site intensity ratio $I_B/I$ and glass transitions ($T_g$) in melt quenched $(Ge_{0.99}Sn_{0.01})_xSe_{1-x}$ glasses. The parameter C des cribes chemical preference of Sn to choose A over B sites. C = 1 implies that Sn chooses to attach in A and local units randomly.

means that $Ge_2(Se_{1/2})_6$ ethane-like unit constitutes the principal
building block of a $Ge_2Se_3$ glass. In our spectra, the observation
of a finite intensity of site B at $x = 0.333$ constitutes, therefore,
the first clear evidence for intrinsically broken Ge chemical order
in a $GeSe_2$ glass. We have developed a model to relate the populations ($N_A$ and $N_B$) of $Ge(Se_{1/2})_4$ and $Ge_2(Se_{1/2})_6$ units in the
glass network to the measured Sn site occupations (intensities $I_A$
and $I_B$) of these units. Let us suppose that $I_B/I$ is a smooth function of x and takes on values of 0 and 1 at $x = x_0$ and $x = x_1$.
Further, let us suppose that the intensities ($I_A$, $I_B$) depend on the
populations ($N_A$, $N_B$) and on the chemical affinity
$C = \exp[(E_B - E_A)/kT_g]$of Sn atoms to attach themselves in respective Ge units. $E_B - E_A$ represents the bond energy difference
in moving a Sn atom from a site B to a site A. On minimizing the
free energy, one can show that

$$\frac{I_B}{I_A + I_B} = \frac{N_B}{C\,N_A/2 + N_B} = \frac{x - x_0}{C(x_1 - x)/2 + (x - x_0)} \cdot \quad (4)$$

The smooth curve through the data points in Fig. 11 is a fit to
Eq. (4), and yields $C = 1.0$, $x_1 = 0.385(4)$. The value of $C = 1$
indicates that Sn atoms choose randomly the available Ge units,
and further that $I_B/I$ equals the fraction of Ge sites in ethane-like
units of the network. We define the degree of broken order (DBO)
in a $GeSe_2$ glass as the fraction of Ge sites in ethane-like units
of the network, i.e., $2N_B/(N_A + 2N_B)$ and find its value to be
$0.16(1)$.

The trend of an increase in the site intensity ratio $I_B/I$ with
x (Fig. 9), particularly in the composition range $x_0 < x < 0.333$,
where $T_g$ of the glasses increases so rapidly, is a remarkable
result. This trend is much too steep to be described by a model in
which the chemical order breaking site B is identified with some
Ge-Ge bonds that are formed at random in an ordered bond network.
This is seen by comparing the observed slope $d(I_B/I)/dx$ at
$x = 0.33$ of $32(2)$ [Fig. 11 and Eq. (4)] with the calculated slope
$d(N_{Ge-Ge}/N)/dx$ of 18 describing the change in the fraction of Ge
sites in Ge-Ge bonds at $x = 0.333$ in an ordered bond network.
Indeed, as the number of Ge-Ge bonds at $x = 0.333$ increases (increased disorder), the slope $d(N_{Ge-Ge}/N)/dx$ can be shown to
decrease from its maximum value of 18 to a minimum value of 4,
for a completely random covalent network. This trend, on the
other hand, is better described as reflecting a rapid growth in the
fraction of Ge sites in $Ge_2(Se_{1/2})_6$ clusters with x. Such sites

correspond to Ge-Ge bonds. According to our data, the fraction $Ge_2(Se_{1/2})_6$ clusters, i.e., $N_B/(N_A + N_B)$, is predicted to vary nearly linearly with x (straight line corresponding to C = 2 in Fig. 9). This linear variation is expected to scale with the scattering strength of the $180cm^{-1}$ feature [shown in ref. 32 to be the benchmark of a $Ge_2(Se_{1/2})_6$ unit in Raman spectra of $Ge_xSe_{1-x}$ glasses]. The overall picture of the network structure emerging from different types of measurements will be discussed in Section IV.

B.    Se Chemical Order

Selected $^{129}I$ emission spectra of ternary $GeSe_{2-x}Te_x$ glasses [23,30] and the elemental chalcogens are displayed in Fig. 12. The central result to emerge from these spectra is that in glasses of $GeSe_{2-x}Te_x$ and $GeS_{2-x}Te_x$, even as x approaches 0, there are two inequivalent I sites, A and B. For example, as can be seen in Fig. 12, a qualitative improvement in the fit to a g-GeSe₂ spectrum results in going from a one-site to a two-site fit. The NQI parameters $e^2QV_{ZZ}^A$, $e^2QV_{ZZ}^B$, $\eta^B$, and $\eta^A$ for $GeSe_{2-x}Te_x$ alloys are shown as a function of x in Fig. 13. These parameters were obtained by standard Mössbauer spectra analysis described in Section II.B, and for $x \to 0$, they are compared in Table III with the elemental solute parameters. Note that the sign of $e^2qQ$ is negative at both A and B sites, suggesting that the parent Te site is twofold coordinated in each instance. There is compelling evidence that the two sites seen in the present glasses do not originate from a nuclear aftereffect, such as bond breaking following nuclear transmutation. To date, such an effect has not been observed in any metallic or semiconducting host. Our observation of a unique and static NQI in the elemental glasses (S, Se), see Fig. 12 and Table III, follows this pattern, and it strongly suggests the absence of a nuclear aftereffect in these semiconducting glasses.

Most recent workers have assumed that the atomic structure of chalcogenide glasses can be described in terms of a chemically ordered CRN as originally proposed by Zachariasen [1]. Our data provide the first direct evidence that this is not the case. Specifically, let us suppose that chalcogen sites are of two types, a chemically ordered site A bonded to two Ge atoms, and site B bonded to a Ge atom and a chlacogen atom. Let the probability that a Te atom occupies an A or B site be $p^A$ or $p^B$, and let the branching probabilities that B site daughter I atoms will choose to attach themselves to the Ge or chalcogen neighbors be $f_G$ or $f_C$. Then, the probabilities $P_{G,C}$ that $^{129}I$ will be bonded to Ge or the chalcogen are given by

$$p_G = p^A + p^B f_G, \tag{5}$$

$$p_C = p^B f_C. \tag{6}$$

Returning to Table I, we see by comparison with the elemental chalcogen parameters ($e_2 Q V_{zz}$, $\eta$, and isomer shift) that for $x \rightarrow 0$, i.e., $GeSe_2$ (and also $GeS_2$), the B site definitely corresponds to an I-chalcogen bond. Further, we note by comparing

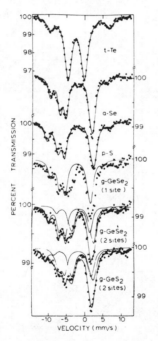

Fig. 12  $^{129}I$ emission spectra of indicated hosts taken from ref. 30. A noticeable improvement in the fit to the spectrum of g-$GeSe_2$ results in going from a one-site to a two-site fit.

A-site parameters in $GeSe_2$ with the A-site parameter in $GeS_2$ that these are really the same site, and identify it with an I-Ge bond. Regardless of the transmutational branching probabilities $f_G$ and $f_C$, Eq. (6) tells us that there must be a substantial probability $p^B$ of the type B Te sites in the glass, i.e., the chemical ordering must be intrinsically broken.

According to Fig. 13, $I_B(x) > I_A(x)$, i.e., $p_C(x) > p_G(x)$, at $x = 0$ and further $p_C(x)$, $e^2QV_{ZZ}^B$, and $\eta^B$ all sharply change with increasing x, especially $p_C$ which is already halved at $x = 0.2$. This suggests several interpretations of the chalcogen environments

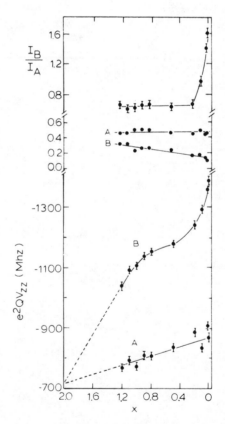

Fig. 13     The observed variation of the quadrupole couplings
            $(e^2Q_gV_{zz})$, intensity ratio $I_B/I_A$, and asymmetry para-
            meter $(\eta)$ for the two sites in $g-GeSe_{2-x}Te_x$ alloys (x)
            of the glass.

and the physical mechanism responsible for the transmutational branching ratio. Because $p_C(0) > p_G(0)$, we conclude that $f_G/f_C$ is small and is probably close to zero, i.e., I prefers to form I-Se(S) rather than I-Ge bonds in the glass. This result is somewhat surprising, because according to Pauling [34] the ionic contribution to the heat of formation of I-Ge bonds should be almost 50 times greater than that of I-Se(S) bonds. We must remember, however, that these bonds are not formed in the vapor or in dilute solution. Instead, they are formed in the melt quenched glass which is 90% as dense as the crystal. In dense covalent networks the Van der Walls repulsive energy between nonbonded lone-pair electrons can be quite large, as is shown by the near equality of bonded and nonbonded interatomic spacings in elemental Se crystals. Thus, at B sites, I-Se(Se) bonds are formed in preference to I-Ge bonds, apparently because in the latter case, the repulsive or steric hindrance nonbonded I-Se(S) interactions overwhelm the ionic energy difference.

The rapid variation in the site intensity ratio displayed in Fig. 13 with x will be discussed in the next section in connection with the structure of $GeSe_2$ glass.

Table III  $^{129}I$ quadrupole coupling ($e^2QV_{zz}$), asymmetry parameter ($\eta$), and isomer shift ($\delta$) deduced from spectra of Fig. 1. $\delta$ is quoted relative to $Na^{129}I$.

| Host | $e^2QV_{zz}$(MHz) | $\eta$ | $\delta$(mm/s) |
|------|------|------|------|
| t-Te | $-$ 397 (2) | 0.70 (1) | 1.16 (1) |
| a-Se | $-$ 1341 (10) | 0.11 (2) | 1.26 (4) |
| p-S | $-$ 1453 (6) | 0.13 (2) | 1.26 (2) |
| $GeSe_2$ A | $-$ 860 (12) | 0.49 (6) | 0.76 (3) |
| $GeSe_2$ B | $-$ 1360 (9) | 0.15 (6) | 1.28 (4) |
| $GeS_2$ A | $-$ 936 (15) | 0.57 (6) | 0.62 (4) |
| $GeS_2$ B | $-$ 1432 (10) | 0.25 (6) | 1.29 (4) |

IV.    STRUCTURE OF GeSe$_2$ GLASS

The network structure of GeSe$_2$ has been the focus of previous Raman and diffraction measurements. In this section, we will show that the present Mossbauer results extend in a significant way the understanding of the morphological structure of this material as deduced from the previous investigations. We shall conclude by showing that the present experiments strongly suggest that GeSe$_2$ glass is an incompletely polymerized tetrahedral network, and is better described in terms of a heterogeneous microstructure consisting of clusters.

A.    Diffraction Experiment

Chemically specific diffraction experiments [35] utilizing EXAFS of the Ge K edge and Se K edge in glassy and crystalline GeSe$_2$ have been observed. The interference function analysis indicates that on an average Ge and Se atoms are coordinated respectively to four Se and two Ge near-neighbors in accord with the 8-N coordination rule. These experiments make it plausible to visualize the GeSe$_2$ glass network in analogy to the crystal to consist predominantly of Ge(Se$_{1/2}$)$_4$ tetrahedral building blocks.

It also appears that the tetrahedral units in the glass, as in the high temperature crystalline phase of GeSe$_2$, may be present in a layered morphology. Both x-ray [36] and neutron [37] structure factor of g-GeSe$_2$ reveal an anomalous sharp diffraction peak at k = 1.1 Å$^{-1}$ (see Fig. 14). Anomalous x-ray scattering techniques [38] using a syncrhrotron source have further shown that the peak at k = 1.1 Å$^{-1}$ corresponds to Ge-Ge correlation. Because this correlation translates into a real space distance of about 6Å, which also happens to be the Ge-Ge interlayer distance in β-GeSe$_2$ [39], it appears plausible that this peak may signify the presence of medium range order in a network which consists of locally layered like structures.

Neutron diffuse scattering experiments on GeSe$_2$ glass and liquid show certain common features. Surprisingly, both structure factors display the anomalous sharp first diffraction peak at k = 1.1 Å$^{-1}$ alluded to above, albeit the width of this peak being broader in the liquid phase. Furthermore, Raman spectra of liquid and glassy GeSe$_2$ both display [40] the so-called A$_1$ companion mode. This mode is generally taken to be signature of medium-range order and will be discussed in more detail next. To summarize, characteristic features of both Raman and diffraction

experiments, thus, clearly indicate that the medium-range
structure of the glass bears some relationship to that of the
liquid. It is plausible to imagine that characteristic molecular
fragments having a local layered-like arrangement of tetrahedral
units persist in both these phases. After all, the glass does
evolve from the liquid and is, in fact, the supercooled liquid.

B.    Vibrational Spectroscopy

Direct confirmation of the tetrahedral nature of the principal building block of a $GeSe_2$ glass network has emerged by de-
coding Raman scattering [41,42] and IR reflectance spectra [43].
These spectra reveal several sharp vibrational modes. Several
of these modes have been positively identified as the normal
modes of nearly decoupled tetrahedral $Ge(Se_{1/2})_4$ units in
$g$-$GeSe_2$. Variation of mode strengths with glass composition in
$Ge_xSe_{1-x}$ glasses, measurements of depolarization ratios [42],
and, finally, comparison of observed mode frequencies with theo-

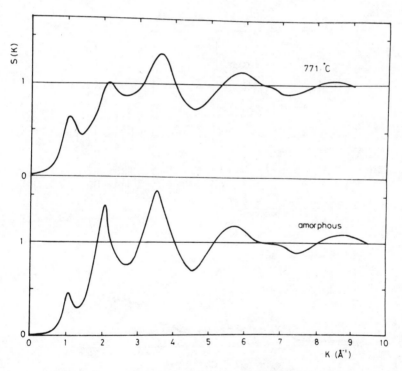

Fig. 14    Neutron scattering structure factor of liquid and glass
$GeSe_2$ (taken from ref. 37) showing the first anomalous
sharp peak at $k = 1.1 Å^{-1}$.

retical estimates of one phonon density of states in model
clusters [45] have served to confirm the microscopic origin of the
modes. Specifically, in the Raman spectra of g-GeSe$_2$(see Fig.
15), the sharp mode at 200 cm$^{-1}$ and the broad one at 304 cm$^{-1}$
are identified as the A$_1$ symmetric breathing mode and the F$_2$ high
frequency scissor mode of a Ge(Se$_{1/2}$)$_4$ tetrahedral unit. In the
spectrum, there are two surprises, however, and these include
(a) a weak mode at 180 cm$^{-1}$ which appears as a shoulder to the
A$_1$ mode, and (b) a narrow and intense mode at 210 cm$^{-1}$, usually
labelled as A$_1^C$, the companion of the A$_1$ mode. The microscopic
origin of the 180 cm$^{-1}$ is identified as the normal mode of nontetrahedral species, specifically A$_{1g}$ mode [42] of an ethane-like
Ge(Se$_{1/2}$)$_6$ cluster. The origin of the A$_1^C$ mode, on the other hand,
has been the subject of some controversy. Nemanich and Solin [1]
were the first to emphasize that the rapid variation of the A$_1^C$ mode
strength in the Se-rich (x > 1/3) phase of the Ge$_{1-x}$Se$_x$ binary requires the presence of large clusters in the glass. In their model,
their clusters were described to be 12-atom rings of Ge$_6$Se$_6$ which

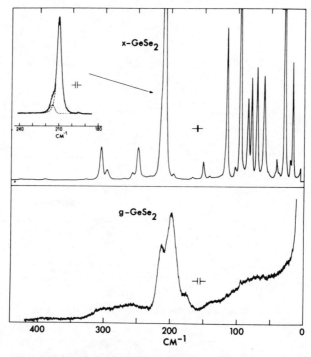

Fig. 15     Raman spectrum of crystalline and glassy GeSe$_2$ taken
            from ref. 46.

are embedded in a perfectly chemically ordered network (Fig. 16a). Bridenbaugh et al. [46], on the other hand, proposed that chemical order is intrinsically broken in the stoichiometric glass. They suggested that two kinds of partially polymerized clusters are present which are either cation-rich or chalcogen-rich. The latter were proposed to be raft-like fragments of the high temperature crystalline form, and were laterally bordered by chalcogen-chalcogen bonds (Fig. 16b). These workers identified the $A_1^C$ mode as a Ge-Se stretch mode localized at the edges of this chalcogen-rich cluster. This particular interpretation of the $A_1^C$ mode is now supported by the recent photostructural transformation [2] studies and vibrational density of states calculations [45], and also by our Mössbauer spectroscopy results [2,23,30,48] which will be discussed next.

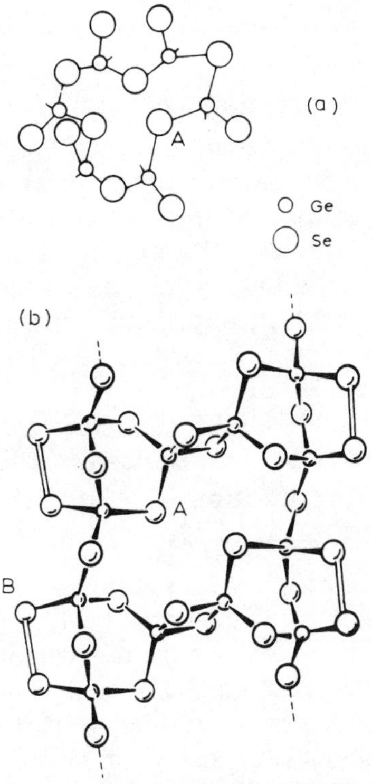

Fig. 16  Elements of medium-range order in g-Ge-Se$_2$ according to (a) CRN model, and (b) outrigger-raft model; (a) shows a Ge$_6$Se$_6$ ring cluster providing for one A site, while (b) shows a Ge$_6$Se$_{14}$ outrigger-raft cluster providing for two chemically inequivalent A and B Se sites.

The heterogeneous character of Ge(Se or S)$_2$ glass network has been beautifully illustrated in a series of recent pressure dependent Raman [49] and optical absorption edge shift [50] experiments. In the Raman experiments on g-GeSe$_2$ for example, Murase et al. observe the scattering strength of the $A_1^c$ mode extrapolate to zero at a pressure of 48 kbar. Apparently, the molecular clusters prevailing at ambient pressure coalesce through cluster surface reconstruction upon application of pressure. The high pressure phase of g-GeSe$_2$ may represent the second clear example of a Zachariasen network, the first example being the ternary glass Ge$_{0.65}$Sn$_{0.35}$Se$_2$, as illustrated by the Mössbauer experiments of Stevens et al. [48].

## C.  Mössbauer Spectroscopy

### (i) Degree of Broken Chemical Order on Cation and Anion Sites Compared

Both the $^{119}$Sn and $^{129}$I experimental results on (Ge$_{0.99}$Sn$_{0.01}$), Se$_{1-x}$, and GeSe$_{2-x}$Te$_x$ glasses, which were presented in Section III, provide conclusive evidence for the existence of symmetry breaking B sites in GeSe$_2$. It is instructive to inquire if the degree to which the chemical order is broken at cation sites and anion sites bears any relationship to each other. This requires that we establish explicitly the ratio of occupation probability of the probe atoms (Sn or Te) to attach themselves in the two chemically inequivalent (A,B) sites of the network. We have already demonstrated in Section III that for the case of Sn, probe atoms randomly select A and B sites of the network. This led us to conclude that the fraction of nontetrahedral Ge sites present in GeSe$_2$ network is 0.16(1).

Our $^{129}$I experiments provide evidence of a high selectivity of the Te atoms to replace the symmetry breaking B sites over the A sites, and this is seen by analyzing the variation of the site intensity ratio $I_B(x)/I_A(x)$ which rapidly increases in the composition range $0 < x < 0.1$. We have built a statistical model [30] in which GeSe$_{2-x}$Te$_x$ alloy glasses for $x < 0.1$ are visualized as made up of strings consisting of single chalcogen strings: Ge-Se-Ge and Ge-Te-Ge, and double chalcogen strings: Ge-Se-Se-Ge, Ge-Se-Te-Ge, and Ge-Te-Te-Ge. The crosslinking of the strings at Ge insures that 8-N coordination rule is satisfied, i.e., Ge is always fourfold and the chalcogen always twofold coordinated. By assigning characteristic energies to these strings

and minimizing the free energy of the stringed network, we have
projected (Fig. 17) the population ratio R of Ge-Se-Te-Ge strings
(B sites) to Ge-Te-Ge strings (A sites) as a function of x. The
calculations show that in the Se-rich phase of the present ternary
this ratio depends on two parameters $\rho$ and $\beta$. We define $\rho$ as the
population ratio of all double chalcogen strings to all single chalcogen strings, while $\beta$ as the probability ratio of Te to select B
sites over A sites. We find from Fig. 17 that to reproduce the
rapid variation of $I_B/I_A(x)$, the parameters $\rho \leq 1/16$ and $\beta \geq 13$.
The finite value of $\rho$ indicates that the stringed network is chalcogen rich, and is characterized by a stoichiometry of
$GeSe_2(1+2\rho)/[(1+2\rho)/(1+\rho)]$. Specifically, the value of $\rho \leq 1/16$
implies that the stringed network has on average stoichiometry
that is Se poorer than $Ge_{17}Se_{36}$. The value of $\beta \geq 13$, on the
other hand, shows that there exists a high preference for Te to
populate B sites over the A sites.

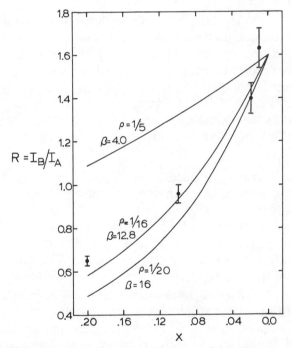

Fig. 17    Plot of R = Ge-Se-Te-Ge strings to Ge-Te-Ge strings in
           $GeSe_{2-x}Te_x$ glasses as a function of x. The parameters
           $\rho = M/N$ = #double chalcogen strings/#single chalcogen
           strings; and $\beta$ describes the chemical preference of Te
           to select B sites over A sites. The data points are taken
           from ref. 23.

It becomes transparent from the above analysis that one can describe the $GeSe_2$ glass to be phase-separated into a Se-rich stringed cluster of $Ge_{17}Se_{36}$ stoichiometry and a compensating Ge-rich cluster of $Ge_2Se_3$ stoichiometry as follows:

$$Ge_{17}Se_{36} + 2\ Ge_2Se_3 + 21\ GeSe_2 \qquad\qquad (7)$$

then an upper limit to the degree of broken Ge chemical order would be $4/(4+17) = 0.19$. Note that all Ge sites in the ethane-like cluster possess locally a nontetrahedral symmetry while those in the Se-rich cluster all possess locally tetrahedral symmetry. We find that this value of $N_B/N = 0.19$ is very compatible with the value of $0.16(1)$ that we deduced earlier from our $^{119}$Sn experiments directly. These results are particularly striking because these have been obtained by two completely independent probes of the chemical order of $GeSe_2$ glass.

(ii)  B Sites are not Defects in a CRN

It is conceptually helpful at this stage to recall general features of the Mössbauer results that rule against assigning B sites to isolated bonding defects (homopolar bonds) frozen in a CRN which consists predominantly of heteropolar bonds. In Section IIIA and B, we indicated that the relative population of B sites seen in both the $^{119}$Sn and $^{129}$I experiments vary rapidly and systematically with glass composition. This rapid variation requires, on account of the law of mass action, that these sites be formed in a cluster.

If B sites were point defects in a CRN, one would expect their thermal population to change with degree of annealing of the network, as defects normally do in a crystal. Guided by these considerations, we performed $^{119}$Sn experiments on melt quenched $Ge_{0.99}Sn_{0.01}Se_2$ and $Ge_{0.99}Sn_{0.01}S_2$ glasses in their virgin (as quenched in water) state and annealed state. The annealed state was achieved by taking the virgin sample to the glass transition temperature for the order of tens of minutes and thereafter cooling to room temperature slowly (over a period lasting 10-15 minutes). The spectra revealed no measurable change in the DBO on account of this thermal annealing. As the temperature of annealing exceeded the crystallization temperature of the glass, first order changes in the spectra could be understood in terms of crystallization of specific phases. These thermal annealing studies reinforce the view that the B sites seen in our spectra form an intrinsic part of the glass network.

(iii)  Sites Intensity Ratio:  Signature of Cluster Size

The proposal of outrigger-raft [46] for the structure of $GeSe_2$ glass has several attractive features that provide a way to understand the Mössbauer results.  In this cluster, the symmetry breaking Ge-Se-Se-Ge strings (providing a source of $^{129}I$ B sites) reside on the outer edges of the cluster in contrast to a Ge-Se-Ge strings (source of $^{129}I$ A sites) that occur in the interior of the cluster.  One may understand the high selectivity ($\beta \geq 13$) of the oversize anion probe (Te) to choose B sites over A sites in terms of a strain mediated process.  Apparently, either in the melt, or in the process of quenching, Se replacement by Te in the interior of the cluster must induce sufficient strain to drive the oversized impurity to cluster surfaces in much the same way that impurities segregate at grain boundaries in polycrystalline materials.

The outrigger-raft model proposal [46] consisting of two corner-sharing chains, on the other hand, is quantitatively incompatible with the degree of broken chemical order deduced from the present Mossbauer experiments.  This is seen by writing the stoichiometry of the two-chain raft as $Ge_6Se_{14}$, which requires that the following stoichiometric relation exist for cluster phase separation

$$Ge_6Se_{14} + 2\ Ge_2Se_3 = 10\ GeSe_2\ . \tag{8}$$

The above equation requires that the degree of broken Ge-chemical order be $4/(4+10) = 0.40$.  Inspection of the crystal structure of $g\text{-}GeSe_2$ reveals that there are specific lateral dimensions (along b axis) at which bordering by Se-Se bonds can be invoked to produce characteristic clusters of progressively reduced Se excess.  One such cluster of $Ge_{22}Se_{46}$ stoichiometry (Fig. 18) consisting of six corner-sharing chains laterally, provides, according to our data, an excellent candidate to be the Se-rich cluster of $GeSe_2$ glass.  The stoichiometry of the reconstructed fragment requires that we write the molecular phase separation as

$$26\ GeSe_2 = Ge_{22}Se_{46} + 2\ Ge_2Se_3\ . \tag{9}$$

According to this equation, the broken Ge chemical order is expected to be $0.154$, and it compares favorably to the value of $0.16(1)$ deduced from our $^{119}Sn$ experiments.

D.    Correlation of Raman and Mössbauer Results

It is instructive to inquire at this stage if one can quantitatively understand features of broken chemical order from the Raman and Mössbauer spectra of GeSe$_2$ glass in terms of the idea of molecular phase separation into a large Se-rich and a small Ge-rich cluster. In Raman spectroscopy, scattering cross sections can vary by order of magnitude from one vibrational mode to another. This, of course, is in sharp contrast to Mössbauer spectroscopy where the absorption cross section ($n\sigma_0 f$; n = area density of resonant nuclei, $\sigma_0$ = nuclear resonant cross section, and f = recoil free fraction) is nearly site chemistry independent, particularly at low temperatures, where f-factors tend to saturate. Furthermore, while scattering strengths in Raman spectroscopy scale as the number of bonds, resonant absorption signal in Mössbauer spectroscopy scales as the number (n) of sites. The molecular phase separation of a GeSe$_2$ glass described by Eq. (9) requires that there be two homopolar Ge-Ge bonds for every 100 heteropolar Ge-Se ones. If one assumes a priori that the Raman scattering cross section of a Ge-Se bond and a Ge-Ge bond is the same, then one finds surprisingly that the observed scattering strength [1] ratio of the 180 cm$^{-1}$ (A$_{1g}$ ethane-like) to the 202 cm$^{-1}$ (A$_1$ mode) of 2% or so (Fig. 15) is in excellent accord with the present model of molecular phase separation.

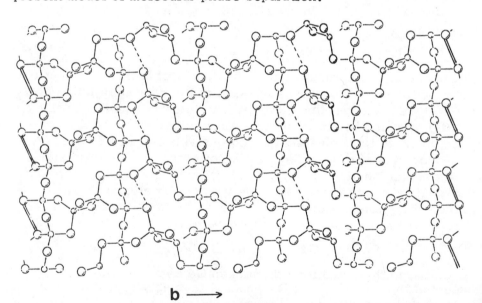

b $\longrightarrow$

Fig. 18    Crystal structure of β-GeSe$_2$ showing provision for laterally terminating the network to form outrigger-rafts containing 2,4, or 6 chains. As the number of chains increases, the Se excess of the rafts decreases and the anticipated broken chemical order reduces.

## V.    CONCLUDING REMARKS

We conclude with several general comments central to the application of Mössbauer spectroscopy as a probe of structure of network glasses. These comments are meant to serve as a guide to the scope and limitations of the present approach.

(i)    Our use of Sn and Te impurity atoms as probes of Ge and Se sites (chemical order) in a $GeSe_2$ glass is contingent upon the probe atoms replacing their lighter isovalent counterparts in the host. If supercooled melts of $SnSe_2/GeSe_2$ and $GeSe_2/GeTe_2$ display a pronounced tendency to phase separate, then clearly the present approach will not serve its intended purpose. As far as we can tell, the evidence at our disposal, from thermodynamics and microscopic measurements [18], strongly suggests that indicated melts when water quenched yield homogeneous single phase bulk glasses that display a single glass transition temperature [47]. This transition, furthermore, is found to monotomically and nonlinearly decrease with composition. The reader is referred to the extensive discussion of this point in ref. [48].

In general, the validity of the present method as a probe of the morphological structure of network glasses requires that probe atoms such as Sn, Sb and Te replace their lighter isovalents Ge, As, Se in host chalcogenide networks.

(ii)    Because of the open 5p shell in the Mössbauer probe atoms Sn, Sb, and Te, the EFG parameters and contact charge density in dense covalent networks can be expected to be largely determined by the distribution of valence electrons. This has the important consequence that the nuclear hyperfine structure is largely determined by nearest neighbor coordination shell. It is for this reason that observation of chemically inequivalent sites in the spectra establishes aspects of short-range order of a glass network.

(iii)    The use of oversized atoms can be a distinct advantage to probe network chemical order or the lack of it, for it may provide the driving mechanism that determines the high selectivity of probe atoms to populate one or more of the inequivalent sites (A,B) of a network. If this turns out to be the case, this, of course, needs to be established. This selectivity could be an asset, particularly if probe atoms have affinity for replacing the symmetry breaking sites of a network. The intensity enhancement of the B sites in $^{129}I$ emission spectra of $GeSe_2$ glass, where $I_B > I_A$ is a spectacular case in point.

(iv)    Compositional variation of site intensity ratios $I_B/I_A$ in the
$^{129}I$ emission spectroscopy and in the $^{119}Sn$ absorption spectroscopy provides important clues on elements of medium-range order
of the network. We have shown specifically that the rapid variation of these site intensity ratios with x requires, because of
statistical considerations, that clustering must occur in the networks. These structure results are nicely complemented by recent
Raman and optical absorption edge measurements [49,50].

(v)    Measurements of T-dependence of Mössbauer recoil-free
fraction f(T) can provide valuable insights in the low frequency
one phonon density of states in glasses. We have, for example,
studied in detail the f(T) of the two $^{119}Sn$ A and B sites in $GeS_2$
glass and find that the characteristic vibrational frequency of
the tetrahedral A sites falls in the domain of the low frequency
$F_2$ scissor mode of tetrahedral $Sn(S_{1/2})_4$ units seen in Raman
vibrational spectroscopy. Aspects of this chemically specific
vibrational spectroscopy of glasses remain to be explored and
these will complement results of IR and Raman spectroscopy on
these materials. A more general, and in some sense more elaborate, method of studying vibrational excitations in glasses make
use of Rayleigh scattering of Mössbauer resonant radiation and
this method has been discussed in ref. [51].

(vi)    Because of the high sensitivity of the present method to
aspects of broken chemical order, it is of interest to inquire
what, if any, is the role of sample preparation on the microstructure of a network glass. Although such questions have been
surely asked in the past, answers on a quantitative level have
been hard to come by. Using the present approach, however, we
are now in a position to quantitatively compare the DBO in a melt
quenched glass with that of an amorphous film of the same composition prepared by vapor deposition, such as evaporation or
sputtering. Preliminary experiments indicate that evaporated $GeSe_2$
films do, indeed, exhibit a substantially higher degree of broken
order than a melt quenched glass. More significantly, the annealing kinetics of DBO in vapor deposited films are notoriously slow.
These results will be discussed in forthcoming publications.

ACKNOWLEDGEMENTS

    During the course of work on glasses I have benefitted
greatly from collaboration and discussions with John deNeufville,
Stan Ovshinsky, Jim Phillips, Peter Suranyi, Michael Tenhover,

and Mark Van Rossum. In addition, I have also had the pleasure
to work closely with Wayne Bresser, Jeff Grothaus, Mark Stevens,
George Lemon and Dave Ruffolo, all of whom have actively parti-
cipated in this research effort on glasses at the University of
Cincinnati. I acknowledge their contribution as well. This effort
is currently supported by National Science Foundation Grant No.
DMR-82-17514.

REFERENCES

1.    W.H. Zachariasen, J. Am. Chem. Soc. 54, 3841(1932);
       M.F. Thorpe in Vibration Spectroscopy of Molecular Solids,
       edited by S. Bratos and R.M. Pick, Plenum, New York
       (1979), p. 341; R.J. Nemanich, G.A.N. Connell, T.M.
       Hayes and R.A. Strèet, Phys. Rev. B 18, 6900(1980).
2.    J.E. Griffiths, G.P. Espinosa, J.P. Remeika and J.C. Phillips,
       Solid State Commun. 40, 1077(1981); Phys. Rev. B 25,
       1272(1982); K. Murase, T. Fukunaga, Y. Tanaka, K.
       Yakushiji and I. Yunoki, Physica 117B and 118B, 962(1983).
3.    P. Boolchand, J. Grothaus, W.J. Bresser and P. Suranyi,
       Phys. Rev. B 25, 2975(1982); P. Boolchand, J. Grothaus
       and J.C. Phillips, Solid State Commun. 45, 183(1983).
4.    S.R. Ovshinsky, AIP Conference Proceedings 31, 67(1976).
5.    J.P. deNeufville (private communication); also see J. Non-
       cryst. Solids 8-10, 85(1972).
6.    J.C. Phillips, J. Noncryst. Solids 34, 153(1979); 43, 37
       (1981).
7.    A. Feltz and H. Aust, J. Noncryst. Solids 51, 395(1982).
8.    R.L. Mossbauer, Z. Physik 151, 124(1958); Naturwissen-
       schaften 45, 538(1958). For review articles on the Möss-
       bauer method used as a probe of glasses, see P.P. Seregin,
       A.R. Regel, A.A. Andreev and F.S. Nasredinov, Phys. Stat.
       Sol. (a) 74, 373(1982) and W. Müller-Warmuth and H.
       Eckert, Phys. Reports 88, 93(1982).
9.    An excellent overview of the metallic glass systems inves-
       tigated can be obtained in Proceedings of the International
       Conference on Amorphous Systems investigated by Nuclear
       Methods Balatoonfured, Hungary, September 1981, pub-
       lished in J. Nucl. Instrum. Methods 199, (1982). A re-
       view of the early oxide work has been given by C.R.
       Kurkjian, J. Noncryst. Solids 3, 157(1970).

10.   A.C. Wright and A.J. Leadbetter, Phy. and Chem. of Glasses
      17, 122(1976).

11.   M.H. Brodsky in Amorphous Semiconductors, Topics in Ap-
      plied Physics, Springer Verlag, (1979), vol. 37.

12.   L. Ley, M. Cardona and R.A. Pollak in Photoemission in
      Solid II, Topics in Applied Physics, Springer Verlag, (1979),
      vol. 27, p. 11.

13.   P.J. Bray, F. Bucholtz, A.E. Geissberger and I.A. Harris,
      Nucl. Instrum. Methods 199, 1(1982).

14.   M. Rubinstein and P.C. Taylor, Phys. Rev. B 9, 4258(1974);
      also see J. Szeftel and H. Alloul, Phys. Rev. Lett. 42,
      1691(1979); J. Szeftel, Phil. Mag. 43, 549(1981).

15.   The Mossbauer effect of the 13.3keV x-ray in $^{73}$Ge first
      observed by Raghavan and L. Pfeiffer, Phys. Rev. Lett. 32,
      512(1974) has a natural linewidth of $6.98 \times 10^{-3}$mm/s. This
      extremely narrow resonance is hard to observe in well-
      ordered crystals because of inhomogeneous line broadening.
      In glasses this resonance is expected to have little use.

16.   N.N. Greenwood and T.C. Gibb, Mössbauer Spectroscopy,
      Chapman and Hall Ltd., London, (1971); G.K. Shenoy and
      F.E. Wagner, Mössbauer Isomer Shifts, North Holland,
      (1978).

17.   G. Weyer, B.I. Deutch, A. Nylandsted-Larsen, J.U.
      Anderston and H.L. Nielsen, J. Phys. 35, 6-297(1974);
      also see J.W. Peterson et al. Phys. Rev. B 21, 4292(1980);
      D.L. Williamson and S.K. Deb, J. Appl. Phys. 54,
      2588(1983).

18.   P. Boolchand and M. Stevens, Phys. Rev. B 29, 1(1984).

19.   P.A. Flinn in Mössbauer Isomer Shifts, North Holland, (1978),
      p. 595; also see J.D. Donaldson and B.J. Senior, J. Inorg.
      Nucl. Chem. 31, 881(1969).

20.   R.M. Sternheimer, Phys. Rev. 84, 244(1951); 95, 736(1954).

21.   G. Czjzek, J. Fink, F. Gotz, H. Schmidt, J.M.D. Coey,
      J.P. Rebouillat and A. Lienard, Phys. Rev. B 23, 2513
      (1981).

22.   A. Coker, T. Lee and T.P. Das, Phys. Rev. B 13, 55(1976);
      22, 2968(1980); 22, 2976(1980).

23.   W.J Bresser, P. Boolchand, P. Suranyi and J.P. deNeufville,
      Phys. Rev. Lett. 46, 1689(1981); P. Boolchand, W. Bresser
      and G.J. Ehrhart, Phys. Rev. B 23, 3669(1981).

24.   C.S. Kim and P. Boolchand, Phys. Rev. B 19, 3187(1979).

25.     Sputtered amorphous films GeS, GeSe and GeTe have been studied by ion implanting $^{129}Te^m$ atoms and recording the $^{129}I$ emission spectrum of such targets by P. Boolchand et al., using the isotope separator at Leuven University, Belgium.

26.     P. Boolchand, W.J. Bresser, P. Suranyi and J.P. deNeufville, Nucl. Instrum. Methods 199, 295(1982).

27.     More precisely, the isomershift of $NaI^{129}$ is taken to be +0.08(3)mm/s with respect to the closed shell configuration I. See ref. 28.

28.     H. deWaard in Mössbauer Effect Data Index, edited by J.G. Stevens and V.E. Stevens, Plenum Press, N.Y., (1975), p. 447. (Covers 1973 literature.)

29.     P. Boolchand, W.J. Bresser and M. Tenhover, Phys. Rev. B 25, 2971(1982).

30.     W.J. Bresser, P. Boolchand, P. Suranyi, J.P. deNeufville and J.G. Hernandez, Phys. Rev. B, to be published.

31.     P. Boolchand, B.L. Robinson and S. Jha, Phys. Rev. B 2, 3463(1970).

32.     G. Lucovsky, R.J. Nemanich and F.L. Galeener, in Proceedings of the Seventh International Conference on Amorphous and Liquid Semiconductors, edited by W.E. Spear, Edinburgh, Scotland, (1977), p. 125.

33.     A. Feltz, K. Zickmuller and G. Pfaff, in Proceedings of the Seventh International Conference on Amorphous and Liquid Semiconductors, edited by W.E. Spear, Edinburgh, Scotland, (1977), p. 133.

34.     L. Pauling, The Nature of the Chemical Bond, Cornell University Press, Ithaca, (1960).

35.     D.E. Sayers, F.W. Lyttle and E.A. Stern, in Proceedings of the Fifth International Conference on Amorphous and Liquid Semiconductors, edited by J. Stuke and W. Brenig, Taylor and Francis Ltd., London, (1974), p. 403.

36.     L. Cervinka and A. Hruby, in Proceedings of the Fifth International Conference on Amorphous and Liquid Semiconductors, edited by J. Stuke and W. Brenig, Taylor and Francis Ltd., London, (1974), p. 431; also see L.E. Busse and S.R. Nagel, Phys. Rev. Lett. 47, 1848(1981).

37.     O. Uemura, Y. Sagara and T. Satow, Phys. Stat. Solid A 26, 99(1974); J. Noncryst. Solids 33, 71(1979).

38.     P.H. Fuoss, P. Eisenberger, W. K. Warburton and A. Bienenstock, Phys. Rev. Lett. 46, 1537(1981).

39.   G. Dittmar and H. Schafer, Acta. Cryst. B 31, 2060(1975).
40.   J.R. Magaha and J.S. Lannin, J. Noncryst. Solids 59-60,
      1055(1983).
41.   P. Tronc, M. Bensoussan, A. Brenac and C. Sebenne, Phys.
      Rev. B 8, 5947(1973); G. Lucovsky, J.P. deNeufville and
      F.L. Galeener, Phys. Rev. B 9, 1591(1974).
42.   G. Lucovsky, F.L. Galeener, R.H. Geils and R.C. Keezer,
      in Proceedings of the Symposium on the Structure of Non-
      crystalline Materials, Cambridge, U.K., edited by P.H.
      Gaskell, Taylor and Francis Ltd., London, (1977), p. 127.
43.   G. Lucovsky, R.C. Keezer, R.H. Geils and H.A. Six, Phys.
      Rev. B 10, 5134(1974).
44.   R.J. Nemanich, S.A. Solin and G. Lucovsky, Solid State
      Commun. 21, 73(1977).
45.   J.A. Aronovitz, J.R. Banavar, M.A. Marcus and J.C. Phillips,
      Phys. Rev. B 28, 4454(1983); T. Fukunaga, Ph.D. thesis
      (unpublished), Osaka University (1982).
46.   P.M. Bridenbaugh, G.P. Espinosa, J.E. Griffiths, J.C.
      Phillips and J.P. Remeika, Phys. Rev. B 20, 4140(1979).
47.   D.J. Sarrach, J.P. deNeufville, and W.L. Hayworth, J.
      Noncryst. Solids 22, 245(1976).
48.   M. Stevens, J. Grothaus, P. Boolchand and J.G. Hernandez,
      Solid State Commun. 47, 199(1983); J.C. Phillips, ibid.
      47, 203(1983).
49.   T. Fukunaga, Proceedings of Optical Effects in Amorphous
      Semiconductors, Snowbird, Utah, August 1984; K. Murase
      and T. Fukunaga, 17th International Conference on Physics
      of Semiconductors, San Francisco, California, August 1984.
50.   B. Weinstein and M.L. Slade, Proceedings of Optical Effects
      in Amorphous Semiconductors, Snowbird, Utah, August
      1984; B. Weinstein et al., Phys. Rev. B 25, 781 (1982).
51.   G. Alebenese and A. Deriu, Riv. Nuovo Cimento 2, 1(1979).

# DISLOCATION MEDIATED PSEUDO-MELTING AT
# SILICON-METAL INTERFACES

B.K. Chakraverty*

Laboratoire d'Etudes des Propriétés Electronique des

Solides, B.P. 166, 38042 Grenoble Cedex, France

## I.    INTRODUCTION

It is known [1] that the silicide formation between silicon and transition, as well as noble, metals occurs at 0.3 - 0.4 of the melting temperature of silicon. It is proposed that the chemical reaction is preceded by an interfacial "melting" process due to defect generation (dislocation or disclination) via the Kösterlitz-Thouless mechanism. This is possible at temperatures lower than the melting temperature due to overall softening of the shear vibration mode of the surface silicon layers, and may lead to the low temperature chemical reaction observed. The softening mechanism of atoms on the (111) surface of silicon due to metal overlayer is suspected to be both elastic and electronic in origin, and leads to the breaking of the strong covalent silicon bonds ($\sim 2\text{eV}/$bond) at low temperatures.

For the purpose of this paper, we shall consider the interface of silicon-metal as two dimensional or pseudo-two dimensional, as shown in Fig. 1 (metal atoms large circles, silicon

---

*Currently visiting Energy Conversion Devices, Inc. An abridged version of the paper was presented at the 17th International Conference on Physics of Semiconductors, San Francisco, August 6-10, 1984

atoms small dots), where the left hand of the figure shows the
two lattices in perfect coincidence or lattice matching, while in
the right hand of the same figure, they are out of registry with each
other, although having the same orientation. Thermal fluctuations,
as we increase the temperature, and stress, as we increase the
thickness of the metal overlayer, would cause progressive loss of
positional order of the silicon atoms, induce covalent bond-
breaking, and perhaps eventual melting of the silicon surface. The
manifestations of loss of positional order for the (111) silicon surface are surface dislocations and disclinations, shown in Fig. 2,
which can be thought as precursors of the melting phenomenon. If
this happens, i.e., dislocations and disclinations form spontaneously and in large quantities at some temperature $T^* < T_M$, where
$T_M$ is the bulk melting temperature of silicon, we shall have a
supply at the interface of weak or broken silicon bonds where
chemical reaction with the metal can be initiated, which will then
proceed to form the silicides (see Fig. 3) which generally have a
large negative formation enthalpy $\Delta H_f$ [2].

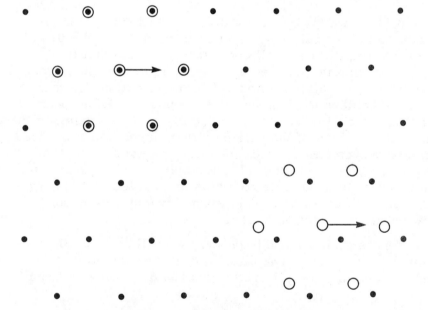

Fig. 1     Schematic representation of an interface. The atoms
           (open circles) on the left are positioned coincident with
           the black dots. The atoms in the right-hand region are
           not in registration with those on the left, as seen by
           lack of registry with the dots. The two regions do have
           the same orientation.

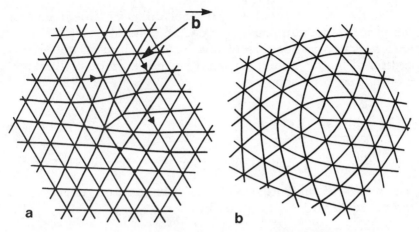

Fig. 2  (a) Schematic representation of a dislocation in a triangular
        two-dimensional lattice.  Such defects are topological de-
        fects in the displacement fields of the solid.  The atoms are
        located at the vertices.  (b) Disclination in a triangular lat-
        tice.  The lattice axis rotates by 60° around the central
        fivefold symmetric site.  There are also disclinations in
        which the rotation is -60° which have senvenfold symmetry
        in the central region.  The large strains in such disclina-
        tions are relieved by the free dislocations.

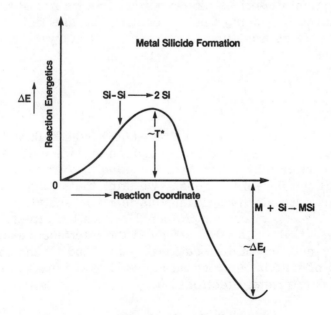

Fig. 3      Metal silicide formation.

The Kösterlitz-Thouless [3] mechanism of two-dimensional lattices propose spontaneous dislocation generation at the melting temperature which is seen as follows. The energy to add a surface dislocation of Burger's vector $\vec{b}$ to a perfect surface of linear size R is (neglecting core energy):

$$U = \frac{Kb^2}{8\pi} \ln(\frac{R}{b}),$$ (1)

where $K = \frac{4\mu B}{\mu + B}$ with $\mu$ and B, the shear and bulk moduli; b is the $|\vec{b}|$, magnitude of the Burger's vector. At non-zero temperature, the free energy F = U-TS is given by:

$$F = \frac{K}{8\pi} b^2 \ln(\frac{R}{b}) - kT \ln(\frac{R}{b})^2$$

$$= (\frac{Kb^2}{8\pi} - 2kT) \ln(\frac{R}{b}).$$ (2)

At low temperature, the free energy for creation of a dislocation is thus very large for a macroscopic system. However, if $T > T^*$, the temperature given by $kT^* = \frac{Kb^2}{16\pi}$ the free energy becomes negative, and hence free dislocations will spontaneously form. Once this occurs, the system will no longer resist shear, since shear stress can be relieved by moving free dislocations as in a fluid. We can thus identify T* as a melting temperature, given by the universal relationship:

$$T^* = \frac{b^2}{16\kappa_B} K(T^*).$$ (3)

Using the values for silicon, at $T^* \sim 1600^\circ K$, the bulk melting temperature will require at the bare silicon surface a softening of the shear modulus $\mu$ by $\sim 50\%$ from its value at room temperature. Since this is not known to happen, we think that a pure silicon surface does not melt via spontaneous thermal generations of dislocations. A metal overlayer profoundly alters the situation in two major ways: (1) the mismatch stress at the interface makes the task of thermal generation of dislocations easier, and (2) the bare shear modulus $\mu_0$ of silicon can also be softened at the surface by coupling through the metal electrons.

These two mechanisms will be briefly treated in Section II, entitled, respectively, elastic mechanism and electronic mechanism. In Fig. 4, we show the perfect (111) surface and the elementary excitation responsible for a surface dislocation-- a five-seven pair pentagon and septagon are shown, which are also known as disclination. These excitations should properly be called "Pandeyons," after Pandey who first showed their importance for the reconstruction of the (111) silicon surface [4].

When formed in large quantities, these elementary excitations can form a grain-boundary, or can coalesce into dislocation loops on the surface enclosing bad or faulted material. We note that these excitations are entirely localized in the surface layer, and one can hence invoke a Kösterlitz-Thouless mechanism for their formation. We also note that the elementary excitation in Fig. 4B is obtained by moving a surface atom into a wrong position, creating no new dangling bonds and forming an extra first silicon neighbor $\pi$-bond, gaining thereby energy to compensate for the distortion involved. We shall now answer the question why they should form in large quantities in the presence of certain metal

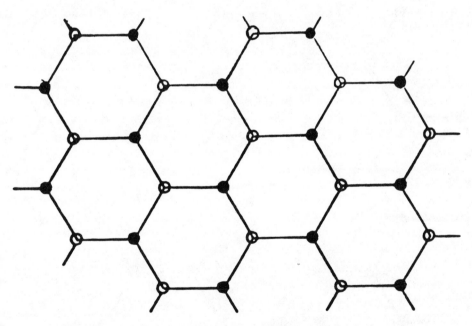

Fig. 4A      Unreconstructed (111) surface.
  ● Outer most atom with one dangling bond/atom.
  o Next sub-layer with no dangling bond.

overlayers. <u>It is to be emphasized that these five- and seven-
sided polygons are the fundamental defect units both in amorphous
solids [5] as well as the metal-silicon interfaces, and are certainly responsible for some of the electronic properties of these sys-
tems</u>, which we shall not discuss here. It is evident that the
metal overlayer disturbs the silicon surfaces, forces bond rearrangement, and a stressed configuration that resembles more and
more at the interface an amorphous solid. In this sense, the
transition of a two-dimensional system to a liquid through the
Kösterlitz-Thouless mechanism is similar to that of a three-
dimensional solid to an amorphous ("spin-glass" like) phase.

## II.     MECHANISM OF SURFACE SHEAR-MODE SOFTENING

### A.     Electronic Mechanism

We shall first look into the softening mechanism of the
base shear modulus $\mu_o$ (at T = 0) in the absence of any dislocations, in the presence of the electrons of the metal layer. We
have earlier postulated [6] that the bulk modulus B of silicon can

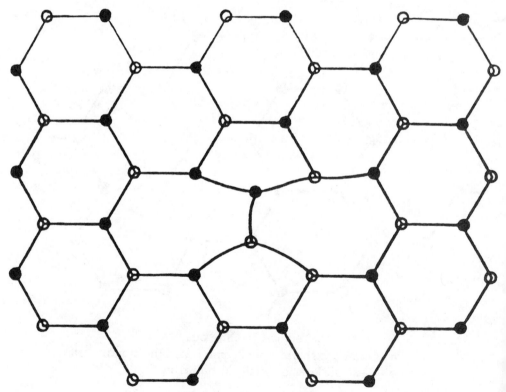

Fig. 4B     (111) surface with disclination pairs.

be decreased in the surface layer by screening due to metal electrons. We think now, however, that the screening mechanism of the plasma oscillations of the ions proposed, using the dielectric constant enhancement of silicon, is unlikely since the static dielectric constant of homopolar semiconductors do not have any wave-vector q dependence as $q \to o$.

The shear constant $\mu$ (or $C_{44}$) of a homopolar semiconductor is a result of near-neighbor and next near-neighbor interaction constants $\alpha$ and $\beta$, respectively, given by [7]:

$$\mu = \frac{\alpha\beta}{a(\alpha+\beta)} , \tag{4}$$

where 4a is the lattice constant of silicon.

For silicon, we have $\alpha = 0.485 \times 10^{12}$ dynes/cm$^2$, $\beta = 0.138 \times 10^{12}$ dynes/cm$^2$ to give a $\mu = 0.79 \times 10^{12}$ dynes/cm$^2$ in agreement with the experimental value. We note that $\alpha$ and $\beta$ are the two second derivatives of the interaction energy between the two silicon atoms evaluated at the equilibrium near-neighbor and next near-neighbor distances.

1.    Modification of $\alpha$

If the metal overlayer affects in any way the near-neighbor interaction energy, i.e., the covalent bond energy $\epsilon_b$, the force constant will be changed. In the simplest tight-binding model of homopolar covalent solid, one makes an sp$^3$ hybrid from an s and p orbital, which can be written as:

$$|h> = \cos\theta\, |s> + \sin\theta\, |p> . \tag{5}$$

The expectation value of the hybrid $< h|H_o|h >$ is :

$$\epsilon_h = \cos^2\theta\, |\epsilon_s| + \sin^2\theta\, |\epsilon_p| , \tag{6}$$

where $\epsilon_s$ is the energy of the s-level, and $\epsilon_p$ is the energy of the p-level. For a sp$^3$ hybrid, $\cos^2\theta = \frac{1}{4}$, $\sin^2\theta = \frac{3}{4}$, with $\epsilon_h = \frac{\epsilon_s + 3\epsilon_p}{4}$. Silicon has two electrons with $\epsilon_s$ and two with $\epsilon_p$, so transforming to four sp$^3$ hybrids costs $\epsilon_p - \epsilon_s$ per atom which is divided equally among the bonds costing a promotion energy $\frac{\epsilon_p - \epsilon_s}{2}$ per bond. The covalent or the resonance energy

between two $sp^3$ orbitals on different atoms is given by
$V_2^o = <h^i|H_o|h^j>$, where $i$ and $j$ are two atoms forming a bond,
and from (5) we get:

$$V_2^o = \cos^2\theta V_{ss\sigma} - 2\sin\theta \cos\theta V_{sp\sigma} - \sin^2\theta V_{pp\sigma}. \tag{7}$$

Similarly, the resonance energy between two hybrids on the same
atom is given by:

$$V_1^o = <h'|H_o|h> = (\epsilon_s - \epsilon_p) \cos^2\theta. \tag{8}$$

This gives us $4V_1^o = \epsilon_p - \epsilon_s$.

The bond energy $\epsilon_b$ is a result of three terms: $\pm V_2$, the
difference between bonding and antibonding levels; $V_1$, the promotion energy; and a third term, the so-called overlap or the increase of the average energy of the $sp^3$ orbital due to nonorthogonality. One can think of the overlap as the excess kinetic
energy of the electrons which keeps the bond from collapsing under
the attractive $\sigma$-bonding energy. The covalent bond energy is
given approximately by [8]:

$$\epsilon_b = -2V_2 + 2V_1 + V_2, \tag{9}$$

where the first term is resonance energy of the two electrons
constituting the bond, $2V_1$ their promotion to be paired, and $V_2$ is
the repulsive overlap nonorthogonality cost. Using a $V_2^o \sim 5eV$
and $V_1^o \sim 1.5eV$, we get a bond energy for silicon $\sim 2eV$ of the
same magnitude as the experimental value.

In the presence of a metal layer, the covalent bond
weakens due to what Pauling [9] has termed fractional order. This
is due to the fact that the p-level of silicon tends to bond with
d-level of the transition or near-noble metals so that the p-electron
spends proportionally less time in the $sp^3$ orbital, and, thereby,
weakening it. Equation (5) is no longer a correct description of the
hybrid. Now, the electron is no longer localized on an $sp^3$ orbital
and has developed a lifetime. Of the matrix elements $V_2$ and $V_1$,
$V_1$ will be most affected, as it involves perturbation of two $sp^3$
orbitals on the same atom due to metal, and will now be given by:

$$V_1 = <h|H|h'>,$$

where H is the perturbed Hamiltonian due to metal layer. To the second order of perturbation, we then have:

$$V_1 = V_1^o - \frac{<h|H'|m><m|H'|h'>}{\epsilon_m - \epsilon_h}$$

or

$$V_1 = V_1^o - \frac{|T_{hm}|^2}{\epsilon_m - \epsilon_h} , \tag{10}$$

where $|T_{hm}|^2$ is the square of the matrix element taking an electron from the covalent bond to metal and vice versa with $\epsilon_m$ a relevant energy level in the metal, say the d-band. We note that Eq. (10) simply translates the fact that there is now an extra channel available to go from one hybrid orbital to another on the same atom, which is through the metal layer. It is this increase of $V_1$, without necessarily any concomitant increase of $V_2$, that affects the bond strength, $\epsilon_b$, if the bond length remains unaltered. On the other hand, a reduced $\epsilon_b$ will induce the bond length to increase, which, in turn, will reduce $V_2^o$ which depends exponentially on the bond length, thereby reducing further $\epsilon_b$ in a self-consistent manner. Any reduction of $\epsilon_b$ will reduce the near-neighbor Keating interaction or force constant $\alpha$, affecting $\mu$ directly.

## 2.    Modification of $\beta$

The second or next near-neighbor interaction cannot be described the same way as that of $\alpha$, because the two non-near-neighbor (6 on the 111 surface) atoms do not form a direct covalent bond. On the other hand, these atoms will cause excitations and de-excitations of metal electrons at the Fermi level between two silicon atoms at positions i and j on the surface. This will give [10] to an interaction energy between these two non-near-neighbor atoms:

$$\Delta E = \sum_m \frac{<0|\emptyset_i|m><m|\emptyset_j|0>}{E_o - E_m} , \tag{11}$$

where $\emptyset$ are the perturbing potential at site i and j; $|0>$ and $|m>$ are the ground and excited states of the electrons in the metal. Equation (11) can be evaluated to give an oscillatory interaction of the form:

$$V_{ij} = A \frac{\cos(2k_F \cdot R_{ij} + \delta)}{R_{ij}^2} \, , \tag{12}$$

where $R_{ij}$ is the distance between two silicon atoms on the surface, and $\delta$ is a phase shift of the metal electrons at the Fermi level due to scattering by the Si atoms. A is the interaction constant:

$$\sim \frac{(\text{Metal-Silicon Coupling})^2}{|E_F^M - E_F^{Si}|} \sim eV.$$

For $\delta = 0$, we find that $V_{ij}$ is attractive for Pd, Au, Ni, Cu, and Mo. On the other hand, if there is strong resonance between metal and silicon, $\delta \approx \pi$, and the interaction is repulsive. As a result of the repulsive interaction, the potential-well of second-neighbor silicon-silicon interaction is flattened out and $\beta$, the curvature of the well, is diminished.

We, thus, see from considerations in sections 1 and 2 that both $\alpha$ and $\beta$ can be profoundaly modified due to the metal layer, especially if there is strong coupling between the two surfaces, and will lead to a shear constant $\mu$ quite different from the bare silicon surface value $\mu_o$.

B.      Elastic Mechanism

In K-T theory, thermal generation of dislocation occurs in pairs of +- dislocations (see Fig. 5) (so that the total Burger's vector $\sum_i b_i = 0$), with approximately energy $\sim 2E_c$, where $E_c$ is the core of one dislocation. With n dislocation pairs, the mutual screening reduces $E_c$, and hence renormalizes K, hence increases in turn n self-consistently, leading to a new K (and hence $\mu$) to 0 at $T > T^*$. The screening argument is fundamental to K-T theory, and one can postulate a renormalized shear constant $\mu_R$ given by:

$$\mu_R = \frac{\mu_o}{\epsilon} \, , \tag{13}$$

where $\epsilon$ is a pseudo-dielectric constant $= 1 + 4\pi\chi$ , the "susceptibility" $\chi$ is:

$$\chi = \Sigma \, A(R) \, n_{+-}(R), \tag{14}$$

where $A(R)$ represents the polarizability of a +- dislocation pair separated by a distance R; $n_{+-}(R)$ the total number of such pairs in the system at T.

In the presence of an overlayer $n(R)$ is profoundly modified because of the stress $\sigma$ exerted by the overlayer on silicon, given by:

$$\sigma = \frac{\mu}{1-\nu} \; \frac{a_M - a_{Si}}{a_{Si}} \; , \tag{15}$$

where $\nu$ is the Poisson ratio, $a_M$ and $a_{Si}$ are the surface lattice constants of metal and silicon.

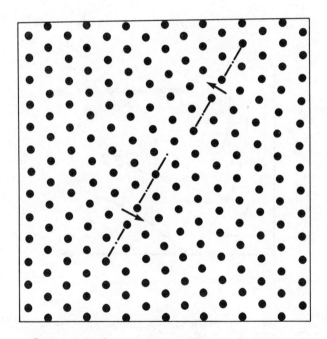

Fig. 5          Pair of dislocations. The dash-dotted lines show the axes of symmetry of the dislocations. The arrows indicate the Burger vectors, and are located so that the center of each arrow coincides with the position of the dislocation according to our definition.

We now have a Peierl's force $\sigma b$ acting on a dislocation pair separated by a distance R, tending to increase R to $R + \Delta R$ which will be resisted by the stiffness or the second derivative of the interaction potential between the two dislocation, so that we have for one pair energy increase:

$$\Delta E = \gamma (\Delta R)^2 - \sigma b \, \Delta R , \qquad (16)$$

which gives, at equilibrium $\overline{\Delta R}$:

$$\overline{\Delta E} = - \frac{\sigma^2 b^2}{4\gamma} . \qquad (17)$$

This decreases drastically pair creation energy self-consistently (see Fig. 6):

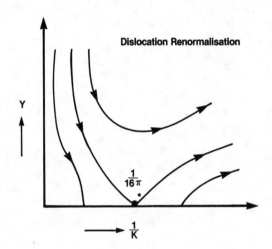

$$Y(T) = \exp\frac{-2E_c}{kT}$$

$$K(T) = \frac{4 \mu B}{\mu + B}$$

$E_c \cdot \cdot$ Core energy of dislocation
$\mu \cdot \cdot$ Bulk shear modulus of silicon
$B \cdot \cdot$ Bulk modulus of silicon

Fig. 6

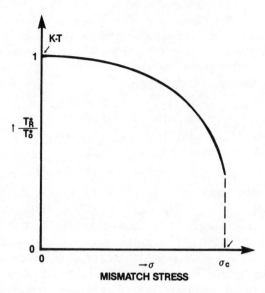

Fig. 7     $T_R^*$, renormalized transition temperature as a function of
           mismatch stress; $\sigma = 0$, $T_O^*$ is Kösterlitz–Thouless point.

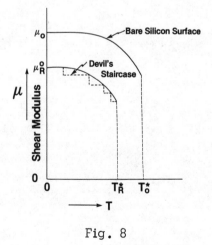

Fig. 8

$$n_{+-} = \exp - \frac{1}{kT}\left[2E_c - \frac{n_{+-}(\sigma^2 b^2)}{4\gamma}\right],$$  (18)

completely analogous to K-T theory, the increased number of $n_{+-}$ increases screening, reduces $\mu$ to $\mu_R$ and gives us a new $T_R^*$ which is shown in Fig. 7 as a function of mismatch stress $\sigma$. This prediction implies critical mismatch stress (i.e., either critical mismatch or thickness of the overlayer).

In Fig. 8, the combined effect of reduction of bare $\mu$ (through metal electrons) and that through dislocation is shown. It is quite possible that the transition to the final chaotic state takes place through a succession of small first-order transitions as in a devil's staircase [11], especially if the disclination pairs could lock into states of immediate order, but that is a matter of conjecture. This would happen if the Kösterlitz-Thouless transition becomes of low order in presence of external stress, as is likely. There has been ample evidence [12,13] that the metal-silicon interface is amorphous or completely disordered. In this paper, we show the physical reasons behind the process of disordering.

ACKNOWLEDGEMENTS

I would like to thank Drs. Iris and Stan Ovshinsky for their wonderful hospitality during my stay at ECD.

REFERENCES

1.      L.J. Brillson, Surf. Sci. Rep. 2, 250(1982).
2.      K.N. Tu and J.W. Mayer, Thin Films, Interdiffusion and Reactions, edited by J.M. Ponte and J.W. Mayer, Wiley, N.Y., 1981, p. 359.
3.      J.M. Kösterlitz and O.J. Thouless, J. de Phys. C6, 1181 (1973).
4.      K.C. Pandey, Phys. Rev. Lett. 47, 1913(1981).
5.      N. Rivier, Phil. Mag. A 40, 859(1979); also S.R. Ovshinsky, J. de Phys. 42, C4-1095(1981).
6.      B.K. Chakraverty, Solid State Commun., to be published.
7.      P.N. Keating, Phys. Rev. 145, 637(1966).
8.      W. Harrison, Phys. Rev. B 27, 3592(1983).

9.  L. Pauling, _The Nature of Chemical Bond_, Cornell University Press, Ithaca, 1960.

10. K.H. Lau and W. Kohn, Surf. Sci. _75_, 587(1978).

11. P. Bak, Rep. Progr. Phys. _45_, 587(1982).

12. R.W. Bene and R.M. Walsey, J. Vac. Sci. Tech. _14_, 925(1977).

13. G.U. Lay, M. Mannerville and R. Kern, Surf. Sci. _65_, 261(1977).

# VIBRATIONAL PROPERTIES OF AMORPHOUS SOLIDS

Gerald Lucovsky

Department of Physics, North Carolina State University

Raleigh, North Carolina 27695-8202

## I. INTRODUCTION

The objective of this paper is to discuss the relationship between the vibrational properties of an amorphous solid and the
local atomic structure. Vibrational spectroscopy is an indirect
structural probe, inasmuch as it requires a theory or model to relate spectral features to specific types of local bonding arrangements. It is complementary to other probes of local atomic structure, such as x-ray or neutron scattering which give direct information about bond-lengths and other aspects of atomic-scale
geometry. We shall concentrate here on two specific types of
amorphous solids: oxide and chalcogenide glasses, and amorphous
silicon (a-Si) based alloys. The vibrational spectra of these materials, and the methods used to obtain information about their atomic
structure, will illustrate techniques of analysis which are applicable to other material systems as well.

Before discussing the various types of vibrational spectra
which are obtained experimentally, it is necessary to develop a
framework for describing local atomic structure in noncrystalline
or amorphous materials. This requires precise definitions of the
concepts of short-range and intermediate-range order (hereafter
designated as SRO and IRO, respectively). We shall use the term
SRO to describe those aspects of the local atomic structure that
are associated with the bonding at each specific type of atomic
site. We, therefore, define SRO to include: (1) bond-lengths, or

nearest neighbor distances; (2) bond-angles, or next-nearest neighbor distances; and (3) the local site symmetry.

In a complementary way, we define IRO to include those aspects of order that include more than three atoms and less than 10 to 20, the following elements of the atomic geometry: (1) dihedral angles, equivalent to specifying third neighbor distances; (2) small rings of atoms, generally the most important include four and six atom bond sequences; and (3) the local topology, whether it is fully three dimensional, layer-like and two-dimensional, or chain-like and one-dimensional.

The first section which follows will deal with network glasses, the second with alloy atoms in a-Si host. We shall present representative spectra, and emphasize the way features in the spectra are related to aspects of both the SRO and the IRO.

## II.    NETWORK GLASSES: OXIDES AND CHALCOGENIDES

Oxide and chalcogenide glasses have received considerable attention because of their importance in a variety of technological applications. In this section of the paper, we shall restrict the discussion to four glasses, v-$GeO_2$, v-$As_2O_3$, v-$As_2S_3$, and v-$GeSe_2$ (we use the prefix v- to designate glasses that can be quenched from a melt, i.e., vitreous solids). The spectra discussed will serve to illustrate more general relationships between spectral character and structure, relationships that apply in a variety of other glass-forming oxide and chalcogenide systems as well.

### A.    Oxide Glasses I: v-$GeO_2$

The IR and Raman spectra of v-$GeO_2$ have been studied by Galeener and Lucovsky [1], and the neutron spectra by Galeener and coworkers [2]. To compare theory and experiment in a quantitative manner, it is usually necessary to analyze a spectrum, leading to a "reduced spectrum" which can then be more easily compared with calculations. In this context, we focus on the IR and Raman spectra, following the observation made by Galeener et al. [2] that the inelastic neutron scattering spectrum gives the same spectral information that is also contained in the depolarized Raman response.

Figure 1 [1,2] gives the IR reflectivity and polarized (HH) and depolarized (HV) Raman spectra for v-$GeO_2$. The IR reflectivity is dominated by features in three distinct spectral regimes, at about $250cm^{-1}$, $550cm^{-1}$, and $850cm^{-1}$. In contrast, the polarized Raman response is dominated by a single feature at about $420cm^{-1}$. The polarized and depolarized Raman spectra also show weak structure at other frequencies. Figure 2 indicates the reduced Raman spectra [1,2] and two dielectric functions, the IR dielectric constant ($\epsilon_2$), and the energy loss function ($-1m\left(\frac{1}{\epsilon}\right)$). The dielectric functions are obtained from the IR reflectivity via a Kramers-Kroenig analysis in which the experimental reflectivity must be analytically extended to both high and low wave number limits. The positions of the peaks in the

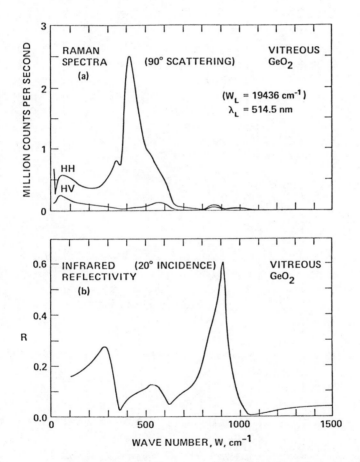

Fig. 1     Vibrational spectra of v-GeO$_2$ [1,2]; (a) polarized (HH) and depolarized (HV) Raman spectra, and (b) IR reflectivity.

IR dielectric constant give the frequencies of IR active vibrations,
analogous to transverse optic modes of crystals (TO phonons),
and the positions of the peaks in the energy loss function give
the frequencies of the longitudinal modes (LO phonons). For purposes of comparison with theory and/or models, we emphasize
the IR active vibrations as defined by the peaks in the dielectric
constant. Galeener and coworkers [2] have shown that the depolarized or HV Raman response replicates to a very large extent
feature in the entire vibrational density of states (VDOS). In
contrast, the spectral features that dominate in the polarized or HH

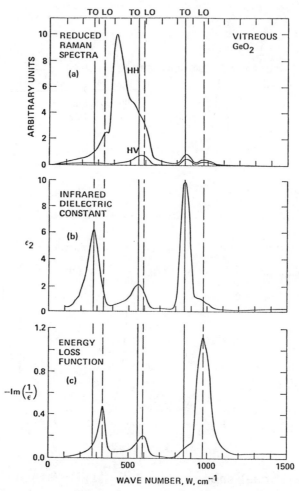

Fig. 2    Reduced vibrational spectra of $GeO_2$[1,2]; (a) Raman
          response functions, (b) IR dielectric constant, and
          (c) energy loss function.

Raman response and are not generally enhanced in the HV spectrum, are derived from vibrations with a high degree of symmetry (defined either with respect to the local site symmetries of the constituent atoms, or to the symmetry of a particular element of the IRO). A more precise definition of symmetry will evolve when we discuss the nature of the atomic displacements of the various IR and Raman active vibrations. At this point of the discussion, we emphasize that the IR and Raman spectra are complementary, each in effect amplifying different features in the VDOS. In this context, we can view each of these distinct spectra as the product of a VDOS, and a matrix element which is qualitatively different for each of the three reduced experimental spectra, the IR dielectric constant, and the polarized and depolarized Raman response functions.

There has been a variety of different approaches to the theory of the vibrational properties of amorphous solids. One class of theories is based on finite clusters either containing a small number of atoms, generally less than 10, or a large number of atoms, generally more than a few hundred. The other class of models is based on an infinite aperiodic structure, the Bethe Lattice. The most important and, indeed, successful cluster models were developed by Bell and Dean [3-5], and were based on ball and stick models for v-$SiO_2$. The construction of the cluster was based on the concept of a covalent random network (CRN), in which each of the constituent atoms satisfies its normal bonding requirements; i.e., fourfold-coordinated Si atoms and twofold-coordinated O atoms. The construction used an $SiO_4$ tetrahedral building block. These units were interconnected through the twofold-coordinated O atoms with the following rules of construction applied:

1.    The bond angles at the Si atoms sites deviated by no more than 10 degrees from the ideal value, 109.47 degrees.

2.    The tetrahedra were corner-connected (for example, as opposed to edge-connected).

3.    The bond angles at the O sites had an average value of 150 degrees and spread of approximately plus or minus 25 degrees.

4.    There was an equal probability for all dihedral angles.

5.    There was no correlation between the bond angle at an O atom site and the magnitude of the neighboring dihedral angles.

The network constructed in this way was space-filling and the application of a force constant model replicated the major features in the IR and Raman spectra. The major problem in this model was the sensitivity of the spectra to the boundary conditions imposed on the atoms at the perimeter of the cluster. For the 500 atom arrays generally used, a relatively large fraction of the atoms are surface atoms which in turn have unsatisfied or dangling bonds. These atoms can be allowed to move freely, or be constrained. In both cases, the spectra are different. In spite of this problem, the modeling of Bell and Dean [3-5] gave a tremendous amount of insight into the nature of the local atomic motions associated with the various IR and Raman active modes. Bell and Dean [3-5] also applied their technique to other glasses that were assumed to have the same atomic scale structure as $v-SiO_2$, $v-GeO_2$, one of the glasses we are discussing, and $v-BeF_2$.

One of the most important results of their vibrational analysis for $v-SiO_2$ and for $v-GeO_2$ is that the major spectral features could be associated with different types of O atom motion which, in turn, are correlated with the local symmetry of that atom. Figure 3 includes the O atom bonding, and defines the three classes of atomic motion (we use $v-GeO_2$ as the example in the ensuing discussion). Motion in the plane of the Ge-O-Ge bond, and along a direction defined by a vector parallel to a line joining the two Ge atoms, is called stretching (or sometimes asymmetric stretching); motion in the same plane, but along the bisector of the Ge-O-Ge bond is called bending (or sometimes symmetric stretching); and finally, an out-of-plane motion, perpendicular to Ge-O-Ge plane, is called rocking. Rather than discuss the assignment of the spectral features in terms of these motions at the present time, we will go on to describe the Bethe Lattice approach, and then reintroduce these types of motions in the context of the local (vibrational) densities of states (L(V)DOS or simply LDOS).

The first application of the Bethe Lattice Method to oxide glasses is due to Laughlin and Joannopoulos, who applied the method to $v-SiO_2$[6,7]. The first paper [6] used two-body central and non-central forces, the so-called Born forces, and the second paper [7] extended the approach to the use of the Valence Force Field (VFF), short-range forces which include both two- and three-body forces. There are two aspects of the work reported in refs. 6 and 7 which have recently been improved upon by Lucovsky, Wong, and Pollard [8], and in this paper we will emphasize their approach.

In the application to v–SiO$_2$, Laughlin and Joannopoulos based their Bethe Lattice construction on an interconnection of SiO$_4$ tetrahedra using a particular value of the dihedral angle that facilitated convergence of their calculation. This arbitrary choice of a structural parameter introduced relatively sharp features into the spectrum, because a single value of the dihedral angle in effect builds aspects of long range order into the Bethe Lattice which are not present in the glass. The second problem relates to the calculation of the IR and Raman spectra. Since the Bethe Lattice "overfills" real space, Laughlin and Joannopoulos found that the

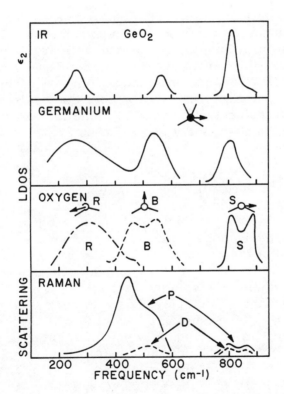

Fig. 3     Calculated spectra for v–GeO$_2$ [8]; (a) infrared dielectric constant, (b) local density of states (LDOS) for the Ge atom, (c) LDOS functions for the O atom, and (d) polarized (P) and depolarized (D) Raman response functions.

IR and Raman response functions diverged when calculated for the entire infinite structure. Therefore, they arbitrarily truncated the calculation for a finite cluster with relatively few atoms, of the order of 20-25. The work reported in ref. 8 remedied both of these deficiencies, and will now be discussed in the context of v-GeO$_2$. The approach includes the following two innovations:

1.   IRO is introduced in a specific way through an average over a set of dihedral angles that are determined by the bonding chemistry of the constituent atoms in their molecular and crystalline forms.

2.   Long-range disorder is introduced through the use of an r-dependent broadening term in the Green's function that decreases the long range correlations between atomic motions, and thereby ensures a convergence of the calculated IR and Raman response functions, even for the infinite Bethe Lattice structure.

The calculations in ref. 8 are based on a use of two- and three-body valence forces in the VFF representation. In addition, the IR response is calculated using both static and dynamic contributions to the IR effective charge, and the Raman response includes both bond stretching and bond bending contributions to the changes in the polarizability. We present the results of these calculations in the context of the LDOS functions for the O and Ge atoms, and then relate the IR and Raman response to these spectral functions.

Figure 3 includes the LDOS for the Ge and O atoms and the calculated IR and Raman response functions, specifically the IR dielectric constant and reduced polarized and depolarized Raman response. The LDOS is a function of the symmetry of the atomic sites. The Ge atoms are in sites with tetrahedral symmetry so that the motion of the Ge atoms is fully degenerate with respect to any set of cartesian coordinates, and all projections of the LDOS are equivalent. Therefore, there is only one LDOS function, and, in this instance, it displays peaks in three distinct regions of the spectrum. The number of peaks in this LDOS function is determined in part by the frequency spectrum of the O atom, and, in particular, the number of discrete features its component spectra display. The symmetry at the twofold-coordinated O atom is significantly reduced from the full cubic symmetry at the Ge atom site, and as a result of this, there are three distinctly different projections of the LDOS, which in turn contribute to features in three distinct spectral regimes. We use a coordinate system that

is referenced to the local symmetry at the O atom site, and has
two of its axes in the plane of the Ge-O-Ge bond. One of these
is along the symmetry axis defined by the bisector of the Ge-O-Ge
bond angle, and the second is parallel to a line joining the two
Ge atoms. The third axis is perpendicular to the plane of the
Ge-O-Ge bond. Motion along these axes corresponds, in the order
defined above, to the bending, stretching, and rocking motions
discussed earlier in the text. The stretching band is at the highest frequency regime, and is completely distinct from the lower
frequency bending and rocking bands which show a small amount
of spectral overlap. We now discuss the way motions of O and/or
Ge atoms contribute to the various features of the IR and Raman
response.

An IR active vibration is characterized by opposite motions
of the two atomic species, in this case, the Ge atoms and O
atoms. If we use the three types of O atom motion, stretching,
bending, and rocking, as a frame of reference, then we expect IR
activity for each of these three mutually orthogonal motions combined with a similar type of oppositely direct Ge atom motion.
This is borne out by the calculation of the IR dielectric function.
The O stretching motion at $860 \text{cm}^{-1}$ combines with Ge atom motion
and produces the highest frequency IR band. In a similar way,
O atom bending motion at $560 \text{cm}^{-1}$ and O atom rocking motion at
$260 \text{cm}^{-1}$ combine with Ge atom motion at the same frequencies,
respectively, and give rise to the other two IR active vibrations.
We can view these three vibrations as "rigid sub-lattice" vibrations in which the two atomic species move along three mutually
perpendicular axes with opposite phases. The IR activity derives
from two sources, an ionic contribution from a static charge
transfer between the Ge and O atoms, and a dynamic contribution
from charge redistribution. The relative activities of the three
vibrations demonstrate that the dynamic contribution dominates in
the high frequency vibration.

All vibrations which are IR active also contribute to the depolarized Raman response. In contrast, contributions to the
polarized Raman response can only come about from vibrations
which preserve some aspect of the local atomic symmetry or some
aspect of the IRO, such as the full symmetry of a ring of bonded
atoms. In the case of $v\text{-GeO}_2$, there is no evidence for any major
feature in the polarized Raman response being due to IRO, and the
major contribution to the polarized Raman response then derives
from preservation of atomic site or local symmetry. Since the Ge

atom is in site with tetrahedral symmetry, there is no displacement of that atom by itself that preserves the full tetrahedral symmetry; hence; the dominant contribution to the polarized Raman response must be derived from O atom motion. Of the three mutually perpendicular motions of the O atoms, only the bending motion along the bisector of the Ge-O-Ge bond angles preserves the full symmetry at the O atom sites; hence, this is the motion that contributes to the polarized Raman response. This argument is confirmed by a comparison of the LDOS and polarized Raman response in Fig. 3. The polarized Raman band clearly derives from O atom bending motion with the dominant peak at $420 cm^{-1}$ being due solely to O atom motion, and the shoulder at $550 cm^{-1}$ possibly involving Ge atom motions as well. Finally, the depolarized Raman response at about $950 cm^{-1}$ is also associated with O atom motion, in this instance, stretching motion which does not preserve the local site symmetry of the O atoms; hence, the depolarized character of the vibration. A comparison between the experimental spectra in Fig. 2 and the calculated spectra in Fig. 3 indicates very good agreement. All features in the experimental spectra are accounted for with the exception of the weak, but relatively sharp feature at about $320 cm^{-1}$.

Galeener [9] has shown that this feature in $v-GeO_2$, and similar sharp, but weak, features in $v-SiO_2$, are associated with the small planar rings of atoms. The polarized character of these vibrations has been shown to be associated with O atom motions that preserve the full symmetry of the various ring configurations. In this context, the relative sharpness of these modes is a result of the spatially correlated motions of all of the O atoms in the particular ring. Atomic motions that preserve the full symmetry of small ring configurations have also been discussed for other oxide glasses, and, in effect, turn out to be the most important aspect of IRO that is easily identified via contributions to the polarized Raman response. In the case of the vibration at $320 cm^{-1}$ in $v-GeO_2$, and the two analogous vibrations at approximately 500 and $600 cm^{-1}$ in $v-SiO_2$, the fraction of O atoms occurring in the ring configurations is small, of the order of 1 at .%. This estimate is made by Galeener [9] through a comparison of the scattering strength in these vibrations and the scattering strength in the dominant Raman peak which is associated with a similar type of O motion. In the case of a number of other oxide glasses, including $v-B_2O_3$ and $v-As_2O_3$, the fraction of O atoms in the ring configurations is significantly larger. We now go on to discuss the

spectra of v-As$_2$O$_3$ in order to illustrate the way that Bethe
Lattice calculations can be applied to glasses in which there is
evidence for a significant degree of IRO.

B.      Oxide Glasses II:  v-As$_2$O$_3$

The vibrational spectra of v-As$_2$O$_3$ have received considerable study [10-12], with the most complete results being described in ref. 11.  The IR and Raman spectra show significantly
more structure than those of v-GeO$_2$ (see Fig. 1 of ref. 11).  The
vibrational spectra have also been the subject of numerous theoretical studies [8,9,12,13].  Rather than discuss each of the
models proposed to explain the spectra, we simply note that all
of the recent attempts to account for the sharpness of the features
and the large amount of structure have invoked a significant degree
of IRO [8,12].  In particular, the recent work of Lucovsky, Wong
and Pollard [8] demonstrated that six-membered rings of alternating
As and O atoms were the primary contributor to the IRO.  The
authors of ref. 8 reached this conclusion by means of a Bethe
Lattice calculation that specifically included aspects of IRO.
They were able to change input parameters, and, in effect, "turn
off" the contributions to the IRO, and thereby demonstrate that the
resultant spectra lost the sharp polarized Raman features and
replicated spectra calculated using a CRN model that included a
random distribution of dihedral angles, and hence no aspects of
IRO [13].

Figure 4 displays the reduced IR and Raman spectra from
refs. 11 and 12, and Fig. 5 gives the results of the Bethe Lattice
calculations [8].  Before discussing these calculated functions,
we highlight the way the parameters used in the calculations were
obtained.  All bonds were assumed to be heteropolar, i.e., As-O
bonds.  This bond-length  and the bond-angles at the O and As
atom sites were obtained from x-ray studies which generated
Radial Distribution Functions.  These studies also confirmed the
coordinations of three and two at the As and O atom sites,
respectively.  In the case of the v-GeO$_2$ calculation discussed
above, the dihedral angle distribution was assumed to be random,
and there were no small ring configurations which would contribute
any aspect of IRO.  In the case of v-As$_2$O$_3$, the dihedral angle
distribution is taken to be non-random, and is determined from the
crystalline polymorphs, Arsenolite and Clauditite I and II.  The
most important aspects of this restricted set of dihedral angles
derive from chemical bonding considerations wherein specific

bonding arrangements come about from a minimization of repulsions between the nonbonding or lone-pair electrons on both the As and O atoms. This distribution of angles includes an important contribution from the six-membered ring configuration in Arsenolite, and, in effect, this contribution to the dihedral angle distribution serves to ·emulate the ring geometry and the associated symmetry properties in the Bethe Lattice calculation. Finally, the IR and Raman matrix elements are obtained from a fit to the vibrational

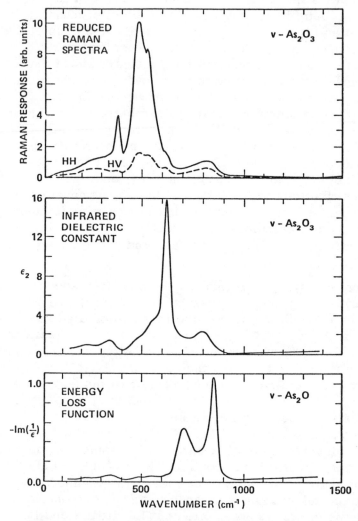

Fig. 4        Reduced vibrational spectra for v-$As_2O_3$ [11]; reduced HH (P) and HV (D) Raman spectra, infrared dielectric constant, and energy loss function.

spectra of the $As_4O_6$ molecule, the building block of the molecular crystal Arsenolite. The only adjustable parameter in the calucu- lation involves the frequency of an inactive vibration in Arsenolite, which, in effect, allowed a variation of one of the three-body VFF forces until a best fit could be obtained.

The results of the calculation appear in Fig. 5. The first thing to be noted is that the correspondence between the calculated reduced IR and Raman spectra (both polarized and depolarized) and the experimental results in Fig. 4 is excellent. This then enables us to specify, with a high degree of confidence, the atomic motions that contribute to each of the dominant spectral features,

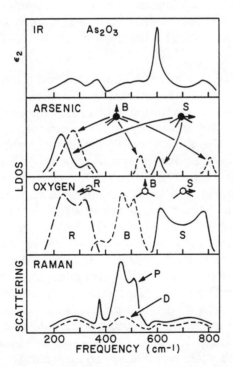

Fig. 5       Calculated spectra for v-$As_2O_3$ [8]; IR response, LDOS functions for As and O atoms, and P and D Raman response functions.

and also to identify which features derive from the IRO as contained in the rings discussed above. Consider first the LDOS spectra. The central portion of the figure gives the projections of the LDOS functions for the As and O atoms relative to their respective symmetry axes. The As atom sits in at a pyramidal site with a three-fold symmetry axis. As such, there are two independent types of As atom motion: in the direction of the threefold symmetry axes which we designate as bending (this definition parallels that used for bending at the twofold-coordinated O atom site), and at right angles to the threefold axes which we designate as stretching (this motion is doubly degenerate). The O atom displays three independent projections of the LDOS that are defined in a completely analogous way to what we have already discussed for the O atoms in v-$GeO_2$. We designate the two LDOS functions for the As atoms as B and S (corresponding respectively to bending and stretching motions), and the three LDOS functions for the O atoms as R, B, and S (corresponding respectively to rocking, bending, and stretching). The As atoms LDOS functions show considerably more structure than the single LDOS function for the tetrahedrally coordinated Ge atoms, while the O atoms show a qualitatively similar set of LDOS functions as compared with the O atom LDOS functions in v-$GeO_2$.

Features in the IR dielectric constant are associated with out-of-phase As and O atom motions, with the relative strength being determined by the particular mixture, which reflects the relative importance of static and dynamic contributions to the IR effective charges. The strongest IR response at about $615 cm^{-1}$ comes from S motions of both As and O atoms. The relative strength of this feature compared to the other IR active modes at $800 cm^{-1}$ (this combines O atom S motion with As atom B motion) and those at 250 and $340 cm^{-1}$ (these combine O atom R motion with As atom S and B motions) shows that the major contribution to the IR response is associated with the dynamic rather than the static component of the IR effective charge. It further demonstrates that the most important component of this dynamic charge is associated with S motion of the O atoms. This component of the motion produces a renormalization of the bonding charge, including any back donation from the O atom nonbonding states into antibonding states on the As atoms. The calculated IR response is relatively insensitive to the IRO and does not change in any significant way if the dihedral distribution is made random. The Raman response is qualitatively different inasmuch as it varies considerably with changes in the dihedral angle distribution.

As in the case of v-GeO$_2$, the major contributions to the polarized Raman response in v-As$_2$O$_3$ are associated with O atom bending motion, which preserves the local symmetry at these sites. However, since the As atom resides at a site of reduced symmetry as well, As atom bending motion also has the necessary symmetry to produce polarized Raman modes.

The dominant polarized Raman feature at 480cm$^{-1}$ is associated with O atom B motion, and is not sensitive to the dihedral angle distribution. In this context, it is the analogue of the v-GeO$_2$ Raman mode at 420cm$^{-1}$. In contrast, the other two polarized Raman modes at 525 and 380cm$^{-1}$ combine both O and As atom B-type motions. The 525cm$^{-1}$ mode combines out-of-phase motions of these atoms and displays a weak dependence on the dihedral angle distribution, whereas the 380cm$^{-1}$ mode combines in-phase B-type motions and displays a much greater dependence on the choice of dihedral angles. These observations are in accord with the "parentage" of these two vibrations. The As$_4$O$_6$ molecule has two polarized Raman modes at 550 and 380cm$^{-1}$ which are associated, respectively, with out-of-phase and in-phase B motions of the As and O atoms. The molecule has four six-membered chain-type rings of alternating As and O atoms, and a dihedral angle which we define as zero degrees. The 380 and 525cm$^{-1}$ vibrations in the Bethe Lattice calculation increase in relative strength as the relative fraction of zero degree dihedral angles is increased. Furthermore, a calculation of the modes of a vibrationally isolated six-membered chain-type ring yields polarized Raman modes at 380 and 525cm$^{-1}$. On the basis of these observations, we assign the modes in v-As$_2$O$_3$ to similar rings of bonded atoms within the glass phase. We estimate that the fraction of atoms in these rings is at least 25%. The specification of a set of dihedral angles that includes a significant fraction of zero degree values then has the effect of emulating the symmetry properties of rings of atoms.

Finally, it should be noted that the correspondence between the calculated spectral functions in Fig. 5 and the experimentally-determined ones in Fig. 4 is excellent. Each spectral feature is reproduced in the calculations with essentially the correct center frequency, effective linewidth and relative amplitude. This detailed correspondence gives us confidence that our descriptions of both the SRO and IRO are accurate in detail.

C.     Chalcogenide Glasses I: v-As$_2$S$_3$ and v-GeSe$_2$

A large number of studies have been performed on chalcogenide glasses, initially those in which the metalloid species
were restricted to As and Ge, and more recently to glasses in
which the metalloid species is Si. We shall not review this work
in any great detail, but, instead, present calculations for v-As$_2$S$_3$
and v-GeSe$_2$ that serve to illustrate the origin of the large differences that occur in changing from O to either S, Se , or Te
atoms. These changes in the character of the spectrum derive
from two sources: (1) the increase in atomic mass from O to S, Se,
or Te, and the accompanying decrease in the ratio of the chalcogenide atom mass to the metalloid atom mass; both of these factors serve to compress the entire spectrum and to change the ratio
of atomic displacements of the constituent atoms; and (2) the decrease in the bond angle at the twofold-coordinated site from an
average value of 125-150 degrees for the oxides to 95-105 degrees
for the chalcogenides; this has the effect of decreasing the relative splitting between the B and S bands of the twofold-
coordinated atom species. There is no simple corresponding set
of general statements that can be made about the IRO. IRO,
particularly in the form of relatively small rings of bonded atoms
(four to six in number), is dictated by specific details of the
local chemistry, and as such must be considered separately in
each system. We first discuss v-As$_2$S$_3$, where there is nothing
in the polarized Raman response to suggest any IRO in the form
of rings, and then discuss v-GeSe$_2$, where there is definite evidence in the polarized Raman response for contributions related
to the IRO.

Figure 6 summarizes the results of calculations for v-As$_2$S$_3$.
These calculations use parameters scaled from a-As$_2$O$_3$, but
instead of using a particular distribution of dihedral angles, assume this to be random. This last choice is dictated by: (1) the
bonding chemistry of various As-S molecules and crystals, and
(2) the fact that the experimental spectra for v-As$_2$S$_3$ do not show
sharp features in the polarized Raman response that are associated
with IRO [14]. The LDOS functions for As and S are qualitatively
similar to those for As and O as shown in Fig. 5 with differences
deriving from the points discussed above. In the case of v-As$_2$O$_3$,
the spectrum extends to about 850cm$^{-1}$, whereas for v-As$_2$S$_3$, it
terminates at 400cm$^{-1}$. This results in part from the increased
mass of the S atom relative to that of the O atom, and also from
a decrease in the strength of the short-range VFF forces. The

As-LDOS functions are qualitatively similar to those in Fig. 5,
with the primary difference being a decrease in the number of discrete B bands, from three to two. The differences overlap between
the B and S bands, which is a direct result of a decrease in the
As-S-As bond angle. This, in turn, has important effects on the
qualitative aspects of the IR and Raman spectra.

The IR dielectric constant displays three features: (1) a
low frequency band centered at about $160cm^{-1}$, (2) the strongest
IR feature at about $325cm^{-1}$, and (3) a shoulder on the dominant
peak extending to about $400cm^{-1}$. The low frequency band is
associated with S atom R-type motions combined with As atom B

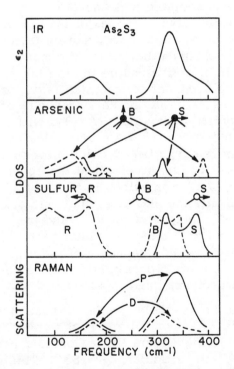

Fig. 6        Calculated spectra for $v-As_2S_3$ [8]; IR response, LDOS
               functions for As and S atoms, and P and D Raman res-
               ponse functions.

and S motions. The dominant feature combines As and S atom
motions, and finally the high frequency shoulder is associated
with S atom S-type motions and As atom B motions. Therefore,
the features in the IR spectrum of v-$As_2S_3$ parallel those in the
IR spectrum of v-$As_2O_3$. The major differences between the
spectra of the oxide and sulfide are (1) in the amount of resolvable structure in all of the bands, and (2) in the sharpness of the
dominant feature. These differences between the two materials
derive more from the change in the mass of the twofold-coordinated
atom than from any aspect of IRO present in v-$As_2O_3$ and absent
in v-$As_2S_3$. The differences between the polarized Raman response
functions of the two materials are much greater, and derive primarily from the differences in IRO in the two materials. There is
only one broad feature at about $350cm^{-1}$ that dominates polarized
Raman response function of v-$As_2S_3$ and it is associated with S
atom B motion. The low frequency feature at about $175cm^{-1}$ in
the depolarized spectrum is associated with a combination of As
atom B and S motion and S atom R motion. The depolarized feature
at about $325cm^{-1}$ is associated with the same S motions of the
As and S atoms that give rise to the major feature in the IR response. The dominance of the S atom S motion feature in the IR
indicates that the mechanism that makes the most important
contribution to the IR effective charge is also the dynamic contribution and that the dominant contribution is associated with
S-type S atom motions. The IR and Raman spectra of v-$As_2Se_3$
are qualitatively similar to those of v-$As_2S_3$, with no features in
the polarized Raman response that are indicative of any significant
degree of IRO. This is supported by calculations based on the
Bethe Lattice approach. Therefore, all of the spectral features in
v-$As_2S_3$ and $As_2Se_3$ can be explained in terms of a structural
model that includes only the SRO at each atomic site. There is
no evidence in any of the spectra for IRO, and in particular for
any effects associated with extended regions with the layer-like
two-dimensional local bonding geometry of the crystals, even
though the two- and threefold coordination of the constituent atoms
mandates a local two-dimensional topology.

In contrast to the As-chalcogenide discussed above, the Ge-
and Si-chalcogenides display features in the polarized Raman
response that indicate significant IRO. Consider first the Ge-
dichalcogenides and, in particular, v-$GeSe_2$. This material has
received considerable study [15-17], and is currently the subject
of much debate and controversy [15-20]. The main issue revolves

around the polarized Raman spectrum and, in particular, the so-
called "companion" Raman line at 220cm$^{-1}$ [15,17,18]. Rather
than state all of the arguments relative to the origin of this feature,
we will present a model for the vibrational properties of v-GeSe$_2$
that is similar to those presented in this work and in ref. 8 for a
number of other oxide and chalcogenide glasses. Figure 7 displays the LDOS functions and the calculated Raman response functions for v-GeSe$_2$ as calculated, using a Bethe Lattice formalism.
The results presented in Fig. 7 are for a calculation in which the
dihedral angle distribution is assumed to be random, and account

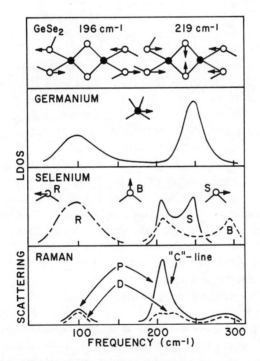

Fig. 7     Atomic displacements of polarized Raman modes for
           four-atom ring configurations for v-GeSe$_2$ [8]. Spectral
           functions: LDOS functions for Ge and Se atoms, and P
           and D Raman response functions.

for all features in the IR dielectric constant and the polarized and depolarized Raman response functions with the exception of the companion line, "C". It is worthwhile to compare the LDOS functions for the Ge and Se atoms in Fig. 7 with the corresponding LDOS functions for Ge and O atoms, respectively, in v-GeO$_2$, as displayed in Fig. 3. The Ge LDOS function for the selenide displays three. This derives from the differences in the LDOS functions of the respective Se and O atoms. The S and B bands for O atoms in v-GeO$_2$ are distinct, whereas in v-GeSe$_2$ the S and B bands of the Se atoms display significant overlap. One major difference between the bands of v-GeSe$_2$ (and also v-GeS$_2$) and the S and B bands of the As-chalcogenides and the As- and Ge-oxides is that in v-GeSe$_2$ (and v-GeS$_2$) the B band extends to higher frequencies than the S band. This accounts for the IR and depolarized Raman features at approximately 315cm$^{-1}$ in v-GeSe$_2$ (and 430cm$^{-1}$ in v-GeS$_2$). The calculation also accounts for: (1) the IR and depolarized Raman modes at 100cm$^{-1}$ in v-GeSe$_2$ and 200cm$^{-1}$ in v-GeS$_2$, (2) the dominant IR modes at 250cm$^{-1}$ in v-GeSe$_2$ and 365cm$^{-1}$ in v-GeS$_2$, and (3) the dominant Raman features at 200cm$^{-}$ in v-GeSe$_2$ and 345cm$^{-1}$ in v-GeS$_2$. There has been considerable debate over the structural origin of the companion lines at 200cm$^{-1}$ in v-GeSe$_2$ and 370cm$^{-1}$ in v-GeS$_2$. The scaling arguments presented in ref. 20 demonstrate that both lines derive primarily from chalcogen atom motion. The relative sharpness of the features suggests some element of IRO. The aruguments given in refs. 8, 19, and 20 support an assignment in terms of rings of atoms. The possible ring configurations are those which occur in the layered crystal form of GeSe$_2$ [20], and the most likely ring is a four-atom configuration that results from an edge connection of GeSe$_4$ tetrahedra. Figure 7 gives the results of calculations based on small, vibrationally isolated clusters that support this assignment [20]. Further support for this assignment comes from the recent studies of Tenhover and his coworkers on v-SiS$_2$ and v-SiSe$_2$, where similar companion features are found in the polarized Raman response [21-23]. The crystalline forms of these Si-chalcogenides contain ribbon-like structures in which SiS$_4$ and SiSe$_4$ tetrahedra are edge-connected and, therefore, have four-atom rings of bonds. The explanation given above is favored over explanations proposed by Phillips and coworkers and based on an "out-rigger raft" structure in which the companion lines result from vibrations involving Se-Se (and S-S) bonds [18].

). Summary

This section of the paper has dealt with oxide and chalogenide glasses. It has been shown that the Bethe Lattice method, using the innovations discussed in ref. 8, can be used o calculate LDOS functions, the IR dielectric constant, and the polarized and depolarized Raman response functions. The glasses f interest fall into two classes, those displaying strong, sharp polarized features that are indicative of IRO, usually due to small four to six) rings of bonded atoms, and those that display only broad features indicative of the absence of small ring configurations in the IRO. The Raman spectra of $v-As_2O_3$, $v-GeS_2$, and $v-GeSe_2$ fall into the first category, and the Raman spectra of $v-GeO_2$, $v-SiO_2$, $v-As_2S_3$, and $v-As_2Se_3$ fall into the second. $v-GeO_2$ and $v-SiO_2$ display weak features indicative of a relatively low concentration of small rings containing about one atomic percent of the O atoms. One of the other materials, not discussed in this paper but displaying IRO identified with a ring component of structure is $v-G_2O_3$ [24,25].

III. ALLOY ATOMS IN a-Si HOSTS

This section of the paper deals with a complementary topic in the vibrational spectroscopy of disordered solids, the local bonding arrangements of alloy atoms in a-Si hosts. Studies of these materials have been stimulated by the applications of a-Si alloys in a variety of technological applications including photovoltaic devices for solar energy conversion, thin-film field effect transistors, and photoreceptors for electrophotography. Lucovsky and Pollard [26] have recently written an extensive review article in this area, and we shall not attempt to discuss the subject in anywhere near the detail discussed in that work. We highlight the main aspects of the physics relating to the interpretation of the spectra, and then give a few examples that serve to illustrate the nature of the spectra and the basis for a structural interpretation.

A. General Considerations of Alloy Atom Bonding

The qualitative behavior of vibrational modes involving alloy atoms in an a-Si host parallels the behavior of alloy atoms in a crystalline host material. The incorporation of alloy atoms produces "optic mode" vibrations in which the atomic displacements of the atoms are localized on the alloy atom and its immediate neighbors. In a crystalline material, the alloy atom is constrained to either substitute for an atom of the host, reside at an interstitial site, or be associated with a complex that includes native defects (usually vacancies). In contrast, in an amorphous solid, an alloy atom can reside at a site in which the local atomic geometry is dictated by the valence bonding requirements of that alloy atom. This is a particular manifestation of the eight-N rule for idealized covalent amorphous solids. The satisfaction of local bonding also establishes the local symmetry at the alloy atom site, and hence determines the character of the vibrational modes involving alloy atom motions. There are two types of modes that are known to occur in Si-hosts: (1) local modes at frequencies in excess of the highest vibrational frequencies of the host material, and (2) resonance modes at frequencies within the frequency spectrum of the host. A third type of alloy atom mode falling within a gap between acoustic and optic mode branches is not possible in a Si-host, since its frequency spectrum does not have a gap between acoustic and optic branches. We will emphasize the local modes, but note that recent experiments and calculations have addressed the questions of resonance modes as well [27-30].

Figure 8 includes the local bonding environments for a number of important alloy atoms in an a-Si host. We have classified the alloy atom sites according to their valence bonding requirements (one, two, three or fourfold coordination), and have given in each case representative examples for specific alloy atoms. Also shown in the diagram are: (1) the local site symmetry, (2) the displacements of the fundamental alloy atom vibrations, and (3) the frequencies of these modes.

Univalent species include hydrogen, as shown in the diagram as well as the halogens, F, Cl, etc. The bonding environment shown in the figure is designated as the monohydride configuration and has two vibrations, both of which are at frequencies in excess of the highest frequencies of the Si host ($500 cm^{-1}$). One of these is a stretching vibration with atomic displacements along the direction of the Si-H bond, and the second is a bending vibration where the atomic displacement of the hydrogen atom is at right angles to the direction of the Si-H bond. The stretching vibrations of D (deuterium), F, and Cl are at frequencies greater than $500 cm^{-1}$, hence they are local modes, whereas the bending vibrations of these atoms are at frequencies in the spectral range of the a-Si host, hence they are resonance modes. There are also other resonance vibrations that involve predominantly Si-atom

notion, but are also localized near the alloy atom sites. These
re discussed in refs. 29 and 30. Bonding configurations involv-
ng more than one H atom attached to the same Si site have been
identified, and their spectral signatures are well understood.
These will be discussed a little later on in this paper.

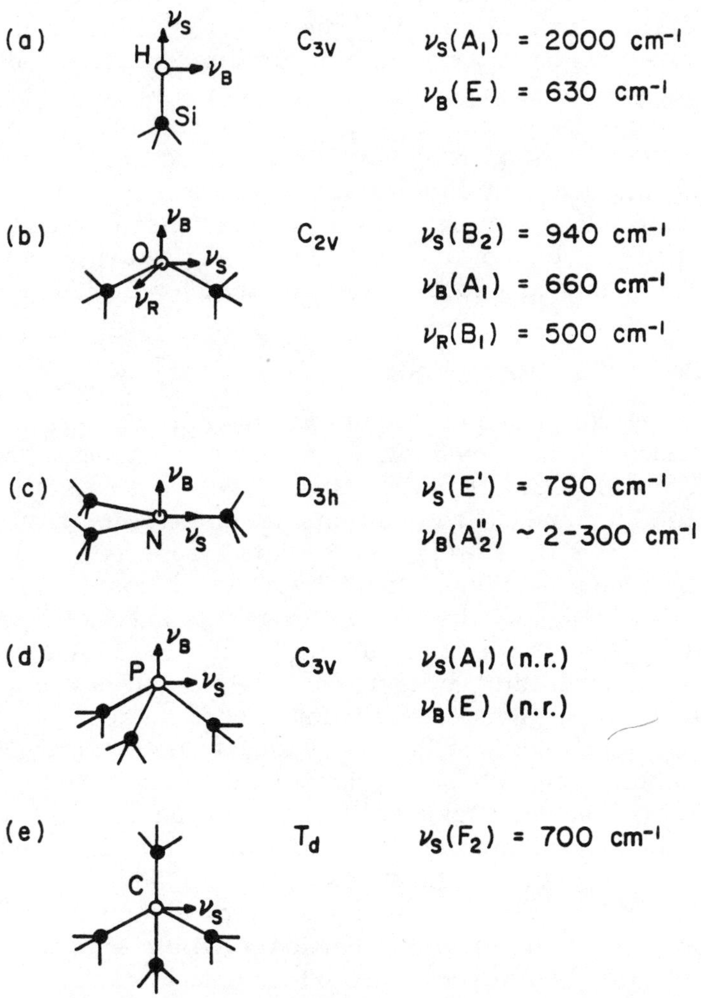

(a)    $C_{3v}$    $\nu_S(A_1) = 2000$ cm$^{-1}$
              $\nu_B(E) = 630$ cm$^{-1}$

(b)    $C_{2v}$    $\nu_S(B_2) = 940$ cm$^{-1}$
              $\nu_B(A_1) = 660$ cm$^{-1}$
              $\nu_R(B_1) = 500$ cm$^{-1}$

(c)    $D_{3h}$    $\nu_S(E') = 790$ cm$^{-1}$
              $\nu_B(A_2'') \sim 2\text{-}300$ cm$^{-1}$

(d)    $C_{3v}$    $\nu_S(A_1)$ (n.r.)
              $\nu_B(E)$ (n.r.)

(e)    $T_d$    $\nu_S(F_2) = 700$ cm$^{-1}$

Fig. 8    Local bonding environments of alloy atoms in a-Si host
materials. The diagram includes the bonding geometry,
the atomic displacement vectors of the symmetry determined localized vibrations, the local site symmetry,
and the frequencies of the symmetry determined vibrations [26].

Part (b) of the figure illustrates the local bonding of the O atom, and includes the three degrees of motion discussed previously in the context of the vibrational properties of v-GeO$_2$. For an O alloy atom in an a-Si host, two of the vibrations are local modes, the stretching and bending vibrations, while the rocking motion gives a resonance mode. Part (c) and (d) of the diagram give the local bonding environments of threefold-coordinated atoms from Volumn V of the periodic table. Nitrogen is unique inasmuch as it bonds in a planar site, while P and As bond in pyramidal arrangements. The stretching vibration of the N atom is a local mode, and the out-of-plane bending is a resonance mode. P and As bonding have not received extensive study and we shall not comment on their vibrational modes. Finally, Column IV atoms reside at substitutional sites with tetrahedral symmetry, and hence have only one type of alloy atom vibration. For the C atom, this is a local mode, whereas for Ge it is a resonance mode.

B.      Theory of Vibrational Modes of Alloy Atoms

The local vibrations of alloy atoms in a-Si host are now well understood through the application of two complementary theoretical techniques: (1) calculations based on clusters of intermediate size, and (2) calculations using the Bethe Lattice approach. Figure 9 illustrates intermediate size clusters (14 and 18 atoms) that have been used to study the local modes of H and F in a-Si hosts [31,32]. The concept of an intermediate size cluster follows from the localization of the atomic displacements of the alloy atom. In this context, we create clusters which include the alloy atom in its normal valence bonding geometry (as in Fig. 8), and then add shells of Si neighbors until a size is reached in which atoms at the perimeter of the cluster remain essentially at rest for the alloy atom local vibrations. This, generally, requires about three shells of Si atom neighbors as shown in Fig. 9. The vibrational properties of H and F have been calculated using VFF forces obtained from molecules and then adjusted to take into account the specific chemical environments in the a-Si host material (see refs. 31 and 33 for details of the empirical procedures employed). In order to further test the concept of "intermediate size," the cluster in Fig. 9(b) was used. It includes four more atoms than that of Fig 9(a); these are arranged in a way that forces all terminal atoms of the cluster into six-membered rings. It rurns out that the local modes calculated using either the 14 or 18 atom clusters are identical; however, there are differences in the resonance modes since these involve a greater participation of the host atoms (see, for example, refs. 29 and 30).

Figure 10 indicates a comparison between calculations of the local modes of the Si-H group using the finite cluster approach and the Bethe Lattice approach. The finite cluster calculation is essentially a calculation of the vibrations of a large molecule. As such, the calculation yields the eigen-frequencies and eigen-vectors of all of the vibrations of the cluster, 3N-6 in all. Our interest is only in the vibrations involving the symmetry deter-mined fundamental modes of the alloy atom, those shown in Fig. 3. The Bethe Lattice calculations extend the number of Si shells indefinitely and employ the same force field as the finite cluster calculations [32]. These calculations yield LDOS functions for the various atoms in the Bethe Lattice. Figure 10 indicates these functions for the H atom, and the first, second, and third shells of Si neighbors for the bonding arrangement shown in Fig. 9(a). The application of these two methods yields complementary infor-

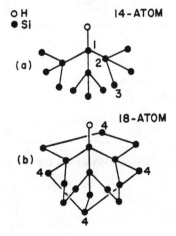

Fig. 9        Cluster configurations for Si-H bonding groups [31].

mation about the vibrational spectra and supports the qualitative
statements made earlier about the spatial localization of the
vibrations. The stretching vibration involves primarily the H atom
and its immediate Si neighbor, whereas the bending vibration involves a displacement of the atoms in the second Si shell as well.

C.     Specific Alloy Systems

Alloy atom bonding configurations have been studied for the
most part using IR absorption spectroscopy. This technique is
particularly well suited to the study of bonds between unlike atoms
where IR activity can result from either a dipole moment due to
static charge transfer (ionic component of the bonding) or from
dynamic charge transfer that occurs during the atomic displacements. In both cases, the moment is generally sufficient to yield

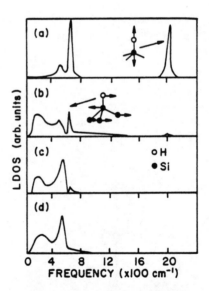

Fig. 10     LDOS functions for the Si-H bonding group; (a) H atom,
            (b) Si atom in first shell, (c) Si atom in second shell,
            and (c) Si atom of the host network. Also included are
            the atomic displacements for the bending and stretching
            vibrations [32].

a detection limit of the order of $10^{16}$atoms/cm$^2$. This translates to approximately 0.1 at .% in a film with a thickness of about one micrometer. Raman scattering spectroscopy is a more powerful tool than IR for studying the properties of the a-Si host network. The results of Raman studies on a-Si, and the theoretical basis for a description of the vibrational properties of the a-Si host are given in ref. 26. The properties of the host depend on the deposition conditions and post-deposition annealing, the central issues revolving around questions of the degree of IRO and the correlations between differences in spectra and differences in the IRO.

The most extensively studied alloy systems to date have been a-Si:H and a-Si:F, primarily because of the interest in these materials for photovoltaic devices [33-35]. Figure 11 includes the various local bonding environments that have been proposed for these univalent atoms. So far, we have mentioned only the monohydride (and monofluoride) geometries in which a-Si atom has only one of the univalent alloy atoms attached which is also spatially remote or isolated from other similar bonding groups. Figure 11 includes other proposed bonding configurations for H and F in a-Si host materials, specifically including situations wherein more than one univalent atom (two or three) are attached to the same Si site, and where these groups may be nearest neigh-

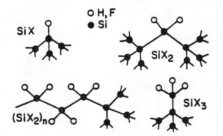

Fig. 11    Local bonding arrangements for H and F atoms in a-Si. Included are monohydride (and monofluoride), dihydride (and difluoride), polysilane (and fluorine substituted polysilane) and trihydride (and trifluoride) configurations [31].

bors in the structure (as in $SiH_{2N}$). Figure 12 indicates IR spectra
for a series of a–Si:H alloys [34]. The absorptions centered at
$2000cm^{-1}$ and $630cm^{-1}$ are the well-known stretching and bending
vibrations, respectively, of the monohydride group [33,34]. Absorptions in the vicinity of $800-900cm^{-1}$ band, in addition to
frequencies of 2090 to $2150cm^{-1}$, are indicative of the attachment
of more than one hydrogen atom to the same Si atom site [33,34].
Figure 13 summarizes the assignments made for the various silicon-
hydrogen bonding arrangements. For a more detailed discussion of
these assignments, and additional references, see the review
article by Lucovsky and Pollard [26].

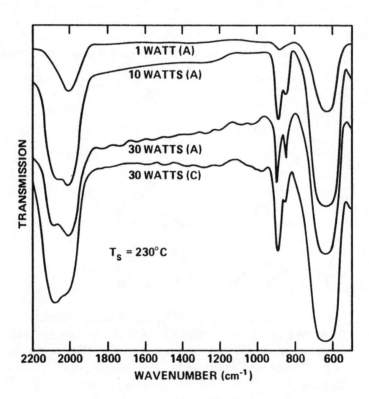

Fig. 12    IR absorption spectra for a–Si:H alloys [34].

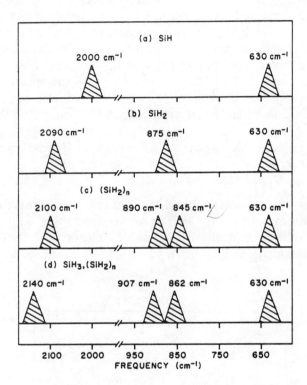

Fig. 13     Assignments for IR active vibrations in a-Si:H alloys [31].

Figure 14 gives IR data for a-Si:F alloys [26,36,37]. Absorption at $830cm^{-1}$ is associated with the stretching vibration of monofluoride bonding groups, while IR absorptions between 850 and $1025cm^{-1}$ derive from bonding arrangements involving more than one fluorine atom attached to the same Si atom site. The spectrum shown in Fig. 14(a) is for a macroscopically homogeneous material. The vibrations between 850 and $1000cm^{-1}$ numbers have been assigned to $SiF_2$ groups. There is still controversy concerning the strong and sharp absorption at about $1015cm^{-1}$. A detailed description of the models proposed and the supporting evidence is given elsewhere [26]. The spectrum shown in Fig. 14(b) is for diphasic material; the absorption at $830cm^{-1}$ is for monofluoride groups in the columns, whereas the absorptions at 950 and $1015cm^{-1}$ are due to F groups in the connective tissue [37]. It is clear that the signatures at 950 and $1015cm^{-1}$ are due to multiple attachment of F atoms to the same Si site, and all possible configurations have been proposed, including $(SiF_{2N})$ and $SiF_2$ groups, as well as $SiF_4$ molecules. The problems involved in making a definitive and an unambiguous assignment for these features are discussed in ref. 26.

So far, we have discussed the general considerations of alloy atom bonding in a-Si, and given as examples the bonding of H and F. We have indicated the spectra signatures that serve to distinguish between one or more alloy atoms being bonded at a

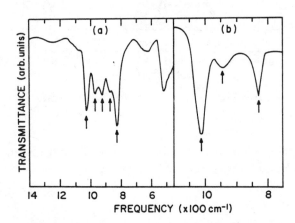

Fig. 14     IR absorption in a-Si:F alloys; (a) homogeneous material, and (b) diphasic material with columnar structure [36,37].

given Si atom site and have indicated the areas within questions remain to be resolved with regard to a definitive assignment. We now go on to discuss some recent work in which the alloy systems are ternary in nature, and in which there is evidence for the attachment of more than one type of alloy atom at the same Si atom site.

Figure 15 gives the IR absorption spectrum of a ternary alloy a-Si:H:O with a H atom concentration of approximately 15 at. % and an O atom concentration of approximately 5 at.% [38]. The figure includes comparisons with the frequencies of the Si-H and Si-O vibrations as identified in studies of the IR absorption spectra of binary alloys of a-Si:H and a-Si:O, respectively. A comparison between the spectrum in the figure and the frequency markers for the Si-H and Si-O modes indicates three features that are not in the spectra of the binary alloys; these are denoted by the markers designated as $v_1$, $v_2$, and $v_3$. There have been additional studies of the a-Si:H:O system [39], as well as studies of ternary alloys in which deuterium has been substituted for hydrogen [40]. These studies have served to identify all the O and O,H related vibrational modes, those displayed in Fig. 15, for

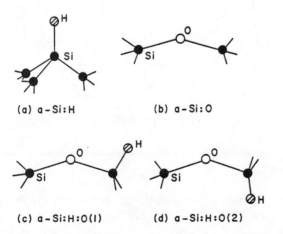

(a) a-Si:H          (b) a-Si:O

(c) a-Si:H:O(l)          (d) a-Si:H:O(2)

Fig. 15    IR absorption in an a-Si:H:O alloy [38].

alloys deposited at substrate temperatures of approximately 250 to 350°C, as well as those deposited onto substrates maintained at lower temperatures [39,40]. The three features mentioned above have been shown to be derived from a bonding site in which H and O atoms are bonded to the same Si atom. Figure 16 includes the local bonding geometry at isolated H and O alloy atom sites, and at sites in which the H and O atoms are bonded to the same Si atom. The arguments for the occurrence of the last type of bonding site derive from: (1) chemically induced shifts in the frequencies of Si-H bond stretching vibration from 2000 to 2090cm$^{-1}$ (due to second neighbor H atoms), and (2) from the occurrence of the new feature at 780cm$^{-1}$. All of these features, as well as a Si-O-Si rocking vibration at 500cm$^{-1}$, grow linearly with increasing O atom concentration [39], and change in predictable ways in ternary alloys in which D atom substitutions have been made [40]. There are two bonding arrangements involving O and H atom bonding that are favored energetically. These are configurations in which

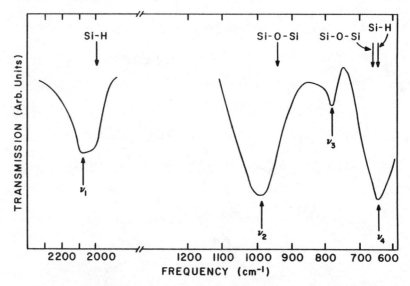

Fig. 16    Local bonding geometry; (a) isolated Si-H site, (b) isolated Si-O-Si site, and (c) and (d) sites with H and O as second neighbors: (c) cis geometry and (d) trans geometry [38,39].

the Si-H bond is in the same plane as the Si-O-Si group. These configurations, both the cis and trans types, minimize the repulsive interactions between the nonbonding electrons on the O atom and the electrons which contribute to the Si-H bond. An analysis of the vibrational properties of cis and trans bonding groups, using the method of clusters of intermediate size and verified by calculations based on the Bethe Lattice approach, establishes the nature of the vibration at $780 \text{cm}^{-1}$. It is a coupled vibration involving simultaneous displacements of both the H and O atoms, and it occurs only for the cis geometry [38,39]. The scaling of the absorption in this band with O atom concentration then establishes that for the O and H atom concentrations studied (0.5 at.% to 15 at.% O and 10-15 at.% H), the fraction of bonds that have a cis conformation is contant [39].

There have been parallel studies of the local bonding in a-Si:H:N ternary alloys [40,41]. These studies have established similar behavior regarding the bonding of both N and H atoms to a common Si atom site. In this case, the simultaneous attachment of the two atoms is identified through: (1) chemically induced shifts of the Si-H and Si-N stretching fequencies, and (2) disorder induced IR activity in a Si atom breathing vibration [41]. There is no distinct vibration that couples H and N atom motions leading to a feature that parallels the $780 \text{cm}^{-1}$ mode in the a-Si:H:O alloys, nor is there a spectral signature that establishes any kind of spatial relationship between the Si-H bond and the plane determined by the N atom and its three Si neighbors [41].

There have recently been studies of a-Ge based alloys in which O has been added to form a ternary system similar to what has been described above for a-Si:H:O [42]. These studies have established qualitatively different bonding arrangements, but nevertheless have given some insight into the origin of the special bonding arrangements of Si,H and O and Si,H and N discussed above. The a-Ge alloy studies have clearly identified the very important role for plasma phase precursors in determining bonding arrangements in the condensed thin films. The alloys of a-Si:H:O, discussed above, have been formed via the glow discharge decomposition of mixtures of $SiH_4$ and $H_2O$ or $O_2$, and show spectra that are essentially the same for either method of O atom introduction. Furthermore, these same films never exhibit an s-OH bonding groups [38,39]. These two observations, coupled with the concentration dependence of the three vibrations associated with Si-O-Si-H bonding sequences discussed above, can be understood in terms of plasma phase interactions that:

    1.   Involve the formation of the precursor molecule, di-siloxane, $O-(SiH_3)_2$.

    2.   Reflect the fact that alcohols of the form $HO-SiH_3$ are not stable and combine to produce disiloxane.

The origin of similar second neighbor relationships between N and H atoms in a-Si:H:N alloys derives from the formation of $N-(SiH_3)_3$ molecules. Since a-Si:H:N alloys condensed on low temperature substrates also show evidence for NH groups, this means that plasma precursors having these groups are also formed, but that condensation on a substrate held above about $300^{\circ}C$ prevents them from remaining intact in the solid phases condensed at the high substrate temperatures [41].

## IV.   SUMMARY

Most of the vibrational modes that have been identified via IR and/or Raman spectroscopy can be understood in terms of the local bonding of the various atomic species present in the compound or alloy. This strong dependence on local bonding geometry, which, in turn, emphasizes the importance of local bonding site symmetry, is the basis for the theoretical techniques that have been used as a framework for discussion of these materials. In addition, features not directly related to SRO, but related to rings of atoms as in several of the glasses, can also be understood qualitatively and quantitatively through extensions of similar theoretical techniques.

## ACKNOWLEDGEMENTS

I would like to thank S.R. Ovshinsky and M.C. Steele for giving me the opportunity to present a lecture at the Institute for Amorphous Studies and for encouraging me in the writing of this paper.

The work reported in this paper was supported in part under ONR Contract N00014-79-C-0133, Air Force Contract F33615-81-C-1428, and SERI Contract XB-2-02065-1.

## REFERENCES

1.    F.L. Galeener and G. Lucovsky, Phys. Rev. Lett. <u>37</u>, 1414(1976).

2.   F.L. Galeener, A.J. Leadbetter and M.W. Stringfellow,
     Phys. Rev. B 27, 1052(1983).
3.   R.J. Bell and P. Dean, Discussions Faraday Soc. 50, 55
     (1970).
4.   R.J. Bell, P. Dean and D.C. Hibbins-Butler, J. de Phys.
     C3, 2111(1970).
5.   R.J. Bell and P. Dean, Phil. Mag. 25, 1381(1972).
6.   R.B. Laughlin and J.D. Joannopoulos, Phys. Rev. B 16,
     947(1976).
7.   R.B. Laughlin and J.D. Joannopoulos, Phys. Rev. B 16,
     2942(1976).
8.   G. Lucovsky, C.K. Wong and W.B. Pollard, J. Noncryst.
     Solids 59 & 60 (1983).
9.   F.L. Galeener, Solid State Commun. 44, 1037(1982).
10.  G.N. Papatheodorou and S.A. Solin, Phys. Rev. B 12, 1741
     (1976).
11.  F.L. Galeener, G. Lucovsky and R.H. Geils, Phys. Rev.
     B 19, 4251(1979).
12.  G. Lucovsky and F.L. Galeener, J. Noncryst. Solids 37,
     83(1980).
13.  D. Beeman, R. Lynds and M.R. Anderson, J. Noncryst.
     Solids 42, 61(1980).
14.  G. Lucovsky, Phys. Rev. B 6, 1480(1972).
15.  P. Tronc, M. Bensoussan, A. Brenac and C. Sebene,
     Phys. Rev. B 8, 5947(1973).
16.  G. Lucovsky, R.J. Nemanich, S.A. Solin and R.C. Keezer,
     Solid State Commun. 17, 1567(1975).
17.  R.J. Nemanich, S.A. Solin and G. Lucovsky, Solid State
     Commun. 21, 273(1977).
18.  J.C. Phillips, J. Noncryst. Solids 43, 37(1981).
19.  R.J. Nemanich, F.L. Galeener, J.C. Mikkelsen, Jr.,
     G.A.N. Connell, G. Etherington, A.C. Wright and
     R.N. Sinclair, Physica 117B/118B, 959(1983).
20.  G. Lucovsky, C.K. Wong and W.B. Pollard, Bull. APS 28,
     327(1983).
21.  M. Tenhover, M.A. Hazle and R.K. Grasselli, Phys. Rev.
     Lett. 51, 404(1983).
22.  M. Tenhover, M.A. Hazle and R.K. Grasselli, Phys. Rev.
     B 28, 5897(1983).
23.  M. Tenhover, M.A. Hazle and R.K. Grasselli, to be
     published.

24.  F.L. Galeener, G. Lucovsky and J.C. Mikkelsen, Jr.,
     Phys. Rev. B 22, 3983(1980).
25.  F.L. Galeener and M.F. Thorpe, Phys. Rev. B 28, 5802
     (1983).
26.  G. Lucovsky and W.B. Pollard, The Physics of Hydrogenated
     Amorphous Silicon II, edited by J.D. Joannopoulos and
     G. Lucovsky, Topics in Appl. Phys. 56, 301(1984).
27.  C. Shen, C.J. Fang and M. Cardona, Phys. Status Solidi
     B 101, 451(1980).
28.  C. Shen, C.J. Fang, M. Cardona and L. Genzel, Phys.
     Rev. B 22, 2913(1980).
29.  E. Martinez and M. Cardona, Phys. Rev. B 28, 880(1983).
30.  W.B. Pollard, to be published.
31.  G. Lucovsky, Springer Series in Solid State Sci. 25, 87
     (1981).
32.  G. Lucovsky and W.B. Pollard, Physica 117B/118B, 865
     (1983).
33.  M.H. Brodsky, M. Cardona and J.J. Cuomo, Phys. Rev.
     B 15, 3556(1977).
34.  G. Lucovsky, R.J. Nemanich and J.C. Knights, Phys. Rev.
     B 19, 2064(1979).
35.  J.C. Knights and G. Lucovsky, Crit. Rev. Solid State Mat.
     Sci. 9, 210(1980).
36.  T. Shimada, Y. Katayama and S. Hirigome, Japan J. Appl.
     Phys. 19, 265(1980).
37.  H. Matsumura, K. Sakai, Y. Kawakyu and S. Furukawa,
     J. Appl. Phys. 52, 5537(1981).
38.  G. Lucovsky and W.B. Pollard, J. Vac. Sci. Tech. A 1,
     313(1983).
39.  G. Lucovsky, J. Yang, S.S. Chao, J.E. Tyler and W.
     Czubatyj, Phys. Rev. B 28, 3225(1983).
40.  G. Lucovsky, S.S. Chao, J. Yang, J.E. Tyler and W.
     Czubatyj, Phys. Rev. B 29, 2302(1984).
41.  G. Lucovsky, J. Yang, S.S. Chao, J.E. Tyler and W.
     Czubatyj, Phys. Rev. B 28, 3234(1983).
42.  G. Lucovsky, S.S. Chao, J. Yang, J.E. Tyler and W.
     Czubatyj, J. Noncryst. Solids 66, 99(1984).

# DENSITY OF STATES IN NONCRYSTALLINE SOLIDS

H. Fritzsche

Department of Physics and The James Franck Institute
University of Chicago
Chicago, Illinois 60637

## I.    INTRODUCTION

The science of semiconductors and semiconductor devices deals with three fundamental questions:

1.    What are the available energy states for electrons in a semiconductor?

2.    How do the electrons in each available energy state respond to externally applied fields?

3.    How can we control the number and distribution of electron energy states so that we can synthesize a semiconductor that has the desirable electronic properties for specific devices?

The last question deals with the task of removing all undesired impurities from the semiconductor material, and with adding those atoms which act as donors or acceptors of electrons that make the semiconductor n-type or p-type, and that change the electrical conductivity, the photoconductivity, or other important properties by many orders of magnitude. It is this variability that makes semiconductors (in contrast to metals and insulators) such versatile substances for solid state devices.

The answer to the second question gives you all the electronic transport properties of the semiconductor material itself, and those of devices. Semiconductor devices are essentially ingenious combinations of n-type and p-type materials, of insulating layers, and of metal contacts.

313

In this paper we consider the first question because, in a way, it is the most fundamental one, and I hope to convey some insight into important aspects of noncrystalline semiconductors. Let us restate the problem: We wish to know (a) the energies of all the electrons in our semiconductor, and (b) all the possible energies electrons can attain after they have been given some additional energy, for instance, by absorption of sunlight or by heating the material.

In order to give you an idea of the difficulty of this problem, let me tell you that there are about $10^{24}$ electrons in 1 cm$^3$ of amorphous (or crystalline) silicon. That is as many electrons as there are sand grains in one million cubic yards of sand--a beach of quite respectable size. In addition to their energies in an equilibrium or unexcited condition, we wish to know all possible higher energies. In my analogy, these correspond to the many states that result from throwing the sand into the air.

There are three principal avenues that may lead to the solution of this problem. The first is to carry out a number of experiments and measurements, and to infer from them the electron energies. A second important way is to use our knowledge of the electron energies in individual atoms and of chemical bonding, and to deduce what the electron energies should approximately be in the solid. The third method involves the theoretical solution of the powerful equations of quantum mechanics and quantum statistics. Despite the enormous numbers of particles, the theoretical method has been successful in crystalline semiconductors. The reason for its success is due to the fact that in crystals the environment of each atom repeats itself with precise periodicity. However, the lack of structural periodicity in noncrystalline materials, the fact that the neighboring of each atom is everywhere different, essentially eliminates the hope for a theoretical solution.

In the following, I shall therefore deduce the electronic states from chemical arguments and from a few experiments.

## II.    CHEMICAL APPROACH

In atoms the available energy states are discrete quantum levels, or energy eigenstates, which are labelled with numbers and letters which signify conserved quantities (quantum numbers) like angular momentum. Only two electrons with opposite spins can occupy a state characterized by a given set of quantum numbers. This is the Pauli exclusion principle. Hence, the 14 electrons of

Si normally occupy the lowest 7 quantum states. Of these, the
two highest occupied ones are labelled 3s and 3p. There are many
unoccupied states at higher energy into which electrons can get
excited by absorbing, for instance, a photon of light of just the
right energy. In a Se atom, the four electrons of highest energy
are in 4p states. We are principally interested in the electrons of
highest energy (the valence electrons) because these are farthest
away from the nucleus and therefore form the chemical bonds between atoms. Imagine that we form a solid by slowly decreasing
the interatomic distance, $\underline{a}$, of a large number, N, of Si or Se
atoms. Figure 1 shows what we expect to happen: the N individual
and discrete atomic energy states will interact and broaden into
bands. As the atomic separation is further decreased, the s and p

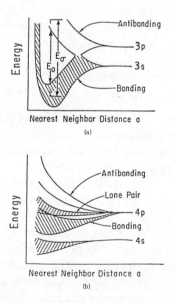

Fig. 1    Sketch of energy bands against near-neighbor distance
          for a tetrahedral semiconductor (a), such as Si, and for
          a lone-pair semiconductor (b), such as Se.

states of Si will mix and lose their identity because covalent bonding occurs with four neighbors in tetrahedral directions. These mixed states are called sp$^3$ orbitals and they form the strong tetrahedral bonds typical for diamond, Si, and Ge. Once formed, it costs energy to pull them apart. Hence, the band of bonding states lies lower than the average energy of the atomic 3s and 3p states. The sp$^3$ orbital can also interact in an opposite manner: by forming antibonding states. These are pushed up in energy and form an antibonding band of states. The equilibrium atomic distance is the one for which the energy of the solid is lowest. Since each Si atom has four valence electrons and forms four sp$^3$ valence bonds, the bonding band is completely occupied according to the Pauli exclusion principle and the antibonding band is empty. All important properties of semiconductors are governed by the electron states near the top of the uppermost filled band (the valence band) and near the bottom of the first empty band (the conduction band) and, of course, by any states which lie in the gap in between on account of defects and other causes, as we shall see shortly.

The reason why other filled bands below the valence band, and other empty bands above the conduction band, are of no concern to semiconductor physics is their large distance in energy. Thermal excitation or optical excitations (infrared, visible, or ultraviolet) can only lift electrons across the gap, which is 1.7eV wide in amorphous Si and 2.2eV wide in Se. The widths of the valence and conduction bands are about 5eV. We, therefore, can forget about the more distant bands of energy states.

It is well known that amorphous Si and chalcogenide glasses are, in many ways, very different noncrystalline semiconductors. The major reason for this was first explained by M. Kastner [1] who pointed out that the uppermost filled band in Se and other chalcogenide glasses is not the bonding band but a band of states formed by electrons which do not participate in bonding (or antibonding). This is illustrated in Figure 1(b). Se has four p-electrons, of which only two are needed for covalent bonding to two neighbors. The remaining two p-electrons remain an unused lone pair. The filled band formed in the solid by these lone-pair electrons is the valence band. This fact is the origin for the rather unusual defect chemistry in chalcogenide glasses [2-5]--a negative effective correlation energy, self compensation, pinning of the Fermi level, and other defect-related phenomena that are absent in a tetrahedral semiconductor such as amorphous Si.

We continue our chemical-intuitive approach, and focus attention on the energy region between the top of the valence band and the bottom of the conduction band, the region of greatest interest to semiconductor physics. Figure 2 sketches the density of electron energy states $g(E)$. This is essentially the distribution of energy states of the CFO model [6] that has been proposed by Cohen, Fritzsche, and Ovshinsky in 1969. Since that time, we have learned many details about specific defect states [7]. Let us make an order of magnitude estimate of the densities of states and describe the main features of Fig. 2 as one moves from the bands progressively deeper into the gap.

Si contains about $5 \times 10^{22}$ cm$^{-3}$ atoms. There are four valence electrons per atom; hence, the valence band and the conduction band contain approximately $2 \times 10^{23}$ cm$^{-3}$ states. Since the bands are approximately 5eV wide, the average density of states (number of states in 1 cm$^3$ and 1eV energy range) is $g(E) = 4 \times 10^{22}$ eV$^{-1}$ cm$^{-3}$.

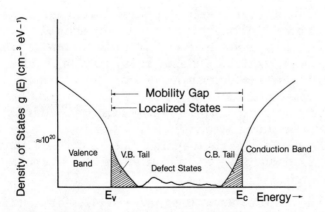

Fig. 2        Band model of a noncrystalline semiconductor. $E_v$ and $E_c$ are the mobility edges of the valence band and of the conduction band, respectively.

Both the conduction band and the valence band have tails of states, and deeper in the gap there are states originating from structural and coordination defects (atoms which are not bonded according to their normal valency) as well as from impurities. Hence, in contrast to crystals, g(E) is nowhere zero.

We believe that the tail states are unavoidable because they are intimately associated with the disorder in the noncrystalline structure. The valence band tail states are covalent bonds that are weaker than normal. This can happen, for instance, when the covalent angle is bent from its equilibrium value, when the bond is stretched due to internal strains, or when some antibonding orbital is mixed in. That occurs, for instance, when atoms bond together in odd-numbered rings containing 5, 7, and 9 atoms (instead of the 6 atoms in cyrstals). Moreover, it was recently found, both theoretically and experimentally, that there are net static charges on some atoms or groups of atoms [8]. These produce potential fluctuations which push states up and down and prevent any sharp feature in g(E). All these effects are expected to produce also a tail of states extending down in energy from the (antibonding) conduction band.

A typical defect-related gap state is a dangling bond, that is, a Si atom bonded to only three nearest neighbors instead of four. Since the dangling bond state is not bonded (and not antibonded), it lies near the gap center. Having one electron only, it is easily detectable by measuring the electron magnetic moment by electron spin resonance (ESR).

The total number of tail states is approximately $10^{-3}$ of the number of states in one band, and the number of defect-related gap states varies between $10^{-5}$ and $10^{-7}$ band states. This sounds like a very small fraction. Indeed, it is, but the requirements for device quality semiconductor materials are very stringent. One wishes, of course, to have as few gap states as possible in order to control the electronic properties by intentional additions of donor and acceptor atoms. Furthermore, the gap states act as recombination centers of photoexcited charge carriers, and thus limit the photoconductivity and the lifetime of electronically injected carriers.

The tail states, as well as the defect-related states, are localized to a region of a few atoms. A charge carrier occupying them has, therefore, no chance of moving away (zero mobility) at low temperatures. Carriers in extended band states, in contrast, have a finite mobility. The demarcation energies $E_v$ and $E_c$ separating extended from localized states thus define a mobility gap.

I have to clarify the term charge carrier which I used in the last paragraphs above. A valence band (or any other band) fully occupied by electrons cannot contribute to electronic conduction. The reason is the following. A motion of an electron in an applied electric field requires that the energy of the moving electron be slightly raised. However, it is clear that when all extended states are already fully occupied, no electron can go into a higher energy states and, hence, conduction cannot occur. That is the reason why semiconductors are good insulators at low temperatures and in the absence of light, even though they contain those incredibly large numbers of electrons. We, therefore, come to the surprising statement that conduction is due to a very minute fraction of electrons, namely, those that are excited either thermally or by light into mobile states near $E_c$ in the conduction band and by the positive holes (missing electrons) left behind near $E_v$ in the valence band. $E_c$ and $E_v$ are referred to as mobility edges. We also talk about electrons and holes occupying localized tail states. They must be taken into account for charge neutrality, but they do not contribute directly to a motion of charge in an applied electric field.

Before we turn to the experimental verification of this density of states model and the experimental determination of specific state densities, let us summarize the important elements of our intuitive picture:

1.      Noncrystalline semiconductors have bands of electron energy states which are extended throughout the material, and hence provide a finite mobility.

2.      The highest filled band (valence band) is separated in energy from the lowest empty band (conduction band) by a mobility gap.

3.      There are localized tail states extending from the top of the valence band and from the bottom of the conduction band into the gap. These localized tail states are weakened or modified band states. They do not conduct--a charge carrier trapped in such a state has zero mobility.

4.      Deeper in the gap there are localized states originating from impurities or from atoms which have either fewer or more covalent bonds than desired by their normal valency.

5.      The mobility edges $E_V$ and $E_C$ separate band states that offer finite mobility from localized states that do not normally permit lateral motion of electrons or holes.

6.      Conduction takes place by thermally or optically excited electrons above $E_C$ or positively charged holes below $E_V$.

III.    EXPERIMENTS

I wish to discuss, in sequence, experiments that yield information about band states, tail states, and then gap states. Before I proceed, however, I must briefly clarify some important aspects of the specific amorphous semiconductor that is used in my examples.

Amorphous silicon has first been prepared by evaporation or sputtering. Films obtained in this manner have no photoconductivity and essentially cannot be doped n-type and p-type. They are useless and uninteresting for device applications. Electron spin resonance measurements show that they have about $10^{20}$ cm$^{-3}$ dangling bond defect states in the gap [9].

The major turning point in our field occurred when Sterling et al. and Chittick et al. [10] prepared a-Si by glow-discharge decomposition of $SiH_4$, and Ovshinsky et al. [11] by glow-discharge decomposition of $SiF_4$; when Spear et al. [12] showed that this new material can be doped; and when Carlson et al. [13] demonstrated the first amorphous solar cell. This new material is so very good and useful because it has hardly any dangling bond defects. It contains about 10 atomic percent hydrogen and/or about 5 atomic percent fluorine which do not allow many Si orbitals to remain without bonding partner; the number of dangling bonds is reduced to about one out of every $10^8$ bonds. Fluorine has an advantage over hydrogen in that it does not escape from the material at high temperatures. In the following, we refer to this material as a-Si:H or a-Si:H:F.

A.      Optical Absorption

Essentially everything we know about energy levels in atoms and molecules has been learned from optical spectroscopy. Although important, these techniques are not quite as powerful in solids because the bands of electron states in solids contain so many closely spaced levels that they cannot be resolved individually by optical excitations. However, gross features of larger groups of levels will certainly be discernible.

In Fig. 3 the band model is plotted on a logarithmic scale in order to emphasize the relative magnitudes of band states, tail states, and gap states. Quantum statistics applies equally to electrons in noncrystalline and crystalline materials. Therefore, there is a Fermi energy $E_F$ below which the energy states are occupied by electrons in equilibrium. Since a photon is essentially absorbed locally, optical excitations between localized states are unlikely because they are separated in space. Essentially no electronic optical excitation can occur unless the photon energy $h\nu$ exceeds a threshold energy that is large enough to lift an electron from $E_v$ to an empty state above $E_F$ or from an occupied state below $E_F$ to $E_c$. These gap state to band state (G-B) transi-

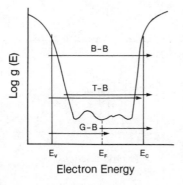

Fig. 3    Optical absorption processes: B-B between extended band states; T-B between tail states and band states; and G-B between gap states and band states.

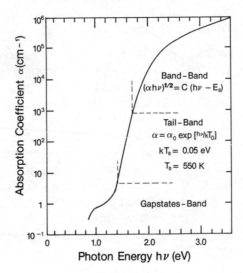

Fig. 4  Absorption spectrum of hydrogenated amorphous silicon.

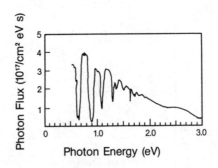

Fig. 5  Solar spectrum of sun as received at noon near the equator.

tions start at the threshold energy for photoconductivity. At higher photon energies $h\nu$, transitions between tail states and band states (T-B) occur, and at even higher $h\nu$ band to band (B-B) transitions take place.

The strength of optical absorption is given by the logarithmic fraction of photons absorbed per cm of material. This absorption coefficient $\alpha$ is proportional to the product of the density of those occupied states and the density of those unoccupied states that can be bridged by the energy $h\nu$ of the incident photon, if the transition probability is the same for all states. This last point is very important and we shall return to it in a moment.

Since the density of states $g(E)$ of band tail and gap states differ so drastically, we expect the absorption spectrum shown in Fig. 4. The strongest absorption is caused by B-B transitions; G-B absorption is weakest and changes with $g(E)$ of the gap states from sample to sample. The T-B absorption region is nearly exponential suggesting that $g(E)$ of the tail states falls off exponentially. The functional forms of $\alpha$ are noted in Fig. 4 for B-B and T-B transitions. These allow the determination of the optical gap $E_O$ and of the exponential decay $1/kT_O$ of the broader of the two band tails.

The absorption spectrum of a-Si:H shown in Fig. 4 differs in important ways from that of crystalline Si. In the latter, the optical transition probabilities are governed by the additional requirement that momentum be conserved. This means that optical transitions can only occur between two energy states that have essentially the same momentum quantum number. Since momentum conservation is related to translational symmetry that is a fundamental property of crystals and, of course, absent in noncrystalline materials, the band-band absorption is much stronger in amorphous than in crystalline Si. As a consequence, $1\mu$m thick a-Si:H solar cells absorb visible light as efficiently as $50\mu$m thick crystalline Si solar cells. This is an important advantage of amorphous materials.

On the other hand, Fig. 4 shows that photons having energies less than $h\nu = 1.7eV$ pass through a-Si:H without getting absorbed. The sun emits lots of these red and infrared photons because it is essentially a 20,000K hot radiator. The solar spectrum at the equator at noon is shown in Fig. 5. If one compares this spectrum with that of Fig. 4, one sees that a great improvement of amorphous solar cells requires the development of a material with a smaller optical gap $E_O$. This might be achieved by alloying Si with Ge or Sn.

We identified the exponential slope of $\alpha$ ($h\nu$ between $\alpha = 10$ and $\alpha = 10^3$ cm$^{-1}$ with the exponential decay parameter $1/kT_o$ of the broader of the two band tails. This may not be entirely true because the optical transition probability may depend on energy $h\nu$ on account of internal potential fluctuations. More research on this question is needed.

The magnitude of the very low absorption region ($h\nu < 1.4$eV) increases with the concentration of gap states. It, therefore, serves as a convenient measure of the quality of the material. As an example, we show in Fig. 6 how the absorption coefficient $\alpha$ in this regime increases with phosphorus doping. One notices that adding phosphorus to a–Si:H not only introduces the desired shallow donor states but also a large concentration of defects.

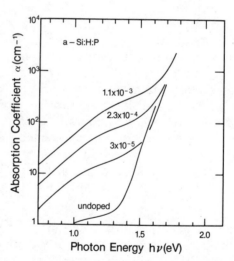

Fig. 6  Absorption in a–Si:H due to defect states produced in the mobility gap by phosphorus doping. The numbers give the doping ratio $PH_3/SiH_4$ in the plasma gas. After W.B. Jackson and N.M. Amer, J. de Physique $\underline{42}$, C4-293(1981).

Different experimental techniques are required to measure absorption coefficients which span the seven orders of magnitude shown in Fig. 4. The most commonly used methods are illustrated in Fig. 7. Direct optical absorption can determine only $\alpha > 1000$ cm$^{-1}$ since the a-Si:H film cannot be made very thick ($d \approx 1$–$3$ $\mu$m). Smaller absorption coefficients are determined by photo-thermal spectroscopic techniques. These work in the following manner. Light absorption causes a small temperature rise $\Delta T$. This in turn warms the medium outside the sample surface. In the photo-thermal deflection method [14–16], a laser beam aimed parallel to the sample surface is bent due to the refractive index gradient caused by the temperature gradient in the adjacent outside medium. Carbon tetrachloride (CCl$_4$) is commonly chosen as the medium

**1. Direct Optical Absorption:**

$$F_t = F_o\, e^{-\alpha d}$$

Limitation: $\alpha d > 0.1$
$d = 10^{-4}$ cm
$\alpha > 1000$cm$^{-1}$

**2. Photo-Thermal Spectroscopy:**

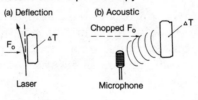

(a) Deflection                    (b) Acoustic

Laser                    Microphone

Limitation: $10^{-4} < \alpha d < 1$

**3. Secondary Photoconductivity:**

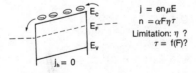

$j = en\mu E$
$n = \alpha F \eta \tau$
Limitation: $\eta$ ?
$\tau = f(F)$?

Fig. 7    Experiments for measuring optical absorption coefficient $\alpha$.

because it has low absorption and a large temperature coefficient of its refractive index. In the photo-acoustic method, $\Delta T$ of the sample heats the gas surrounding the sample at the chopping frequency of the incident light beam. The resultant acoustic wave is then detected by a sensitive microphone. Both photo-thermal methods are calibrated by direct optical absorption above $\alpha = 1000$ cm$^{-1}$.

Photoconductivity is a less reliable method for determining $\alpha$ unless one has independent means for determining the quantum yield $\eta$, that is, the fraction of photoexcited carriers per absorbed photon and the dependence of the carrier lifetime $\tau$ on the photon flux F.

B.        Determination of Tail States Densities

Optical absorption methods have verified the gross features of our density of states model shown in Figs. 2 and 3. These methods, however, always yield the product of the initial and the final density of states and, in addition, involve the optical transition probability. We now turn to methods which yield the individual density of states $g(E)$ of either the valence band tail or the conduction band tail.

1.        Nonlinear Light Dependence of Photoconductivity

When a-Si:H is exposed to a uniformly absorbed light of F photons/cm$^2$ sec., one usually finds that the photoconductivity $\sigma_p$ increases nonlinearly with F as

$$\sigma_p = \text{const.} \; (\alpha F)^\gamma \qquad (1)$$

with an exponent $0.5 \leq \gamma \leq 1$. We assume that photoconduction takes place in the bulk of the film and not in an interface space charge layer and that surface recombination can be neglected. Uniform absorption can be achieved by choosing $h\nu$ such that $\alpha d < 1$.

According to Rose [17], a power law dependence of $\sigma_p$ on F as in Eq. (1) is to be expected when $g(E)$ depends exponentially on energy as

$$g(E) = g_o \exp (E/kT_o). \qquad (2)$$

That is what we expect to occur in the tail state region. The argument proceeds as follows. Let us assume that electrons are the dominant photocarriers with concentration n. The photoconductivity is then

$$\sigma_p = en\mu \tag{3}$$

with

$$n = \alpha F\tau. \tag{4}$$

Here $\mu$ = electron mobility above $E_C$ and the electron lifetime $\tau$ is given by

$$\tau = 1/bN_r. \tag{5}$$

The recombination coefficient $b = \nu_o N_C$ is related to the attempt of escape frequency $\nu_o$ and the effective density of states $N_C$ at $E_C$ because of detailed balance. In general, $\underline{b}$ may have different values for different recombination centers but the argument of Rose [17] assumes that b is constant.

The concentration of recombination centers, $N_r$, includes all states (multiplied by their occupation probability $f(E)$) between the quasi-Fermi levels for electrons $E_{Fn}$ and for holes $E_{Fp}$. If the density of states is very small in the gap but rapidly rising in the tail according to Eq. (2), it is a good approximation to write

$$N_r = kT \ g(E_{Fn})$$
$$= g_o \ kT \ \exp(E_{Fn}/kT_o). \tag{6}$$

By definition $E_{Fn}$ and n are related through

$$n = N_C \exp(E_{Fn}/kT) \tag{7}$$

when we choose $E_C = 0$ as reference energy. Substituting (7) into (6) yields

$$N_r = g_o \ kT \left(\frac{n}{N_C}\right)^{T/T_o}, \tag{8}$$

and combining (8), (5), and (4) yields the final result

$$\sigma_p = \text{const.} \ (\alpha F)^{T_o/(T+T_o)}. \tag{9}$$

A measurement of $\gamma = T_o/(T+T_o)$ determines, therefore, the steepness $1/kT_o$ of the exponential tail at the energy $E_{Fn}$ below $E_C$. By changing the temperature T and/or the light intensity F, one can change $E_{Fn}$ and, hence, probe the exponential slope of the density of states as a function of energy. By this method it has been found [18] that $T_o \approx 300K$ for $E_C - E_{Fn} < 0.3eV$ and $T_o \approx 1000K$ for $E_C - E_{Fn} > 0.35eV$.

Two arguments against this simple explanation have to be resolved: (i) at higher light intensities F, recombination occurs both in the conduction band tail and in the valence band tail. In this case, Eq. (6) will no longer be a good approximation, and (ii) the assumption of a constant , energy independent recombination coefficient b may not be justified.

2.      Time-of-Flight Experiment

This experiment has been discussed in detail by M. Kastner [19]. I shall limit myself to a brief summary of the principal features. As sketched in Fig. 8, excess carriers, both electrons and holes, are created near the front surface of the amorphous semiconductor by a flash of light of sufficiently high photon energy $h\nu > E_o$ that most of the light is absorbed within a surface layer that is thin ($\sim$ 1000Å) compared to the thickness L of the semiconductor. At the same time, a voltage V is applied between the front and the back surface that causes either the excess electrons or the excess holes to drift through the sample depending on the sign of V. Ideally, the front contact should be blocking to the drifting carriers and nonblocking to the others. A uniform drift field V/L is maintained if (i) the dielectric relaxation time $\tau_D = \epsilon/\sigma$ is long compared to the transit time $t_T$, (ii) the amount of injected charge is small compared to Q = CV, the

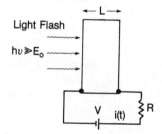

Fig. 8  Sketch of time-of-flight experiment for measuring drift
         mobility.

charge on the electrodes, and (iii) V is applied with or just before the light flash. Because of current continuity, a drift current $i(t)$ can be measured in the external circuit. A plot of log i against log t shows two distinct slopes as sketched in Fig. 9, where $T_O$ is the already familiar parameter of the exponential slope $1/kT_O$ of the tail state density, either of the valence band tail when holes are drifting or of the conduction band tail for electrons.

As explained in detail by Orenstein-Kastner [20] and Tiedje-Rose [21], the decay of the drift current before the break at $t_T$ arises from the fact that each time carriers are re-emitted from shallow tail states, a fraction of them gets stuck in deeper tail states from which a re-release takes a somewhat longer time. The drift current falls off faster with time after the transit time $t_T$ because the total number of excess carriers decreases as an increasing number reach the back electrode. From the transit time one obtains the mobility for either electrons or holes:

$$\mu_d = L^2/Vt_T . \tag{10}$$

The temperature dependence of $\mu_d$ can be fitted to a chosen model for the $g(E)$, the density of the band tails, provided that carriers are not lost to deep traps and that tunneling conduction between tail states can be neglected.

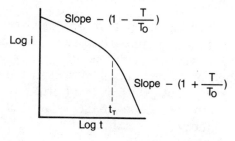

Fig. 9 Sketch of dispersive transport drift current measured in a time-of-flight experiment. The transit time is denoted by $t_T$.

3.        Discussion of Tail States Densities

The experiments discussed above agree well with the exponential energy dependence of $g(E)$ in the tail state regions as expressed by Eq. (2). The parameters obtained from time-of-flight experiments on a-Si:H are shown in Table I, where we have also included values for the band mobilities $\mu_o$. The nonlinear intensity

Table 1.    Tail State Parameters for a-Si:H.

|            | $kT_o$ (eV) | $T_o$ (K) | $\mu_o$ cm$^2$/Vs |
|------------|-------------|-----------|-------------------|
| electrons  | 0.025       | 300       | ~ 15              |
| holes      | 0.042       | 500       | ~ 0.7             |

dependence of the steady state photoconductivity has also yielded $T_o = 300K$ for electrons (although I have noted some reservations about its interpretation). From the exponential regime of the absorption spectrum between $\alpha = 10$ and $10^3 cm^{-1}$ one finds $kT_o = 0.05eV$, a value that is somewhat larger than that for the valence band tail, $kT_o = 0.042eV$, obtained from time-of-flight measurements.

Before leaving this topic, I should mention a few unresolved problems regarding the tail states. The interpretation of the experiments assumed the same attempt-to-escape frequency $\nu_o$ for all states in a tail. If this parameter depends on energy $\nu_o(E)$ then one actually measures the exponential slope of the product $\nu_o(E) \, g(E)$ [22].

The second problem is that we do not know the magnitudes of $g_o$ and of $g(E_c)$ or $g(E_v)$. Figure 10 shows three possible density of states curves and positions of the mobility edge. It is difficult to measure the slopes and magnitudes of $g(E)$ in the vicinity of $E_c$. A recent attempt to construct $g(E)$ from magnitude measurements (capacitance) in the gap and slope measurements ($\gamma$ and time of flight) led to unreasonably large values of $g_o$ for the conduction band tail [18].

Finally, there remains the question how the tails change with temperature. The analysis of drift mobility measurements which are carried out over a temperature range of about 200°C yield values for $kT_O$, the logarithmic slopes of the tail state distributions, by assuming that $T_O$ is independent of T. However, the slope of the exponential part of the absorption edge (Urbach edge) is interpreted to be the slope of the valence band tail; and this Urbach edge has been found to be temperature dependent, changing from $kT_O = 0.055eV$ to $0.070eV$ between 100 and 300K [23]. This raises the question of whether optical transitions and thermal equilibrium transport properties are related to the same density of states function. Optical transitions occur so rapidly that atomic positions cannot adjust to a change of occupancy of the electronic states by an absorption or emission of a photon (Franck-Condon principle). In contrast, the local atomic environment has time to relax during the slower processes occurring during transport measurements. This relaxation caused by electron-phonon interactions increases with the degree of localization, and, hence, the binding energy or the depth of the state. It, therefore, is unlikely that one and the same density of states curve applies to these two sets of measurements.

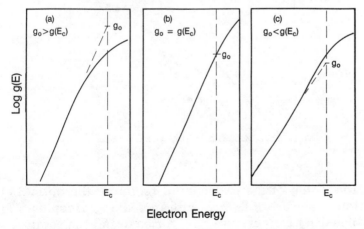

Fig. 10    Three possible forms of the density of states near the mobility edge.

The width of the tail states can be broadened by increasing the disorder, as for instance, by using improper preparation conditions that yield high densities of dangling bonds. An example of this is shown in Fig. 11. It is, of course, most desirable to go the other way and prepare the best amorphous semiconductor. In doing so, it is not clear at present whether there is an intrinsic limit to the steepness of the tails of states that further material improvements cannot surpass.

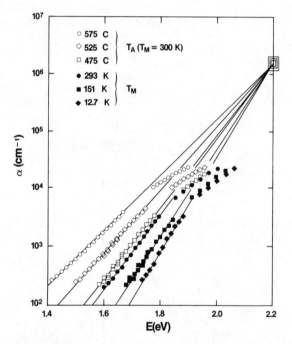

Fig. 11   Broadening of the exponential region of the optical absorption spectrum of a-Si:H with increasing disorder. $T_A$ = annealing temperature, $T_M$ = measuring temperature (after G.D. Cody et al., reference 23).

C.        Determination of Gap State Density

The localized states deeper in the gap are associated with bonding defects--atoms whose nearest neighbor bonds are not satisfied according to their chemical valency.

These are, for instance, dangling covalent bonds such as a threefold instead of fourfold coordinated Si atoms or one of the many coordination defects or impurity complexes discussed by Adler [24]. In solar cell grade a-Si:H only a small fraction, of the order of $10^{-8}$, of the Si atoms have such a defect. Their detection is not a trivial task.

The low energy regime of the photothermal deflection spectra (such as shown in Fig. 6) as well as the magnitude of the electron spin resonance signal yield concentrations of gap state defects between $10^{15}$ and $10^{18} cm^{-3}$. However, these measurements yield only the total concentration of defect states and not their distribution in energy, that is, $g(E)$. This more detailed information is obtained from a variety of techniques, including field effect [25,26], capacitance-voltage measurements [27,28], deep level transient spectroscopy (DLTS) [29,30], space charge limited currents (SCLC) [31,32], and thermally stimulated currents (TSC) [33]. I shall mention the advantages and drawbacks of a few of these methods, and then compare results obtained by them.

1.        Field Effect

When the amorphous semiconductor film is used as one plate of a parallel plate capacitor, it can be charged positively or negatively by applying to the other electrode a gate voltage $V_G$. The charge density $\rho(x)$ in the semiconductor is distributed according to the Poisson equation

$$\frac{d^2 V(x)}{dx^2} = \frac{4\pi e}{\kappa_{sc}} \rho(x) \tag{11}$$

and the bands in the semiconductor space charge region are bent as shown in Fig. 12. $\kappa_{sc}$ is the dielectric constant of the semiconductor. $V(x)$ is determined by Eq. (11) and in turn determines the space charge density

$$\rho(x) = \int_{-\infty}^{+\infty} dE\, g(E)\, [f(E-eV(x))-f(E)]. \tag{12}$$

Here, f(E) is the Fermi distribution function. The band bending
causes an increased current flow in the space charge accumulation
region yielding the field-effect conductance.

$$G = (G_o/t) \int_0^t dx \exp [eV(x)/kT]. \tag{13}$$

By measuring this conductance G between the source and drain
electrodes as a function of the gate voltage $V_G$, one can obtain
the desired density of states g(E) by iteratively solving Eqs. (11),
(12), and (13). The usefulness of the field effect method was
first demonstrated by its success in proving that hydrogenated [34]
and fluorinated [35] glow discharge deposited a-Si:H have a greatly
reduced g(E) compared to evaporated or sputtered a-Si, and that the
lowest g(E) is achieved at deposition temperatures near 260°C.
Moreover, the reversible increase in g(E) caused by light exposure
(Staebler-Wronski effect) [36] and the effect of annealing and
hydrogen effusion on g(E) was observed by means of field-effect
measurements [37].

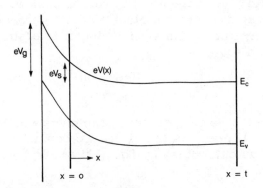

Metal  Oxide  Semiconductor

Fig. 12   Band bending potential V(x) in a MOS field effect transis-
          tor. $V_G$ = gate voltage; $eV_s$ = surface potential; $E_v$ and $E_c$
          are the mobility edges of the valence band and of the
          conduction band, respectively.

When deep level transient spectroscopy measurements (DLTS) later yielded smaller values for g(E) than field-effect measurements, one suspected that the latter method is unreliable because of the following difficulties: (i) the field effect deals with current flow in the first 50-100Å thick layer that may be more defective and have a larger g(E) than the bulk material, and (ii) some of the charge carriers induced by the gate voltage in the amorphous semiconductor film may be stuck in interface states which are neglected in the present field effect analysis.

Besides these factors, there are two additional difficulties that have not been mentioned sufficiently: (iii) at large gate voltages, the band bending creates a quantum well at the interface to which most of the current flow is constricted [38]. It is most likely that quantization normal to the interface (x-direction) localizes states near the mobility edge and thus pushes the mobility edge to energies beyond $V(x)$. Lastly (iv) when one determines the density of states g(E) from the equations above, starting from Eq. (13), one assumes that the conductivity prefactor $\sigma_0$ is independent of the energy separation $E_C - E_F$ (between the mobility edge and the Fermi level), that changes with $V(x)$ in the space charge layer. This, however, is contrary to the observation that $\sigma_0$ decreases exponentially with $E_c - E_F$ according to the Meyer-Neldel rule [38]

$$\sigma_O = \sigma_{OO} \exp M(E_c - E_F) \qquad (14)$$

where the constant M is approximately $(21\pm3)eV^{-1}$. If the effects (iii) and (iv) are included in the analysis of the field effect, one obtains [39] densities of states that are smaller by more than a factor 10, and hence much closer to those determined by DLTS. This actually suggests that the first two problems (i) and (ii) mentioned above may not be as serious as expected. Nevertheless, we do not know enough about transport in a quantum well and about the proper Meyer-Neldel relation to carry out a reliable quantitative analysis of the field effect data.

2.      Capacitance Method

The space charge distribution $\rho(x)$ of Eq. (12) is governed by, and hence contains, the information of the density of states g(E). The total charge of a diode of area A is

$$Q = A \int_O^\infty \rho(x)\, dx. \qquad (15)$$

The change of this charge resulting from a change in the junction potential $\delta V_S$ defines the junction capacitance C

$$C = \delta Q/\delta V_S = \frac{\epsilon A}{<x>} \tag{16}$$

where $<x>$ is the first moment of the charge distribution

$$<x> = \int_0^\infty x \delta\rho\,(x)\;dx/\int_0^\infty \delta\rho\,(x)\;dx. \tag{17}$$

Figure 13 shows the difference in $\delta\rho\,(x)$ when the measurements are carried out (a) by changing $\delta V_S$ very slowly (d.c. method), and (b) by changing $\delta V_S$ at a frequency $\nu$. In the latter case, there is a regime close to the junction in which the states near $E_F$ are too far removed from the nearest conducting band to change their occupancy within the time $1/\nu$. The emission rate $r_e$ of a state separated by $\Delta E$ from the conduction band edge $E_C$ is

$$r_e = \nu_0 \exp\,(-\Delta E/kT) \tag{18}$$

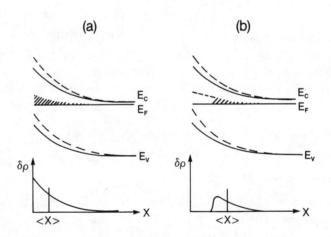

Fig. 13  Change $\delta\rho$ of the space charge density as a function of position as the surface potential is altered by an applied voltage; (a) under d.c. condition; (b) under a.c. condition.

where $\nu_o \simeq 10^{12}/s$ is the attempt-to-escape frequency. As a consequence, all states for which

$$\Delta E > kT \, \ell n \nu_o/\nu \tag{19}$$

are unable to release their electrons at the measuring frequency $\nu$. For this reason $\delta_\rho(x)$ is finite only at larger x where $E_F$ is closer to $E_c$. For the same $\delta V_S$ one finds $<x>$ to increase, and thus C to decrease with increasing frequency $\nu$. By combining V-C and C-T-$\nu$ techniques, one should be able to distinguish spatial from energy variations of $g(E)$.

Even though the capacitance method can conveniently distinguish samples having high and low values of $g(E)$, it is not able to distinguish clearly between different shapes or energy dependencies of $g(E)$. Especially any peak or structure in $g(E)$ cannot be resolved by this method (or by the field effect method) because two integrations are involved in the analysis. These integrations tend to smoothen out any features in $g(E)$. This can easily be verified by working backwards and comparing the C-V (or field effect) data calculated from a variety of choices for $g(E)$.

3.      Deep Level Transient Spectroscopy

DLTS is one of several thermally stimulated transient spectroscopies. These have in common that first a nonequilibrium distribution of carriers over the localized states is established at a low temperature by a pulse of light or voltage, and then the current or capacitance is measured as the sample returns to equilibrium while it is being heated at a constant rate. As the temperature is raised, states of increasing energy depth $\Delta E$ will be able to emit carriers (see Eq. (18)) and return to equilibrium. Hence, any change in current or capacitance at a certain T is related to structure in $g(E)$ at a given E. The thermally stimulated current (TSC) or capacitance (TSCAP) methods differ from the DLTS technique in that a trap filling pulse (light or voltage) is applied once at low T in the case of TSCAP and TSC, whereas in DLTS such a pulse is applied repetitively as T increases and the transients to equilibrium are followed by synchronous detection. The repetitive application of the excitation pulse enhances the signal-to-noise ratio, and offers the additional advantage that DLTS spectra can be taken at constant T or by scanning T up or down.

Figure 14 shows a typical example of density of states curves g(E) obtained by DLTS [40]. Here, we have a series of phosphorus-doped n-type a-Si:H samples containing various amounts of compensating boron impurities. The curves exhibit detailed structure which was shown to be consistent with the DLTS data analysis. A common feature of all g(E) curves determined by DLTS is the pronounced minimum about 0.4eV below $E_C$.

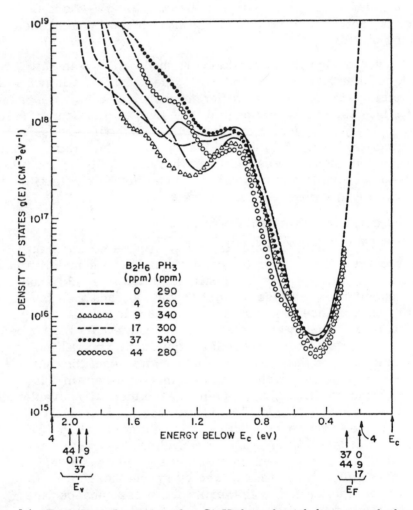

Fig. 14   Density of states of a-Si:H doped with boron and phosphorus at various concentrations. Measurements were made by DLTS (after reference 40).

The reproducibility of the DLTS results is astonishing and better than that of the other methods that are used for determining g(E). Unfortunately, different measuring techniques are employed on samples prepared under various conditions in different laboratories. Furthermore, different methods are best suited for doped and undoped material. Field effect measurements, for instance, can be interpreted best when $E_F$ is near the gap center and the material is intrinsic. Capacitance and DLTS measurements, on the other hand, require a Schottky depletion region, and therefore doped material with $E_F$ close to one or the other mobility edge (usually close to $E_c$).

The DLTS method is the only one that yields the deep minimum in g(E) about 0.4eV below $E_c$. We do not understand the reason for this discrepancy. The other techniques, such as space charge limited currents, capacitance-voltage, and field effect, yield roughly $g(E) \simeq 10^{16} cm^{-3} eV^{-1}$ in the energy range between 0.5 and 0.9eV below $E_c$ and a raise of g(E) beyond that range.

## IV.     CONCLUDING REMARKS

We have made great progress in developing techniques for measuring the density of electronic states in amorphous semiconductors and in understanding the gross features of the state distributions. We still lack the ability to identify the energies and concentrations of specific defects or impurities. The reason for this is the fact that the electronic states form a continuous distribution in the gap without distinct features. This lack of discreteness of the energy levels of the defects might be caused by spatial heterogeneities and internal potential fluctuations. One might have to achieve a control over these features that affect the medium-range order first before deeper insight into the nature of the defect states can be gained.

ACKNOWLEDGEMENTS

I gratefully acknowledge many stimulating and clarifying discussions with S.R. Ovshinsky, my colleagues at Energy Conversion Devices, as well as my students and associates.

REFERENCES

1.      M. Kastner, Phys. Rev. Lett. $\underline{28}$, 355(1972).
2.      R.A. Street and N.F. Mott, Phys. Rev. Lett. $\underline{35}$, 1293 (1975).
3.      M. Kastner, D. Adler and H. Fritzsche, Phys. Rev. Lett. $\underline{37}$, 1504(1976).
4.      M. Kastner and H. Fritzsche, Phil. Mag. $\underline{B37}$, 199(1978).
5.      H. Fritzsche and M. Kastner, Phil. Mag. $\underline{B37}$, 285(1978).
6.      M.H. Cohen, H. Fritzsche and S.R. Ovshinsky, Phys. Rev. Lett. $\underline{22}$, 1065(1969).
7.      D. Adler, Sci. Am. $\underline{236}$, 36(1977).
8.      L. Ley, J. Reichardt and R.L. Johnson, Phys. Rev. Lett. $\underline{49}$, 1664(1982).
9.      M.H. Brodsky and R.S. Title, Phys. Rev. Lett. $\underline{23}$, 581 (1969).
10.     H.F. Sterling and R.C.G. Swann, Solid State Electronics $\underline{8}$, 653(1965); R.C. Chittick, J.H. Alexander and H.F. Sterling, J. Electrochem. Soc. $\underline{116}$, 77(1969).
11.     S.R. Ovshinsky and A. Madan, Nature $\underline{276}$, 482(1978).
12.     W.E. Spear and P.G. LeComber, Solid State Commun. $\underline{17}$, 1193(1975).
13.     D.E. Carlson and C.R. Wronsky, Appl. Phys. Lett. $\underline{28}$, 671(1976).
14.     W.B. Jackson, N.M. Amer, A.C. Boccara and D. Fournier Appl. Opt. $\underline{20}$, 1333(1981).
15.     A.C. Boccara, D. Fournier, W.B. Jackson and N.M. Amer Opt. Lett. $\underline{5}$, 377(1980).
16.     A.C. Boccara, D. Fournier and J. Badoz, Appl. Phys. Lett. $\underline{36}$, 130(1980).
17.     A. Rose, Concepts in Photoconductivity and Allied Problems, Robert E. Krieger Publishing Co., N.Y.(1978).
18.     C.-Y. Huang, S. Guha and S.J. Hudgens, Phys. Rev. $\underline{B27}$, 7460(1983).
19.     M. Kastner, this volume.
20.     J. Orenstein and M. Kastner, Phys. Rev. Lett. $\underline{46}$, 1421 (1981).
21.     T. Tiedje and A. Rose, Solid State Commun. $\underline{37}$, 49(1981).
22.     J. Orenstein, M. Kastner and V. Vaninov, Phil. Mag. $\underline{B46}$, 23(1982).
23.     G.D. Cody, T. Tiedje, B. Abeles, B. Brooks and Y. Goldstein, Phys. Rev. Lett. $\underline{47}$, 1480(1981).

24.     D. Adler, J. de Physique 42, C4-1(1981); also article
        in this volume.
25.     A. Madan, P.G. LeComber and W.E. Spear, J. Noncryst.
        Solids 20, 239(1976).
26.     N.B. Goodman and H. Fritzsche, Phil. Mag. B42, 149
        (1980).
27.     M. Hirose, T. Suzuki and D.H. Döhler, Appl. Phys.
        Lett. 34, 334(1979).
28.     T. Tiedje, C.R. Wronski and J.M. Cebulka, J. Noncryst.
        Solids 35-36, 743(1980).
29.     J.D. Cohen, D.V. Lang and J.P. Harbison, Phys. Rev.
        Lett. 45, 197(1980).
30.     D.V. Lang, J.D. Cohen and J.P. Harbison, Phys. Rev.
        B25, 5285(1982).
31.     W. den Boer, J. de Physique 42, C4-451(1981).
32.     K.D. Mackenzie, P.G. LeComber and W.E. Spear, Phil.
        Mag. B46, 377(1982).
33.     L. Vieux-Rochaz and A. Chenevas-Paule, J. Noncryst.
        Solids 35-36, 737(1980); W. Fuhs and M. Milleville,
        Phys. State. Solid B98, K29(1980); J. Dijon, Solid
        State Commun. 48, 79(1983).
34.     W.E. Spear and P.G. LeComber, J. Noncryst. Solids
        8-10, 727(1972).
35.     A. Madan, S.R. Ovshinsky and E. Benn, Phil. Mag.
        B40, 259(1979).
36.     D.L. Staebler and C.R. Wronski, Appl. Phys. Lett. 31,
        292(1977); J. Appl. Phys. 51, 3262(1980).
37.     N.B. Goodman, Phil. Mag. B45, 407(1982).
38.     H. Fritzsche, Solar Energy Mat. 3, 447(1980).
39.     M.H. Tanielian, N.B. Goodman and H. Fritzsche, J.
        de Physique 42, C4-375(1981).
40.     P. Cullen, J.P. Harbison, D.V. Lang and D. Adler, J.
        Noncryst. Solids 59-60, 261(1983).

# PROBLEMS RELATING TO THE ELECTRONIC STRUCTURE
# OF AMORPHOUS SEMICONDUCTORS

Morrel H. Cohen

Exxon Research and Engineering Company

Annandale, New Jersey 08801

## I.      INTRODUCTION

This paper deals with several problems related to the theory of amorphous semiconductors, some of which have remained unsolved despite the magnitude of the effort devoted to basic and applied research on these materials. I review recent progress, attempt to identify the essential remaining difficulties, and summarize new results of my own and those of my collaborators, as the case may be. The topics covered are band tails, the Lloyd model, shallow impurity levels, deep levels, excitons, electronic transport, and optical absorption. Several of these problems have been covered in recent publications [1-3], from which most of this work has been extracted.

## II.     EXPONENTIAL BAND TAILS

An exponential relationship between the absorption coefficients and the photon energy, known as the Urbach rule, was first enunciated in 1953 for the observed optical absorption edge in AgBr [4]. In many insulators and crystalline semiconductors [5] similar tails have subsequently been observed in the optical absorption. More recent experiments on optical absorption coefficients [6] and on the trap density of individual bands [7] have yielded experimental evidence for exponential tails in the densities of states (DOS) of amorphous semiconductors. Usually

attributed to disorder, i.e., defects, impurities and thermal disorder, the physical origin of this essentially universal behavior is
still not well understood despite extensive experimental [5,6] and
theoretical [8-14] in recent years.

Various attempts [15-17] have been made to calculate the
low-energy tail of the density of states of a disordered system.  In
all these attempts it has been recognized that the density of localized states is several orders of magnitude smaller than the density
of extended states, implying that the former arise from special
atomic configurations.  Consequently, one is forced to abandon
mean-field-like or CPA-like theories and even numerical simula-
tions to obtain the density of localized states.  They arise from
potential fluctuations on wavelength scales of at least several
atomic distances due to a variation in the physical parameters of
the disordered system on the same scale.  It is clearly established [15] that wells of size $\lambda$ in which the minimum kinetic energy of
localization is proportional to $\chi_d/\lambda^2$ gives a DOS N(E) which behaves as:

$$\ln N(E) = \ln (N_o) - (\frac{4}{4-d})^2 (\frac{4-d}{d})d/2 \frac{\chi_d^{d/2}}{2W^2L^d} |E|^{(2-d/2)} \tag{1}$$

where $N_o$ is the pre-exponential, $\chi_d = d \pi^2\hbar^2/2m$, L is the correlation length of the potential fluctuations and is of atomic size, d
is the space dimensionality, and $W^2$ is the variance of the random
potential.  $|E|$ is measured from the bottom of the conduction band.
The energy dependence in Eq. (1) can be understood as follows:
The factor $|E|^2$ comes from the amplitude of the potential fluctua-
tion which becomes Gaussian on the length scale $\lambda \gg L$, the factor
$|E|^{-d/2}$ comes from the spatial extent of the potential fluctuations
$\lambda^d$, which scales as $|E|^{-d/2}$ when one assumes that the kinetic
energy $|E|$ of confinement of the wave function for any dimension
goes like $1/\lambda^2$.

The above treatments, which give the exact one-dimensional
(1-D) asymptotic behavior for the DOS, give an $|E|^{1/2}$ dependence
in the logarithm of the DOS for 3-D in disagreement with the linear
dependence seen experimentally.  Another puzzling feature is the
hierarchy of magnitudes of the characteristic energies:  for example,
from the broadening of the main peak in the absorption coefficient
in passing from the crystalline [6] to the amorphous [6] (a) form of

silicon, one estimates W to be of order eV. The quantity $E_c - E_1$ can be estimated [7] to be of order tenths of eV in a-Si, $E_c$ being the mobility edge and $E_1$ being the energy below which the exponential behavior occurs. $E_o$, the slope of the exponential tail [6,7], is of order hundredths of eV in a-Si.

The discrepancy between theory and experiment for the DOS in the 3-D case derives from the assumption that the energy needed to localize a wave function to a potential well of size $\lambda$ goes as $1/\lambda^2$ in a disordered system. This assumption ignores the short-wavelength fluctuations of the random potential, i.e., it ignores the effects of disorder on the energy of localization. To obtain the energy change $\Delta E$ due to constraining an extended state of energy $E'$ into a volume of linear dimension $\lambda$ in the presence of disorder, we observe that $\Delta E$ equals (within a numerical factor) the shift of the energy due to changing the boundary conditions from periodic to antiperiodic. Thouless [18,19] has shown that this shift can be expressed in terms of the density of states per unit volume $N(E')$ and the dimensionless conductance $g(\lambda, E')$. Thus, we obtain for the energy of localization $\Delta E$ in a three-dimensional material:

$$\Delta E = \frac{B_g(\lambda, E')}{\lambda^d N(E')} \qquad (2)$$

with $d = 3$, where B is a constant. For lengths much higher than interatomic distances we can use the Vollhardt and Wolfle [20] expression for $g(\lambda, E') = \frac{1}{\pi 3}(1 + \frac{\pi}{2} \frac{\lambda}{\xi(E')})$ in three dimensions, where $\xi$ is the correlation length which characterizes the spatial extent of the amplitude fluctuations of extended states of energy $E'$ above the mobility edge. As E increases above $E_c$, these amplitude fluctuations become less severe and finally disappear near an energy $E_u$ at which $\xi$ has reached $\xi_o$, a length of atomic size. At $E_u$, $\Delta E = \chi_3/\lambda^2$ for $\lambda \gg \xi_o$ because of the uniformity of the wave function. Comparing this with (2) gives $2\pi^2 \chi_3 \xi_o N(E_u)$ for B. For $E_c < E' > E_u$, it is possible that $\xi_o \ll \xi$, when $g(\lambda) \to 1/\pi^3$ and $\Delta E \simeq 2\chi_3 \xi_o/\pi \lambda^3 (N(E_u) \simeq N(E'))$. This value of $\Delta E$ is much smaller than $\chi_3/\lambda^2$ because it costs much less energy to compress a highly fluctuating eigen function than a uniform one.

It is easy to show from the Lloyd and Best variational [16] principle that the optimal choice of $E'$ is $E' \simeq E_c$ so that $\lambda \ll \xi$ for any $\lambda$, which enables us to use $\Delta E \simeq 2\chi_3 \xi_o/\pi \lambda^3$. By repeating any of the previous calculations of the Lax and Halperin type for finding the DOS in the tail with this new expression for $\Delta E$ we get:

$$N(E) \sim \exp\left[-\frac{16X_3\xi_O}{3\pi W^2 L^3}|E|\right] \tag{3}$$

in 3-D, where $|E|$ is measured from the mobility edge $E_C$. The inverse slope $E_O$ of the exponential is given by:

$$E_O = \frac{3\pi W^2 L^3}{16X_3\xi_O} \cong \frac{1}{16\pi}\frac{W^2}{\left(\frac{\hbar^2}{2m\xi_O^2}\right)} \tag{4}$$

where we have assumed that $L \simeq \xi_O$. Clearly $E_O$ is only a small fraction of $W$.

For the argument leading to Eq. (3) to hold, the potential well, of extent $\lambda$ and of depth $|E| + \Delta E$, must be isolated. Isolation is guaranteed in 3D if the probability that such a potential fluctuation occur be less than $10^{-2}$. Thus, the smallest value of $|E|$ for which this condition can be satisfied, $E_C - E_1$, is given by:

$$(E_C - E_1) = (1.2 \text{ to } 2.3)\frac{\pi W^2 L^3}{2X_3\xi_O} = \frac{(1.2 \text{ to } 2.3)}{6\pi}\frac{W^2}{\left(\frac{\hbar^2}{2m\xi_O^2}\right)}. \tag{5}$$

Note that $E_C - E_1$ is of order $10^{-1}$ W, while $E_O$ from Eq. (4) is of order $10^{-2}$ of W. The ratio $|E_1|/E_O \simeq (3.2 - 6.2)$ is in agreement with experiment [7].

Our arguments for the exponential behavior of the DOS are very general and can be applied to any type of disordered system whether the disorder is of structural, compositional, or thermal origin. The actual values of $E_O$ and $E_1$ we obtain for a-Si are in agreement with experiment [6,7], but the true significance of the above results is their essentially universal applicability.

In summary, we find that the physical picture of band-tail states embodied in the Halperin-Lax theory is essentially correct.

Below an energy $E_1$, localized states have a physical extent governed by the size of the potential fluctuation localizing them. The fact that Halperin and Lax and their successors obtained the wrong energy dependence of the density of states in $2 + \epsilon$ dimensions derived solely from their use of the incorrect scaling between energy of localization and size of the localized state. The Halperin and Lax arguments do not apply to localized states with energies $E_c > E > E_1$. In that energy range, the long-wavelength potential fluctuations merge, and the physical extent of the localized states is determined by the localization length, which diverges at $E_c$.

## III. THE LLOYD MODEL

Consider an electron moving in a tight-binding s-band the Hamiltonian for which is:

$$H_{\ell m} = G_{\ell} \Delta_{\ell m} = V_{\ell m}. \tag{6}$$

The sites $\ell$, m are periodically arranged so that with $\epsilon_p = 0$, H is that for a crystal. In a disordered system, the $\epsilon_p$ are randomly distributed. In the Lloyd model [21], the $\epsilon_{\ell}$ are randomly distributed according to identical Lorentzian distributions:

$$P(\epsilon_p) = \frac{1}{\pi} \frac{\Gamma}{\epsilon_{\ell}^2 + \Gamma^2} \tag{7}$$

the total density of states is given by:

$$N(E) = \frac{1}{\pi} \text{ tr Im} G(E) = \frac{1}{\pi} \text{ tr Im} <G(E)>. \tag{8}$$

In (8) the energy E has a vanishingly small negative imaginary part inside the Green's function $G(E)$, but is taken to be real inside $N(E)$. The average over the ensemble of all possible values of the $\epsilon_p$ is indicated by angular brackets. The average is, explicitly:

$$<G(E)> = \Pi_{\ell} \int_{-\infty}^{\infty} \frac{de_{\ell}}{2\pi i} \left( \frac{1}{\epsilon_{\ell} - i\Gamma} + \frac{1}{\epsilon_{\ell} + i\Gamma} \right) G(E, (\epsilon_{\ell})). \tag{9}$$

It is carried out by contour integration over $\epsilon_1$, first, holding the remaining $\epsilon_2 \ldots \epsilon_N$ on the real axis. For $\epsilon_2 \ldots \epsilon_N$ real, one easily shows that the only singularity G has in the complex $\epsilon_1$

plane is a pole $z_1$ such that

$$\text{Sgn Im } z_1 = \text{sgn Im E.} \tag{10}$$

The contour is closed in the opposite half plane with the result that $\epsilon_1$ is replaced by $-i\Gamma$ sgn E in $G(E)$. This procedure is iterated for each $\epsilon$ in turn with the final results:

$$<G(E)> = G_O(E + i\Gamma \text{ sgn Im E}) \tag{11}$$

$$N(E) = \int dE' N_O(E') \frac{1}{\pi} \frac{\Gamma}{(E-E')^2 + \Gamma^2} \cdot \tag{12}$$

In (11) and (12) the subscript o indicates crystalline quantities. Equation (12) shows that disorder introduces a Lorentzian broadening of the density of states in the Lloyd model.

The basic results (11) and (12) hold also when:

$$H_{\ell m} = H_{\ell m}^O + \epsilon_{\ell m} \delta_{\ell m} \tag{13}$$

where $H_{\ell m}^O$ is any arbitrary Hermitian matrix and the $G_\ell$ are as before. $G_O$ and $N_O$ then refer in (11) and (12) to quantities relating to H.

## IV.    SHALLOW IMPURITY LEVELS

The generalization of the Lloyd model to the case of arbitrary unperturbed Hamiltonian (13) makes it possible to examine a sequence of problems not yet well understood. One can ask what doping with donors or acceptors does to the band tails of a disordered semiconductor. Consider the conduction band of a crystalline semiconductor doped with one hydrogenic donor. There will be a set of discrete levels below the conduction band edge. We now ask what happens when disorder is introduced. We use the Lloyd model and choose $H^O$ as that of the doped crystal. One immediately sees from (12) that there are two cases. When $\Gamma$ is less than $E_D$, the donor binding energy, there will be peaks corresponding to the lower donor levels, with halfwidth of order $\Gamma$, superposed on the tail of the conduction band. These donor states are resonant with the localized tail states and hybridize weakly with them. When $\Gamma$ exceeds $E_D$, the donor levels merge into the conduction band tail, adding a shoulder to the density of states provided $\Gamma$ is not too large. Such a shoulder has been observed, e.g., by Cody [6] in p-doped a-SiH$_x$.

## V.    DEEP LEVELS

Structural defects or certain impurities can introduce deep levels into the mobility gap of a disordered semiconductor in regions where the density of states would otherwise be very small. One can examine such states via the Lloyd model by taking as $H^0$ the Hamiltonian for a crystal with a single deep level in the gap. Equation (12) then shows us that in the disordered semiconductor the level broadens out into a Lorentzian line of halfwidth $\Gamma$.

The Lloyd model is, of course, very unrealistic, and one does not expect the density of states to be representable as a simple convolution such as (12) in any actual material. Nevertheless, for purposes of estimation or for sketching out theories, it is convenient to consider a generalized version of (12):

$$N(E) = \int N_0(E')g(E - E')dE, \tag{14}$$

where $g(E - E')$ is a normalized broadening function. Given exponential tails, it is clear that $g(E - E')$ should, at least asymptotically, be of the form:

$$g(E - E') = \frac{1}{2E_0} e^{-|E-E'|/E_0}. \tag{15}$$

Consideration of a deep level via (14) shows that the deep-level shape should be that of the broadening function in a disordered semiconductor. An exact treatment would replace $\Gamma$ in (12) by the imaginary part of the self energy, an energy-dependent quantity, and would include the real part in the definition of $N_0(E)$.

## VI.    EXCITONS

Another subject about which little is understood for amorphous semiconductors is the theory of excitons. In molecular materials, excitons are typically of Frenkel type with electron and hole nearly always simultaneously present on a given molecule. The exciton is well separated from the electron-hole continuum and may be regarded simply as an isolated elementary excitation. The influence of disorder on the exciton band is formally the same as on an electron or hole band, and is well understood in principle. In amorphous semiconductors, on the other hand, the entire electron-hole manifold must be considered.

Excitons are typically manifested in optical properties, which are derivable from the microscopic dielectric function. The latter contains as an essential ingredient the electron-hole propagator:

$$G(h\omega^+) \equiv [h\omega + i\delta - H_{eh}]^{-1} \tag{16}$$

where $\omega$ is the angular frequency of the light, $\delta \to \delta^+$, and $H_{eh}$ is the electron-hole Hamiltonian:

$$H_{eh} = H_e + H_h + V_{eh}. \tag{17}$$

For a crystal, $H_e$ is the Hamiltonian of an electron in the conduction band, $H_h$ is the Hamiltonian of a hole in the valence band with a zero of energy chosen to give the correct band gap, and $V_{eh}$ is the electron-hole interaction. We can generalize the Lloyd model to this two band case simply by adding to the crystalline $H_e$ and $H_h$ random single site energies:

$$H_e = H_e^0 + \epsilon_c \tag{18}$$

$$H_h = H_h^0 - \epsilon_v \tag{19}$$

where $\epsilon_\alpha$, $\alpha = c$ or $v$, has the matrix elements $\epsilon_{\alpha\ell m} = \epsilon_{\alpha\ell}\delta_{\ell m}$ and the $\epsilon_{\alpha\ell}$ are all independently distributed according to:

$$P_\alpha(\epsilon_{\alpha\ell}) = \frac{1}{\pi} \frac{\Gamma_\alpha}{\epsilon_{\alpha\ell}^2 + \Gamma_\alpha^2}. \tag{20}$$

One can once again prove that the poles in G are in one-half of the complex $\epsilon_{\alpha\ell}$ plane or the other, but not in both, and thus obtain the results:

$$<G(E)> = G_0(E + i(\Gamma_c + \Gamma_v) \text{ sgn Im } E) \tag{21}$$

$$N(E) = \int dE' N_0(E') \frac{1}{\pi} \frac{\Gamma_c + \Gamma_v}{(E-E')^2 + (\Gamma_c+\Gamma_v)^2}. \tag{22}$$

Thus, the exciton band is broadened by convolution with a Lorentzian the width of which is the sum of the valence and conduction band-tail widths. Consequently, the exciton levels are smeared out before the donor or acceptor levels. If $2(\Gamma_c + \Gamma_v)$ is less than the exciton binding energy, exciton peaks will persist, otherwise

not. Even when $2(\Gamma_c + \Gamma_v)$ exceeds the exciton binding energy
sufficiently that there is no trace of exciton structure left in the
optical absorption, the far tail in the latter can still be dominated
by excitons. These are excitons, however, which overlap the
smeared electron-hole continuum and are therefore resonant with
it.

Consider now a-SiH$_x$. Let us use instead of a Lorentzian
broadening function an exponential broadening function. Equation
(20) for the density of states in the electron-hole manifold would
become:

$$N(E) = \int dE' N_O(E')g(E - E') \tag{23}$$

$$g(E') = \frac{1}{2E_O} e^{-|E-E'|/E_O} \tag{24}$$

$$E_O = E_{oc} + E_{ov}. \tag{25}$$

It is a fact that there is a shoulder in the optical absorption when
donors are present, but not in their absence [22] so that, as consistent with (20)-(23), the exciton levels are smeared to a greater
extent than are the donor levels. An exponential tail persists in
the optical absorption over a range of energies substantially larger
than the sum of valence and conduction band-tail widths [22]. Our
present considerations would suggest that such optical absorption
is primarily due to excitons. However, Cody et al. [23] have
shown that the optical absorption inferred from the photoconductivity by ignoring excitons is the same as that observed directly.
This inconsistency has a simple explanation. The Lloyd formalism
gives for the density of states in the tail of the electron-hole
manifold a sum of contributions from the broadened exciton bands
and the tail of the broadened electron-hole continuum. That is,
the exciton bands are resonant with the broadened electron-hole
continuum. An exciton of energy E must decay into an electron and
a hole with a lifetime of order:

$$\tau_{ex}(E) = \frac{2\pi}{h} M^2 N(E)v \tag{26}$$

where $N(E)$ is the contribution to the density of states from the free
electron-hole pairs. M must be of the order of the exciton binding
energy, but N is decreasing exponentially, v is an appropriate
volume of interaction. For example, for $M \simeq 0.1$ eV,

$N(E) \simeq 10^{17}/eV/cc$, and $v \simeq (10\text{Å})^3$, $\Gamma$ is about $10^{10} sec^{-1}$. Thus, the excitons would have decayed on the time scale of the photo-conductivity experiment.

Turn to $As_2S_3$, Higashi and Kastner [24] have inferred from luminescence studies the existence of two excitation processes, one of which has an abrupt onset at a certain excitation energy. They propose that the abrupt onset is associated with the genera-tion of independent electrons and holes in extended states which then propagate to the luminescent centers and excite them. In this interpretation, the onset energy becomes the mobility gap in the single particle spectrum. The value of the mobility gap so inferred, however, is about a tenth of an eV lower than that in-ferred from transport measurements, and the discrepancy is larger than the combined uncertainties. The discrepancy is readily un-derstandable if it is recognized that excitons dominate the optical absorption sufficiently below the mobility gap in the single parti-cle spectrum, that the excitons will have a mobility edge about the exciton binding energy below the mobility gap, and that ex-tended excitons above their mobility edge can propagate to and excite the luminescent centers before decaying themselves into localized electrons and holes.

## VII.    ELECTRONIC TRANSPORT

Transport by independent electrons can be treated exactly in arbitrarily disordered materials in the absence of Coulomb inter-actions, electron-phonon interactions, and spin-flip scattering [3]. In the particular case of a disordered semiconductor in which the Fermi energy $E_F$ lies closer to one mobility edge, $E_c$, say that in the conduction band, the d.c. conductivity $\sigma$ has the form [3]:

$$\sigma = \sigma_o e^{-E_\sigma/k_BT} \tag{27}$$

with    $E_\sigma = E_c - E_F$  (28)

and    $\sigma_o = \Gamma(v + 1)\sigma_{MIN}\dfrac{a}{\xi_o}\left(\dfrac{k_BT}{E_u - E_c}\right)v.$  (29)

In (29), $\sigma_{MIN}$ is Mott's minimum metallic conductivity [25],

$$\sigma_{MIN} = \frac{1}{2\pi^2} \frac{e^2}{\hbar} \frac{1}{a}, \tag{30}$$

a is the atomic separation, $E_u$ is the energy above which amplitude fluctuations in the extended states is not significant [3], $\xi_o$ is the value at $E_u$ of the correlation length of the amplitude fluctuations [26] and is of order atomic separations [19], $\nu$ is the exponent governing the divergence of the correlation length at the mobility edge [26], and $\Gamma(\nu + 1)$ is the gamma function. $\nu$ has unity as an exact lower bound [3], but field-theoretic calculations suggest that its actual value may be unity [27]. In the following, we shall leave the value of $\nu$ free, supposing it to be slightly greater than unity. The corresponding exact expression for the thermopower is [3]:

$$S = \frac{-k_B}{e} \left[ \frac{E_\sigma}{k_B T} + (\nu + 1) \right]. \tag{31}$$

Note that the conductivity activation energy is not universal; it can vary with structure, composition, disorder, and doping. The quantity $e|S|/k_B$, however, differs from $E_\sigma/k_B T$ only by the universal quantity $\nu + 1 \cong 2$. The pre-exponential $\sigma_o$ falls short of being universal only through the disorder dependence of $E_u - E_c$ and $\xi_o$.

This behavior is sometimes observed. For example [28], p-type amorphous silicon (a-Si) doped with boron has a value of $\sigma_o$ of 4.5 $\Omega^{-1}cm^{-1}$ in samples for which $E_\sigma$ varies from 0.20 to 0.49 eV. A value of 4.5 $\Omega^{-1}cm^{-1}$ is consistent with Eq. (29). We have shown [29] that the exponential tail in the density of states starts at an energy $E_1$ such that $|E_1 - E_c| = 10 E_o$, where $E_o$ is the width of the exponential tail, $E_o \simeq 43$ mV for the valence band of a-Si [7]. We have also argued [29] that $|E_u - E_c| \cong |E_1 - E_c|$. Thus, for a value of 10Å for $\xi_o$, $\sigma_o$ should be about 4.4 $\Omega^{-1}cm^{-1}$, as observed [28]. In the conduction band, however, $E_o$ is about 25 mV [7], which corresponds to a value of $\sigma_o$ of 12 $\Omega^{-1}cm^{-1}$, also observed but at lower temperatures. There are large discrepancies in the value of $\sigma_o$ for the conduction band and needs more careful study [30].

Nevertheless, dramatic discrepancies occur between these predictions and the observations. Instead of being quasi-universal and of order 1-10 $\Omega^{-1}cm^{-1}$, $\sigma_o$ can be orders of magnitude larger and can correlate with $E_\sigma$ [31]:

$$\ln\sigma_o = A + B E_\sigma.$$                                                (32)

Eq. (31) is then violated [32]. Activation of the mean mobility is observed at higher temperatures with a shift at lower temperatures to the weak temperature dependence expected above [33]. For p-type materials the sign of the Hall effect is almost always negative and can shift with temperature [34]. Since the model is exact, these discrepancies can be eliminated only by putting the interactions back into the theory. The electron-phonon interaction is the most important one [1], introducing a strongly energy- and temperature-dependent mobility tail [35]. Thus, transport may occur through localized states, and a much more detailed model of the density of states is required, as well as of the mobility tail itself.

The density of states $\eta(E)$ is exponential with width $E_o$ below an energy $E_1$ with [29]:

$$|E_1 - E_c| \simeq 10 E_o.$$                                              (33)

A suitable model for the density of states is:

$$\eta(E) = A \sqrt{E - E_1 + E_o/2}, \qquad\qquad E > E_1 \qquad\qquad (34)$$

$$= A \sqrt{E_o/2} \, e^{-(E_1-E)/E_o}. \qquad\qquad E < E_1 \qquad\qquad (35)$$

The mobility is taken to be:

$$\mu(E) = \frac{1}{2\pi^2} \frac{e\nu}{\hbar\xi_o} \frac{(E-E_c)^{\nu-1}}{(E_u-E_c)^\nu} \frac{1}{n(E_c)}, \qquad\qquad E > E_c + \frac{\nu-1}{\alpha} k_B T \qquad (36)$$

$$= \frac{1}{2\pi^2} \frac{e\nu}{\hbar\xi_o} \left(\frac{(\nu-1)k_B T}{\alpha e}\right)^{\nu-1} \frac{1}{(E_u-E_c)^\nu} \frac{1}{n(E_c)} e^{-\alpha(E_c-E)/k_B T} \qquad (37)$$

$$E < E_c + \frac{\nu-1}{\alpha} k_B T$$

Eq. (37) introduces into the mobility an exponential tail arising from phonon-assisted hopping between localized states for $E < E_c$ and between the resonances in the extended states for $E_c < E < E_c + \frac{\nu-1}{\alpha} k_B T$. We take $\alpha$ to be a linearly decreasing function of temperature,

$$\alpha = 1 + b \ (T_\alpha - T).$$ (38)

The exact expression for the conductivity from which (27)–(29) were derived now contains an energy integrand which is sharply peaked about the energy, $E^*$. $\sigma$ can thus be evaluated by expansion around the peak with the result:

$$\alpha = \sqrt{\frac{2\pi}{|I''(E^*)|}} \ n(E^*) e\mu(E^*) e^{-(E^*-E_F)/k_BT}$$ (39)

where

$$\ln I(E) = n(E)\mu(E)f(E).$$ (40)

Note that $(E^*-E_F)$ need not equal $E_\sigma$ because $\mu(E^*)$ can be temperature dependent.

There are three distinct temperature regimes. The first is a low-temperature regime in which $T < T_\alpha$ so that $\alpha > 1$. In this case, $E^*$ always lies above $E_c + \frac{\nu-1}{\alpha}k_BT$, in the region of extended states unaffected by the electron-phonon interaction. The transport properties are given by Eqs. (27)–(29), the Hall effect is normal and the mean mobility is weakly temperature dependent. The temperature is simply too low for the electron-phonon interaction to play any role in the transport properties.

The second regime occurs at intermediate temperatures for which $\alpha < 1$ and:

$$T_\alpha \frac{bT_o}{bT_o - 1} > T > T_\alpha$$ (41)

where

$$T_o = E_o/k_B.$$ (42)

A condition for this regime to exist is that $bT_o$ exceed unity, which puts a lower limit on b. The conductivity is given by:

$$\sigma = \tilde{\sigma}_o e^{-E_\sigma/k_BT}$$ (43)

where

$$\overset{\infty}{\sigma}_O = \frac{1}{\sqrt{(\nu-1)e}} \frac{1}{[b(T-T_\alpha)]^{3/2}} \sqrt{\frac{k_BT}{E_C-E_1 + E_O/2}} \sigma_O e^\gamma \tag{44}$$

with $\sigma_O$ given by (29) and $\gamma$ by:

$$\gamma = b(E_C-E_1 + E_O/2)/k_B \tag{45}$$

and where

$$E_\sigma = E_C-E_F-bT_\alpha(E_C-E_1 + E_O/2). \tag{46}$$

The factors preceding $\sigma_O$ differ little from unity when combined so that $\overset{\infty}{\sigma}_O$ is approximately $\sigma_O e^\gamma$. Because $bT_O$ exceeds unity, $\gamma$ is of order 10. Thus, in this intermediate transport regime, the pre-exponential can be three or four orders of magnitude higher than is possible in the low temperature regime. Moreover, under certain conditions, the Meyer-Neldel [32,36] rule, Eq. (32), holds. From (46), $\overset{\infty}{\sigma} \cong \sigma_O e^\gamma$, and (45), we have:

$$E_\sigma = E_C-E_F + k_BT_\alpha \ln\frac{\overset{\infty}{\sigma}_O}{\sigma_O}. \tag{47}$$

The conditions are that changes in the material leave $E_C-E_F$ unaffected while significantly affecting $E_C-E_1 + E_O/2$. We have shown that for disorder potentials small compared to the band width, which is the case for amorphous semiconductors, $E_1$ varies very much more rapidly than $E_C$. The Meyer-Neldel rule thus holds in the intermediate regime when either $E_F$ is pinned or its changes are linearly related to those of $E_C-E_1 + E_O/2$. The mobility is now activated. The thermopower differs from (31) by the mobility activation energy divided by $k_BT$. Finally, the Hall effect for p-type materials can now be negative because transport is by phonon-assisted hopping [37].

In the third regime, $T > \frac{bT_O}{bT_O - 1} T_\alpha$, $E^*$ does not exist. All states with $E < E_1$ become important for transport. The mobility in the tail has become so large that no mobility gap exists and the material no longer behaves as a conventional semiconductor. This

high-temperature regime is presumably important in liquid semiconductors on their way to a metal-nonmetal transition.

Thus, the transport regimes of interest for amorphous semiconductors are the low and intermediate temperature regimes described above. These two regimes of our simple model contain essentially all of the behavior which have been observed thus far,
and are capable of giving a good quantitative account of the data.
Of course, there is competition with Mott hopping [25] at $E_F$ which,
depending on the magnitude of $\eta(E_F)$, can lead to a transition from
Mott hopping into either the low- or the high-temperature regime.

One sees clearly from the above analysis that the focus of
theoretical activity should shift toward understanding the interplay
of disorder and the electron-phonon coupling in determining the
mobility tail.

## VIII.    OPTICAL ABSORPTION

Cody et al. [6] have found that in amorphous $SiH_x$ alloys
the optical absorption has the Tauc-Mott algebraic form [25] at
high energies with an apparent band gap $E_G$ and an exponential
tail with a decay energy $E_0$ at lower energies. As the H content
is varied or the material treated in varying ways, $E_G$ and $E_0$ may
vary but the pre-exponential changes in such a way that extrapolation of the tails to high energy leads to a common intersection
at an energy $E_i$. A linear relation exists between $E_G$ and $E_0$ such
that $E_G$ approaches $E_i$ as $E_0$ is extrapolated to zero. The temperature dependence of $E_0$ is substantial. The intrinsic disorder (structural) and the temperature-induced disorder apparently give comparable contributions to $E_0$. These correlations and regularities
do not hold for doped samples. Similar but less universal results
are found for amorphous Ge.

The behavior in the algebraic domain can perhaps be understood by the use of a generalized form of Bloch functions [1]. The
existence of a common intersection of exponential tails is a common occurrence, although perhaps with multiple explanations.
However, the correlation among $E_G$, $E_i$, and $E_0$ is striking and no
explanation exists, nor is there one for the magnitude and temperature dependence of $E_0$.

## ACKNOWLEDGEMENTS

I am grateful to Professors H. Fritzsche, D. Adler, and M. Kastner and to Drs. T. Tiedje, G. Cody, and P. Persans for discussion of various points important to this paper. I would also like to acknowledge the contributions of Drs. C.M. Soukoulis and E.N. Economou to the work summarized herein.

## REFERENCES

1.     M.H. Cohen in <u>Melting, Localization, and Chaos</u>, edited by R.K. Kalia and P. Vashishta, New York: Elsevier, 1982, p. 125.

2.     C.M. Soukoulis and M.H. Cohen, J. Noncryst. Solids <u>66</u>, 279 (1984).

3.     M.H. Coehn, E.N. Economou and C.M. Soukoulis, J. Noncryst. Solids <u>66</u>, 285 (1984).

4.     F. Urbach, Phys. Rev. <u>92</u>, 1324 (1953).

5.     D.J. Dunstan, J. Phys. C <u>30</u>, L419 (1982).

6.     G.D. Cody, T. Tiedje, B. Abeles, B. Brooks and Y. Goldstein, Phys. Rev. Lett. <u>47</u>, 1480 (1981); G.D. Cody, The Optical Absorption Edge of a-Si:$H_x$ in Amorphous Silicon Hydride, edited by J. Pankove (Vol. 21B of <u>Semiconductors and Semimetals</u>, Academic Press, 1984, p. 11 and references therein.

7.     T. Tiedje, J.M. Cebulka, D.L. Morel and B. Abeles, Phys. Rev. Lett. <u>46</u>, 1425 (1981).

8.     M.V. Kurik, Phys. Stat. Sol. (d) <u>8</u>, 9 (1971) and references therein.

9.     D. Redfield, Phys. Rev. <u>130</u>, 916 (1963); J.D. Dow and D. Redfield, Phys. Rev. B <u>5</u>, 594 (1972).

10.    J.J. Hopfield, Comm. Solid State Phys. <u>1</u>, 16 (1968).

11.    H. Sumi and Y. Toyozawa, J. Phys. Soc. Japan <u>31</u>, 342 (1971).

12.    J. Skettrup, Phys. Rev. B <u>18</u>, 2622 (1978).

13.    M. Schreiber and Y. Toyozawa, J. Phys. Soc. Japan <u>51</u>, 1528, 1537, and 1544 (1982).

14.    S. Abe and Y. Toyozawa, J. Phys. Soc. Japan <u>50</u>, 2185 (1981).

15.    B.I. Halperin and M. Lax, Phys. Rev. <u>148</u>, 722 (1962); ibid <u>153</u>, 802 (1967).

16.    P. Lloyd and P.R. Best, J. Phys. C <u>8</u>, 3752 (1975).

17. J. Ziman, Models of Disorder, Cambridge: Cambridge University Press, 1979, p. 487.
18. D.J. Thouless, Phys. Rev. Lett. 39, 1167 (1977).
19. M.H. Cohen, E.N. Economou and C.M. Soukoulis, Phys. Rev. Lett. 51, 1202 (1983).
20. D. Vollhardt and P. Wolfle, Phys. Rev. Lett. 48, 699 (1982).
21. P. Lloyd, J. Phys. C 2, 1717 (1969).
22. B. Abeles, C.R. Wronski, T. Tiedje and G.D. Cody, Solid State Comm. 36, 537 (1980).
23. G.D. Cody, C.R. Wronski, B. Abeles, R.B. Stephens and B. Brooks, Solar Cells 2, 227 (1980).
24. G.S. Higashi and M. Kastner, Phys. Rev. Lett. 47, 124 (1981).
25. N.F. Mott and E.A. Davis, Electronic Processes in Non-Crystalline Materials, Oxford: Clarendon Press, 1979.
26. Y. Imry, Phys. Rev. Lett. 44, 469 (1980).
27. S. Hikami, Phys. Rev. B 24, 2671 (1981).
28. T. Moustakas, P. Maruska, C. Roxlo and R. Friedman, to be published.
29. C.M. Soukoulis, M.H. Cohen and E.N. Economou, Phys. Rev. Lett. 53, 616 (1984).
30. T. Moustakas, private communication.
31. D.E. Carlson and C.R. Wronski in Amorphous Semiconductors, edited by M.H. Brodsky, New York: Springer-Verlag, 1979, p. 287.
32. H. Fritzsche, Solar Energy Materials 3, 447 (1980).
33. T. Moustakas and W. Paul, Phys. Rev. B 16, 1564 (1977).
34. P. Nagels in Amorphous Semiconductors, edited by M.H. Brodsky, New York: Springer-Verlag, 1979, p. 113.
35. H. Mueller and P. Thomas, Phys. Rev. Lett. 51, 702 (1983).
36. W. Meyer and H. Neldel, Z. Tech. Phys. 18, 588 (1937).
37. D. Emin, Phil. Mag. 35, 1189 (1977).

# NONEQUILIBRIUM TRANSPORT PROCESSES IN AMORPHOUS

# AND OTHER LOW CONDUCTIVITY MATERIALS

Heinz K. Henisch

Materials Research Laboratory, Department of Physics
Pennsylvania State University
University Park, Pennsylvania 16802

## I.    INTRODUCTION

It has to be admitted that macroscopic transport theory, in the sense in which it will here be discussed, is less fundamental than band theory or the theory of electron scattering. However, let us consider how our insights are gained. We conduct some kind of an experiment, and we have designed that experiment in a particular way because we have certain expectations of what the outcome should be. True, some experiments are purely exploratory, but in most cases we have definite expectations. If these are fulfilled, then we feel that the system is well understood. If they are not fulfilled, then we have a few choices. We can invoke the possibility of experimental errors, e.g., through malfunctioning equipment. Alternatively, we could question the model. We could propose, for instance, that mobility that is usually regarded as constant is not in fact constant. It is clear in some such way we can fit any data to a model. This is how physics has always progressed, but if this procedure is to be useful, it is very necessary that we have the correct expectations to begin with.

The equations given in Table I summarize the transport relationships for semiconductors containing two types of charge carriers: electrons and holes. Electron concentrations are traditionally called n, and hole concentrations p, and the two types of charge carriers are in dynamic equilibrium. The product of n and p is, therefore, constant. Graphically expressed, this means that n

361

Table I.   Transport Relationships

(i)     <u>Charge Carriers:</u>

Semiconductors (in general) contain <u>two</u> kinds of charge carriers:

Electrons (-ve)    of concentration n
Holes       (+ve)    of concentration p

"n-type" material: $n > p$;   "p-type" material: $p > n$

(ii)    <u>Electron-Hole Product:</u>

The concentrations n and p are in dynamic equilibrium with one another; carrier generation (e.g., thermal) balanced by carrier recombination.

Simple thermodynamics (corresponding to the chemist's Law of Mass Action) leads to:

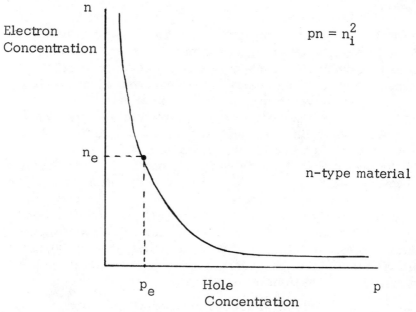

$$pn = n_i^2$$

n-type material

(iii)   <u>Mass Action:</u>

All points on this curve satisfy the Law of Mass Action, but only one (say, $n_e$, $p_e$) guarantees neutrality. In this situation, any charge density $(n_e - p_e)e$ is compensated by ionized impurities.

nd p are in the relationship of a rectangular hyperbola to one nother, as shown in the figure in Table I. Although all points on hat rectangular hyperbola satisfy the Law of Mass Action, only ne point satisfies the neutrality condition. At that point, the carier concentrations are $n_e$ and $p_e$. The semiconductor is further escribed by certain constants: the electron and hole mobilities n, $\mu_p$, and the diffusion constants $D_n$ and $D_p$, connected with one nother by Einstein's relationship. Any current flowing will conist of two parts. One part is carried by electrons, $J_n$, another art is carried by holes, $J_p$. The electron current $J_n$ also has two omponents, one is due to the field $J_{nF}$, one is due to the difusion $J_{nD}$. The same is true for the hole current (see Table II).

## Table II. Transport Parameters

i) <u>Mobility and Diffusion Constants</u>:

Both types of carriers are associated with carrier mobilities and diffusion constants.

| | | |
|---|---|---|
| Electrons | $\mu_n$ | $D_n$ |
| Holes | $\mu_p$ | $D_p$ |

Einstein's relationship: $D = \mu kT/e$

ii) <u>Current Densities</u>:

$$J \quad = \quad J_n \quad + \quad J_p \quad = \quad \text{Total Current Density}$$

$$J_n \quad = \quad J_{nF} \quad + \quad J_{nD} \quad \left.\right\} \quad \text{Field and Diffusion}$$

$$J_p \quad = \quad J_{pF} \quad + \quad J_{pD} \quad \left.\right\} \quad \text{Component}$$

iii) <u>Characteristic Time Constants</u>:

$\tau_0$ = Time constant governing the decay of p-$p_e$ (in n-type material) under conditions of neutrality, called the <u>carrier lifetime</u>.

$\tau_D$ = Time constant governing the decay of excess charge, called the <u>dielectric relaxation time</u>.

$\tau_D = \rho\epsilon$, where $\rho$ = resistivity, and
$\epsilon$ = permittivity.

Microscopic transport theory claims a rare but totally unrealistic privilege: to be able to fix the constants of the material
more or less arbitrarily. Of course, this is not actually correct.
We cannot, in fact, change the doping content without affecting
the mobility and the lifetime; everything depends on everything
else. Macroscopic transport theory, as such, does not know
about this, which means that models must be handled with great
care. Lastly, there are two important time constants. One of
them $(\tau_0)$ is the familiar minority carrier lifetime, which governs
the rate at which a nonequlibrium concentration decays; it is
called the carrier lifetime. The other $(\tau_D)$ is the dielectric relaxation time, and is just as important, even though it is very often
ignored. It governs the decay of the charge, not the decay of
free carrier concentration. Phenomena that are governed primarily
by carrier concentration are controlled by $\tau_0$; phenomena which
are primarily governed by space charge are controlled by $\tau_D$.

II.     THE CONCEPT OF RELAXATION SEMICONDUCTORS

Macroscopic transport theory asks no questions about the
crystalline state of the material. It asks only about the transport
of relationships and their logical consequences. However, what
we have to say is especially relevant to amorphous semiconductors,
as will be seen. Virtually all transport theories handled in textbooks assume that the dielectric relaxation time is negligibly
small. This is a leftover from the historical fact that the transistor was invented on germanium, and germanium and silicon do
indeed have dielectric relaxation times that are usually much
smaller than $\tau_0$. That is why we do not have to worry about free
carrier space charges in germanium or in silicon--they decay very
rapidly, and most researchers simply forget about them. Moreover, we have all been brought up to believe that if we consider a
series of materials beginning with the semiconductors and going
on to higher and higher resistivities, to semi-insulators and to insulators, nothing new is going to happen--just less and less of
the old. However, some years ago, van Roosbroeck [1-3] made a
sensational announcement at a semiconductor conference in Bos-
ton, claiming that the traditional expectation was incorrect. In
the transition from semiconductors to insulators, there is a particular boundary of which new transport properties begin to appear,
namely, where $\tau_0 = \tau_D$. A material for which $\tau_0 > \tau_D$ was hence-
forth to be called a lifetime semiconductor, i.e., a semiconductor
dominated by the lifetime. A material in which $\tau_0 < \tau_D$ was to be

called a <u>relaxation semiconductor</u>. Most amorphous semiconductors come under that latter heading at a sufficiently low temperature, and many are in this category even at room temperature and above, not because they are amorphous, but because they are, on the whole, rather poor conductors.

Van Roosbroeck's suggestions were not immediately popular, and there were many researchers in the field who hoped that the new ideas would soon go away. However, macroscopic transport theory is rather like living in a dictatorship: everything that is not forbidden is compulsory. There is no way of bypassing the transport requirements, short of upsetting some very fundamental laws of electricity and magnetism.

## II. NONEQUILIBRIUM PROCESSES AND TRANSPORT EQUATIONS

This discussion is evidently concerned with nonequilibrium phenomena, because lifetimes and dielectric relaxation times have no relevance as long as equilibrium prevails. Disequilibrium can be produced in a variety of ways, by illumination (as in a solar cell), by magnetic means, or by passing a current through a contact. In what follows, for traditional reasons, we analyze the system which involves passing a current through a contact. Let us suppose that we have a semiconductor, assumed to be n-type. There are two contacts at the end. Nothing can stop us from defining a quantity $\gamma_0$, which denotes the ratio of the minority carrier current to the total current as it prevails in the bulk of the material. $\gamma_0$ is, thus, a constant for the material where equilibrium prevails, but the current composition is bound to be different at the two contacts. Let us first see what the consequences would be of having a $\gamma_1$ and a $\gamma_2$ which differ from the $\gamma_0$ (see Table III). $\gamma_1 > \gamma_0$ means that minority carriers are coming in at a rate greater than required for bulk transport. We call that "minority carrier injection" and this phenomenon is, of course, the basis of transistor action. Next, $\gamma_1 < \gamma_0$ means that minority carriers are coming in at a rate smaller than that required for bulk conduction. This means that bulk conduction sucks out minority carriers from the contact region and creates an electric analog of a vacuum. This process has traditionally been called "minority carrier exclusion." At the other electrode, there are also two processes. $\gamma_2 > \gamma_0$ means that minority carriers are taken out at a rate greater than the replacement rate, and this is called "minority carrier extraction." $Y_2 < Y_0$ corresponds to "minority carrier accumulation." At each contact, one or the other of these situations must prevail, one contact being biased forward, and one reverse. We do not consider the case where $Y_1 = Y_2 = Y_0$.

At every boundary, we are faced with a nonequilibrium of carriers. None of the present arguments actually concerns the contacts as such, except insofar as every contact can be characterized by a certain current composition ratio. How various current composition ratios come about is a whole complex of problems that, for the moment, we will leave undiscussed. What matters here is that $Y_1 \neq Y_0$ and $Y_2 \neq Y_0$, and we are looking for the logical consequence. To do this, we can set up the transport equations. The macroscopic transport equations are widely accepted. What one can argue about is the boundary conditions, and these pose some tricky problems.

Table III.    Characterization of the four nonequilibrium situations which can result from the passage of a current through an interface.

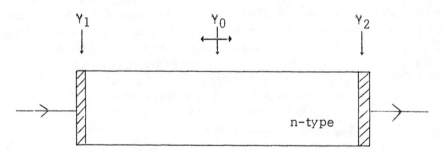

$Y_0$ = current composition ratio in the homogeneous bulk

$$Y_0 = \frac{p_e \mu_p}{p_e \mu_p + n_e \mu_n} \quad \text{for n-type material}$$

$Y_1 > Y_0$ = minority carrier <u>injection</u>

$Y_1 < Y_0$ = minority carrier <u>exclusion</u>

$Y_2 > Y_0$ = minority carrier <u>extraction</u>

$Y_2 < Y_0$ = minority carrier <u>accumulation</u>

$Y_1 = Y_0 = Y_2$ = highly unlikely

To summarize, we have a total current consisting of electrons and holes (see Table IV). The electron current consists of
a field component and a diffusion component--the hole current,
likewise. Then, there are two continuity relationships. The rate
of change of the electron concentration in the steady state is zero,
of course. It is given by the divergence of the electron current
plus a recombination term. The same must be true for the hole
continuity equation, and in a steady state, the two recombination
terms must obviously be the same, because it is ultimately only
with a hole that an electron can recombine. Table IV reflects the
case with traps. We also need boundary conditions. On one side
we are going to assume that the boundary conditions are given by
equilibrium; whatever may be happening must gradually decay into
the bulk. On the other side, where the injection (or extraction)

Table IV.  The standard macroscopic transport equations for a
system with carrier traps. One-dimensional analysis.

i)     $J = J_n + J_p$ = Total Current Density

ii)    $J_n = J_{nF} + J_{nD}$ $\left.\begin{array}{c} \\ \\ \end{array}\right\}$ Field and Diffusion
       $J_p = J_{pF} + J_{pD}$      Components

iii)   $\dfrac{dn}{dt} = \dfrac{1}{e} \, \mathrm{div}(J_n) + R_n$

       $\dfrac{dp}{dt} = \dfrac{1}{e} \, \mathrm{div}(J_p) + R_p$   Continuity Relationships

       where $R_n$, $R_p$ are recombination terms.

       In the steady state: $(dn/dt) = (dp/dt) = 0$
       and $R_n = R_p = R$.

iv)    $dF/dx = [(p - p_e)e - (n - n_e)e + (M - M_e)e]/\epsilon$
             = [Poisson's equation],
       where  M   = concentration of holes in traps
       and    $M_e$  = equilibrium concentration of holes in traps.

v)     Boundary Conditions:

       (a) Right-hand side: conditions in the undisturbed bulk.
       (b) Left-hand  side: conditions at the contact variable.

takes place, the situation is more complicated. For purpose of the calculations that follow, $x = 0$ is actually defined as the point (inevitably present in any injecting system), where the electric field is zero; that is an excellent place in which to start, if one does not want to talk about the barrier itself. Whatever happens within the barrier is beyond the scope of anything here discussed.

## IV.    MODES OF APPROACH

These equations are not generally contested, but there is one problem: they cannot be solved in closed form. There are only three ways of coping with such a situation. One can make approximations, and thereby simplify the equations until they can be analytically solved. This is what used to be done. In that case, how do we know whether the answer is right? The only safe way in which one could find this out would be to compare the answer with the exact solution, and that solution is, by supposition, unavailable. Such an intuitive approach is often right, but can be disastrously wrong. The hope is always that our intuitive approximation will leave the essential physics intact, and that it will affect only the accuracy. However, the truth is that it often affects the physics. Whole groups of phenomena can be ruled out by simply making one plausible assumption, namely, the neutrality assumption, which is often nearly right, but not quite. The small departures from it are exceedingly important, which means that one has to be extremely careful. Even plausible assumptions, which are permissible on their own, are not necessarily right when made in conjunction with other plausible assumptions.

What are the other possibilities? A second possibility is to linearize the equations. That is an algebraic way of approximating them by neglecting higher order terms. One may then be able to solve the equations analytically, but such solutions will be valid only in the framework of what we call a "small signal theory"--i.e., one that is limited to very small currents, very small departures from the equilibrium. Such models have their uses, because such solutions show how various parameters enter into the consequences, but their total scope is very limited. For some phenomena, e.g., extraction, they are not useful at all.

Thirdly, there is numerical computation, and even that is not simple in practice. Before embarking on it, it is desirable to examine one more stage of intuitive reasoning, which turns out to be totally in harmony with the computer solutions. For that pur-

pose, consider the diagram on the left of Fig. 1. Here we have
our rectangular hyperbola, P   N = constant. (The lower case
symbols p and n refer to actual concentrations; the upper case
symbols P and N to concentrations normalized to $n_e$, the majority
carrier equilibrium concentration in bulk.) The starting point is
$N_e$, the equilibrium concentration of electrons, $P_e$ the equilibrium
concentration of holes. Let us consider that we are dealing with
a small bit of material so that concentration gradients can be
ignored. We shall also suppose that, by one external means or
another, an increase of the hole concentration by $\Delta P_0$ has been
brought about. What would happen? We would now have a positive space charge. In a traditional material, like germanium, in
which the dielectric relaxation time is very small, that extra positive charge will almost immediately attract an extra negative
charge consisting mainly of free electrons of equal magnitude into
the neighborhood. Then the extra electrons and extra holes (an
equal number) will recombine, in a time corresponding to the life-

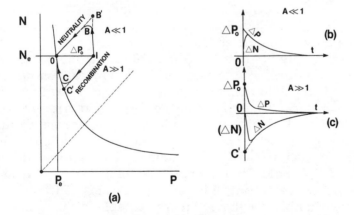

Fig. 1      Minority carrier augmentation in a homogeneous space
            element. $A = \tau_D/\tau_0$. (a) Relationship between electron
            and hole concentration. (b) Time dependences. $A \ll 1$,
            lifetime semiconductors. (c) $A \gg 1$, relaxation semicon-
            ductor. N = normalized n; P = normalized p.

time. Throughout that event there will be more majority carriers in the system than there are in normal times, and this is the behavior that we traditionally associate with a lifetime semiconductor.

Now let us consider something non-traditional, a dielectric relaxation time greater than the lifetime. Again, let us contemplate an increase in the hole concentration by $\Delta P_o$. Once more, there is a positive space charge, but the recombination time is now very small. Very quickly, recombination will take us to the rectangular hyperbola, because that curve defines the locus where no further recombination can take place. This is because the recombination rate depends directly on the product P X N. However, at this stage, the system is not yet neutral. It regains neutrality by creeping back to the starting point along the rectangular hyperbola, and that will happen roughly within a dielectric relaxation time. Thus, the behavior is essentially different in the two cases; the majority carrier concentration is here diminished. (There is also a trivial boundary case, namely, that in which the dielectric relaxation time and the lifetime are equal.)

This intuitive model gives us some useful insights, but its quantitative scope is very limited, mainly because all gradients had to be neglected. Only a computer solution can take all the facts into account.

V.     COMPUTER SOLUTIONS

One solves the differential equations by means of a matrix method, here involving up to 1500 x 1500 elements, but the detailed procedure does not concern us now. Figure 2 shows some concentration and field contours for a moderate relaxation case (injecting boundary at x = 0), with the majority carrier depletion predicted by Fig. 1 clearly in evidence. However, the diminished concentration of majority carriers does <u>not</u> lead to a higher local resistivity, as was at one time believed. The fact that the normalized field F is below the normal bulk field (here F = 1) shows that the effective resistivity is actually smaller. Diffusion currents are responsible for that. Figure 3 displays the results under similar conditions, but for a case more strongly in the relaxation regime. It will be seen that majority carriers are now totally depleted within a certain region. At the same time, a recombination maximum makes its appearance, and does so for good reason. On the left, there are no majority carriers, and on the right, no mino-

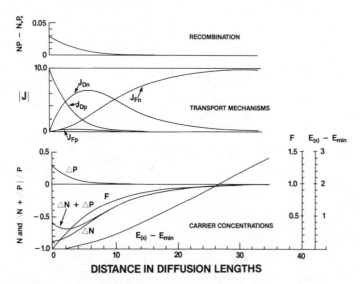

Fig. 2    Typical concentration and field contours for a moderate
          relaxation case. Note the diminished field E and the
          diminished majority carrier concentration ($\Delta n < 0$) with-
          in the injection region. No traps. E = electron field;
          V = local potential. N, P, $\Delta N$, $\Delta P$ normalized to the
          bulk concentration of majority carriers $n_e$. $J_{nF}$, $J_{pF}$
          normalized field current densities. $J_{pD}$, $J_{nD}$ normalized
          diffusion current densities.

rity carriers; the only place in which recombination is possible is one where both exist. Accordingly, there must necessarily be a recombination maximum. As the current changes, this maximum shifts along the semiconductor.

The computer can also calculate the total voltage-current relationship between x = 0 and some place where equilibrium is deemed to prevail. Truly ohmic behavior is represented by the long straight line in Fig. 4. The computed behavior of the system described is indeed linear over a phenomenal region, despite the fact that we are dealing here with a high rate of injection (indeed

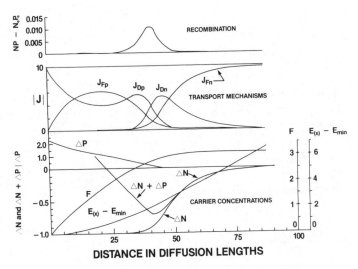

Fig. 3    Data similar to those in Fig. 2, but for a more pro-
          nounced relaxation case. Note total majority carrier
          depletion and the recombination maximum. No traps.
          Symbols as for Fig. 2.

here with $\gamma = 1$). In contrast, most researchers associate linearity with the absence of injection, and this is quite wrong. Whether something is injecting or not, whether something is of low resistance or not, whether something is linear or not (meaning obedient to Ohm's law) are completely different questions, with independent answers. Figure 4 also expresses the fact that the computed currents (for a given voltage) are <u>higher</u> than they would be without injection (total resistance <u>lower</u>).

Of course, computations of this kind can (and should!) be compared with actual measurements. A fine set of such measurements is available for GaAs, by Illegems and Queisser [4]. It reflects the extended linearity very well, but differs from the predictions made above. As a result of injection, the total resistance was <u>higher</u> than normal, not lower than normal as suggested by Fig. 4. This difference can be accounted for by including trapping centers in the analysis.

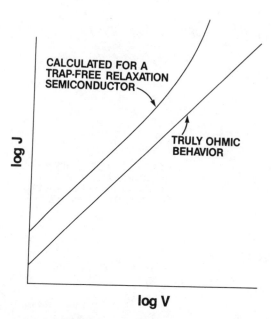

Fig. 4     Schematic representation of ohmic and non-ohmic voltage-current relationships; effective resistivity diminished.

## VI.    SYSTEMS WITH TRAPS

For purposes of the present argument, we shall assume that the semiconductor contains minority carrier traps in substantial concentration. The above analysis can then be repeated, making allowance for the charge density in traps in Poisson's equation. The resulting voltage-current relationships are shown in Fig. 5, and there is a simple explanation for this higher-resistance-than-

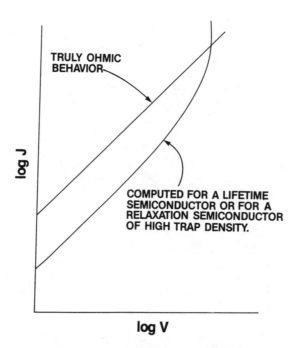

Fig. 5     Schematic representation of ohmic and non-ohmic voltage-current relationships; effective resistivity increased at low current densities.

ormal behavior, in terms of the relationship in Fig. 6. This
hows, as a consequence of hole injection, a $\Delta p$ contour decaying
lowly into the bulk. At the same time, a positive space charge
ill exist in traps, and this could be much greater than $e\Delta p$ at
very point. In turn, this positive space charge will attract an
ncrement $\Delta n$ of free electrons greater than $\Delta p$ at any point. There
re, therefore, two carrier concentration contours, one for holes
nd one for electrons, with the electrons having the greater difusion constant and the greater gradient. The electron diffusion
pposes the current. In order to keep the current constant, we
herefore need (and find by computer) a higher-than-normal electric
ield. This is equivalent to saying that the effective local resis-
ivity has increased, as the experimental results demand. In this

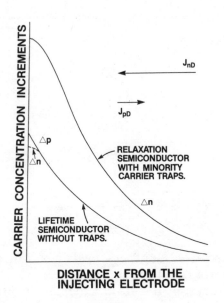

Fig. 6    Electron and hole concentrations as a function of dis-
          tance from the hole-injecting boundary.

way, theory and experiment can be brought into excellent agreement, on relaxation and lifetime materials alike. Figure 7 shows
the local field maximum, and its disappearance at high current
densities. Indeed, whereas the field maximum is more prominent
in a semiconductor with traps than in one without, it is not necessary to have a <u>relaxation</u> semiconductor to observe this form of
behavior. Thus, all the predictions have been qualitatively and
quantitatively confirmed by experiments Ge, with trapping centers
induced by neutron bombardment.

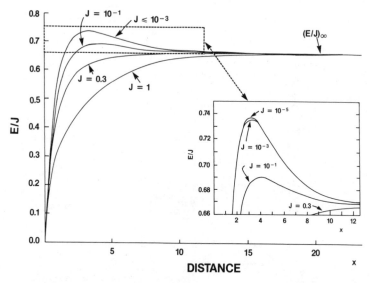

Fig. 7     Normalized field contours for various (normalized) cur-
           rent densities. Lifetime semiconductor. Note gradual
           disappearance of the field maximum at higher current
           densities. Distances expressed as multiples of the
           Debye length.

VII.    NONEQUILIBRIUM SITUATIONS WITHOUT CONTACT
        EFFECTS

        It is possible to produce nonequilibrium effects without any
role of nearby contacts, e.g., in the course of a Hall effect
measurement (see Fig. 8). Textbook Hall theory assumes that the
specimen is infinite in the direction of the Hall field. In fact, of
course, it is always finite. The two kinds of carriers which under
the influence of the horizontal field drift in opposite directions.
Under the influence of the magnetic field, the Lorentz force causes
them to move in the <u>same</u> direction, because of opposite charge
and opposite velocity. As soon as the magnetic field is switched
on, a stream of electron-hole pairs begins to drift from bottom to
top; the magnetic field acts as an electron-hole pump.

        What happens at the top surface? The carriers have to re-
combine, and do so at a rate that depends on the state of the surface. It is crudely described by a surface recombination velocity.
However, the arrival rate has nothing to do with the state of the
surface; that depends on the magnetic field. Therefore, the arrival
and recombination rates are not matched. As a result, one must
envisage near the top surface a cloud of electron-hole pairs which

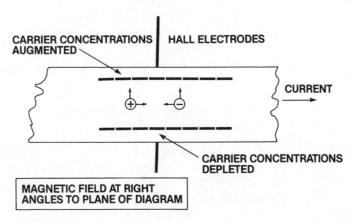

Fig. 8       Schematic representation of a two-carrier Hall effect
             system.

have failed to recombine, which are kept in the steady state by recombination on the one hand, and by back diffusion on the other. Similarly, there is a cloud of missing electron hole pairs at the bottom surface where the opposite happens. Because both regions involve space charges, the field between the top and bottom surfaces cannot be constant. Therefore, the voltage measured between top and bottom cannot be simply interpreted in terms of a Hall field. A computation very similar to that described above is necessary to assess the situation. Its results show not only that the measured Hall voltage is in danger of quantitative misinterpretation, but that the sign of the voltage can be opposite to that expected. Needless to say, all these dangers increase as the specimen becomes smaller in the Hall direction. Sign reversals are known, and some very subtle theories have been devised to account for them. These theories may indeed be right, but before we seriously employ them, we must look at what our normal expectations are on the basis of the macroscopic transport relations. We see that in this case, as in the case of carrier injection, the full transport analysis (which can only be carried out by numerical computation) leads to some new and unexpected results. It goes without saying that such an analysis is similarly important in the context of solar cell design.

REFERENCES

1.    R.W. van Roosbroeck, Phys. Rev. 123, 474(1961).
2.    R.W. van Roosbroeck and H.C. Casey, Jr., Proc. 10th Int.
        Conf. on Physics of Semiconductors, Cambridge, MA.,
        Div. of Technical Infor., U.S. Atomic Energy Com., 1970.
3.    R.W. van Roosbroeck and H.C. Casey, Jr., Phys. Rev. B
        5, 2154(1972).
4.    M. Illegems and H.J. Queisser, Phys. Rev. B 12, 1443
        (1975).

For further reading, with special reference to publications in this field originating from the Pennsylvania State University:

J.-C. Manifacier and H.K. Henisch, Phys. Rev. B 17, 2640(1978).
J.-C. Manifacier and H.K. Henisch, Phys. Rev. B 17, 2648(1978).
J.-C. Manifacier and H.K. Henisch, Nature 272, 521(1978).
S. Rahimi, J.-C. Manifacier and H.K. Henisch, J. Appl. Phys.
    52, 6273(1981).
C. Popescu and H.K. Henisch, Phys. Rev. B 11, 1563(1975).

C. Popescu and H.K. Henisch, J. Phys. Chem. Solids 37, 47 (1975).

H.K. Henisch and C. Popescu, Nature 257, 363(1975).

C. Popescu and H.K. Henisch, Phys. Rev. B 14, 517(1976).

J.-C. Manifacier, H.K. Henisch and J. Gasiot, Phys. Rev. Lett. 43, 708(1979).

Y. Moreau, J.-C. Manifacier and H.K. Henisch, Solid State Elect. 25, 137(1982).

Y. Moreau, J.-C. Manifacier and H.K. Henisch, Solid State Elect. 24, 883(1981).

H.K. Henisch, J.-C. Manifacier, R.C. Callarotti and P.E. Schmidt, J. Appl. Phys. 51, 3790(1980).

G. Rieder, J.-C. Manifacier and H.K. Henisch, Solid State Elect. 25, 133(1982).

S. Rahimi and H.K. Henisch, Appl. Phys. Lett. 38, 896(1981).

G. Rieder, H.K. Henisch, S. Rahimi and J.-C. Manifacier, Phys. Rev. B 21, 723(1980).

J.-C. Manifacier and H.K. Henisch, J. Phys. Chem. Solids 41, 1285(1980).

J.-C. Manifacier and H.K. Henisch, Proc. Int. Conf. on Amorphous and Liquid Semiconductors, Cambridge, MA., August 1979; also in J. Noncryst. Solids 35-36, 1105(1980).

T. Botila and H.K. Henisch, Physica Status Solidi A 38, 331(1976).

# THE PECULIAR MOTION OF ELECTRONS IN AMORPHOUS SEMICONDUCTORS

Marc A. Kastner

Department of Physics, Center for Materials Science
and Engineering and Research Laboratory for Electronics
Massachusetts Institute of Technology
Cambridge, Massachusetts 02139

## I.    INTRODUCTION

In the early 1900's, physicists described a conducting material as a box filled with electrons. The current in the metal was understood to be carried by the electrons drifting in an applied electric field. This works in the following way: if the field is E and the charge on the electron is q, then the force on the electron is qE and the electron experiences an acceleration qE/m, where m is its mass. The electron does not accelerate indefinitely, however. Rather, it slows down because it scatters from the atoms in the metal or from impurities. This scattering acts like a frictional force which keeps the electrons from moving faster and faster, and, instead, they reach a constant drift velocity $v_d$. The relationship between $v_d$ and the acceleration is

$$v_d = \frac{eE}{m} \tau \tag{1}$$

where $\tau$ is the time between collisions (called the scattering time), or, equivalently, the time it would take for the electron to stop if the field were turned off. Equation (1) shows that the drift velocity is proportional to the field and the proportionality constant is called the mobility, $\mu$. The current carried by a single electron is proportional to $qv_d$ and the current density (current per unit area crossing any surface in the material) is $nev_d$, where n is the number of electrons per unit volume.

The surprising thing about this very simple model of conductivity was that it worked too well. One can measure the mobility (as I will show later), and one often finds $\mu > 1000$ cm$^2$/V-s. (The units of mobility are velocity divided by electric field.) This is very large as you can see in the following way. The electrons are actually moving very rapidly in all directions because of their thermal energy; they are like atoms in a gas being bounced around by thermal agitation. This motion is random, however, so that, in the absence of an applied electric field, they have no net average velocity. The drift velocity describes the small average motion of the atoms in an applied field. From Eq. (1) and the definition of mobility, we can calculate $\tau = \mu m/e$. Now, the thermal velocity of the electrons $v_{th}$ is found by equating their kinetic energy, $mv_{th}^2/2$, with the thermal energy kT where k is Boltzmann's constant, which gives $v_{th} \simeq 10^7$ cm/s. But $v_{th}\tau$ is the distance an electron moves before it gets scattered. So knowing $\mu$, we can calculate this distance $\ell$ called the mean-free path. For $\mu = 10^3$cm$^2$/V-s, $\ell$ is about 1000Å. This is very large. The atoms in a solid are only a few Ångstrom apart, so, in a good crystal, the electron passes thousands of atoms before it gets scattered. This was a deep mystery for classical physics: why doesn't the electron scatter every time it comes near an atom? Why isn't $\ell \simeq 3$Å instead of $10^3$Å? The answer came with the invention of quantum mechanics. F. Bloch in 1933 showed that an electron in a periodic solid (a crystal) will not scatter from every atom but only from those that are unusual: defects, impurities, or atoms that are out of place because they are vibrating as a result of thermal excitation.

I like to think of this by making an analogy to the game we all played as children. We would try to walk down the sidewalk without stepping on the cracks between the cement blocks. If you lived in a relatively new suburb, this was easy. All the blocks were the same size; so, by adjusting your stride you could saunter down the sidewalk and never hit a crack. This is what the electron does in a perfect crystal. It adjusts its motion so that it is not scattered by the regularly-spaced atoms. On the other hand, if you lived in the city where the sidewalks had been put down at different times and repaired many times, the cracks could be spaced almost at random, and, because you would constantly have to readjust your stride, you could never go fast. This is what happens to an electron in an amorphous material--it bumps into every atom because the distances between atoms are not regular. The mean

free path is a few Å instead of $10^3$Å, and the mobility is about
1cm$^2$/V-s instead of $10^3$. It is ironic that the mobility of electrons
in amorphous materials which have been studied actively for only
about 20 years is what physicists predicted for <u>all</u> solids before
the advent of quantum mechanics.

There is an even more important result of disorder than the
way it affects the mean free path. Disorder causes localized
states. This means that when electrons are in certain states, they
are trapped some place and cannot move from that spot. The
simplest example of a localized state happens at an impurity in
a crystalline semiconductor. Think of a perfect crystal (of Si,
say), and add one extra electron. There is no reason why that
electron should sit on any particular atom in preference to the
other atoms. So, the electron will have equal probability of being
found anywhere in the crystal. But now suppose that one of the
atoms is exchanged for an ion with a positive charge (say a P$^+$ ion
for a Si atom). Then, the electron will have lower energy if it
moves in an orbit around the ion staying close to the positive
charge, just the way an electron orbits around the proton in a
hydrogen atom. Because the electron is bound to the positive
ion, the probability of finding it far from the ion is very small.
That is why these states are called localized. The electron is
bound to the ion, so that, even if we apply a weak electric field,
it cannot get away. This means that electrons in localized states
cannot support electrical conduction.

In an atom, we usually think about the states as discrete --
separated relatively far apart in energy. But in solids there are so
many atoms so close together that there are states at almost every
energy. The states form a continuous distribution, and it is not
so useful to talk about a state at one particular energy. Rather,
we talk about the number of states in a small range of energies and
we call this the density of states in that range. The density of
states in a perfect crystal falls to zero (with a square-root
dependence on the energy) at some energy $E_v$ , remains zero for
a range of energies called the band gap, and then rises again
(with the same dependence) at a higher energy $E_c$. Almost all the
states below $E_v$ are filled, and they are called the valence states
(or the valence band of states) because they are the highest-filled
energy levels. Almost all the states above $E_c$ are empty, and they
are called the conduction states (or the conduction band of states)
because the small number of electrons in these states are the ones
that conduct electricity in the semiconductor.

Now in this perfect crystal the band gap is completely empty of states. But, of course, in the real world there are no perfect crystals. All crystals have defects. If we found a slightly defective or slightly impure crystal, it would have discrete energy levels in the gap corresponding to localized states like the ones just discussed. Now imagine making the material very defective, or very impure, or amorphous. This will create very many localized states at many energies. There will be so many, in fact, that they will merge with the conduction and valence bands as shown in Fig. 1. But all the states will not be localized. Electrons at sufficiently high energy must be able to move about. Therefore, there will be special energies separating localized from extended (delocalized) states near the edges of the valence and conduction bands. There cannot be localized and extended states at the same energy because then an electron could always get away from the localized state by leaking into the extended one and it would not be truly localized.

In this introduction, I have summarized the basic concepts which are used to describe the electronic properties of amorphous semiconductors. Much more detailed and sophisticated discussions can be found in review articles or the book by Mott and Davis [1]. In the next section, I will discuss some experiments which have been particularly useful in testing these ideas and providing detailed information about amorphous semiconductors.

## II.    TRANSIENT PHOTOCONDUCTIVITY

The first technique I will describe is called time of flight (TOF). A sandwich-type sample is prepared between two metal electrodes. The top one is very thin so that light can pass through the metal and be absorbed in the semiconductor. A voltage is applied across the sample and then the front surface is flashed with light for a short time. We use light of very high energy so that it is strongly absorbed by the sample. This absorption takes place because the energy in the light is used to excite electrons from the valence band to the conduction band. Since the absorption is strong all the electrons are excited near the front surface. Of course, for every excited electron there is an empty state in the valence band. These holes behave just like electrons except that they are positively, instead of negatively, charged. Once separated from each other by the light, both electrons and holes can carry current. However, the electric field (if it points the right way) pulls electrons into the sample and holds the holes near the top electrode.

First, think about what would be observed if this experiment were carried out on a crystalline semiconductor. The electrons would move across the sample with the drift velocity because of the electric field. As long as the electrons are moving the current in the external circuit will be constant. The number of electrons and their drift velocity does not change. However, when the last

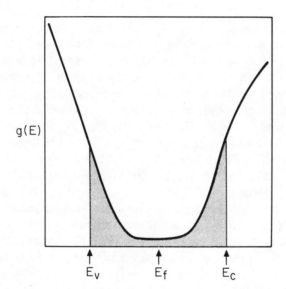

Fig. 1     Schematic density of states in an amorphous semicon-
           ductor. Shaded area represents tails of localized states.
           In this case, the conduction-band and valence-band
           tails overlap near mid gap. The density of states at
           the Fermi energy $E_F$ has been exaggerated for clarity.
           In most models, it is orders of magnitude smaller than
           the densities at the mobility edges $E_V$ and $E_C$.

electrons reach the back of the sample, the current will stop flowing. The time at which the current goes to zero is the transit time across the sample. It is easy to measure this time between the light flash and the time at which the current ceases. Dividing the length of the sample L by this transit time $\tau_t$ gives the drift velocity. Since we know the size of the electric field E, we get the mobility

$$\mu = v_d/E = L/E\tau_t \tag{2}$$

and this technique can be used to measure mobilities in crystalline semiconductors. Because of the way in which the mobility is measured, this is called the time-of-flight technique.

However, if we carry out the very same experiment with a sample of an amorphous semiconductor, an amazing result is found. Instead of finding a constant current until the transit time, the current decreases from the very beginning, as soon as the light flash is complete. This certainly cannot be happening because at the shortest times some of the electrons are traversing the sample. Their mobilities would have to be gigantic. It must be that the current is decreasing because the average drift velocity is decreasing. Because the field is constant, this means that the average mobility decreases with time. This phenomenon was first appreciated by workers at Xerox [2-4]. Upon seeing this behavior, they realized that it was nearly impossible on a plot of i vs t to determine the transit time, but that on a plot of log i vs log t things looked simpler. At first, the current decays as a power law $t^{-1+\alpha}$ (with $\alpha = 0.5$ for a-As$_2$Se$_3$) which looks like a straight line on a double logarithmic plot. At a special time, the power law decay becomes steeper (roughly $t^{-1-\alpha}$), and it is natural to suppose that the special time is the transit time. The theorists at Xerox worked out how such behavior could come about. Their explanation was quite formal, however, and it was clear that a simpler one would be helpful. Such a simple explanation was found by J. Orenstein and me [5], and, independently and almost simultaneously, by Tiedje and Rose [6]. In the next section I will introduce the simple explanation of dispersive transport, the process which is characterized by a mobility that decreases with time.

III.   MULTIPLE TRAPPING MODEL OF DISPERSIVE TRANSPORT

Begin by thinking about a near-perfect crystal with just one energy level in the gap, not too far from $E_c$, at which there are localized states. If electrons are trapped in these states, they cannot move, whereas within the conduction band they can constantly support the flow of electricity. What will happen when electrons are excited into the conduction band by light? At first, the current will be high because all the electrons are free to move, but, as time proceeds, more and more electrons will become trapped in the localized states, and the current will decrease. Only the thermal agitation of the atoms can excite the electrons out of their traps into the conduction band so that, initially, the rate of electron trapping will be greater than the release rate. Thus, even with a single trapping level, the current in the TOF experiment decreases with time at the shortest times. However, this decrease would be exponential with time $[\exp-(t/t_0)]$ because there is only one trapping rate. Furthermore, the current would not continue to decrease with time, but would level off at a value determined by the balance of trapping and release at the single level. This would be hard to see in a linear plot of i vs t. However, on a plot of log i vs log t, there would be an abrupt drop in the current at the trapping time (inverse of the trapping rate). Of course, the current is actually decreasing at the shortest times. However, for time scales much shorter than the trapping time, the decrease is so small that the decrease looks like a step on the double-logarithmic plot. After the trapping time, the current is small because only the small fraction of the electrons that are thermally excited out of the trapping level into the conduction band is free to move.

If we can understand what happens with a single level, we can understand what happens with two levels. At first, when the electrons are excited into the conduction band, the same thing happens as before. The electrons are captured by the two sets of traps. At first, the fact that there are two traps instead of one is irrelevant. The current drops to a lower value at the trapping time. However, in this case, the lower value is determined by the rate of release from the shallower trap resulting from thermal excitation. The thermal release rate from the deeper trap is much slower so the deep one does not influence the current at all just before the trapping time. However, as time proceeds, the electrons are thermally excited out of the shallower level and may be trapped in the deeper level. Of course, they may also be retrapped into the shallower level, but if that happens they get released again a short time later. On the other hand, if they are trapped in the deeper level, they remain trapped for a much longer time. The release from the shallow level limits the number of electrons in the conduction band which, in turn, limits the trapping into the deeper level. Therefore, at a time roughly equal to the inverse of the release rate from the shallower level almost all the electrons trapped in the shallower level will be released and trapped in the deeper level. Since the release rate from the deeper level to the conduction band is much slower, there will be far fewer electrons to conduct and the current will drop to a still lower value. Thus, for two levels the log-log plot will have two steps.

If we can understand what happens with one level and with two levels, we can understand what happens with an infinite number of discrete levels. The log i vs log t plot will have a staircase with an infinite number of steps   The first step occurs when the electrons are trapped at random into any of the levels. The second step occurs when carriers are released from the shallowest level and trapped at random into any of the deeper levels. At the time when electrons are released from the next shallowest level, they will be trapped into still deeper levels and the current will drop to a still lower value. Of course, at any time, electrons are being trapped at levels shallower than the one from which they are being released, but the release rates from these are so fast compared to the release rate from deeper levels that the electrons spend very little time in the shallow levels, and these do not, therefore, limit the number of electrons which take part in conduction. We need not consider these levels in determining how the current changes with time. If the levels were not spaced far apart in energy, but, rather, were distributed quasi-continuously, the decrease of the current with time would approach some power law--a straight line on the double logarithmic plot. This is the basic idea needed to understand the decrease of the average mobility with time. Now we need to make this more quantitative.

Each of the steps in the staircase, except the first, occurs when carriers are released from one of the trapping levels. That is, the step for the ith level occurs when the release rate for that level $\nu_i$ is equal to the inverse of the time of observation, i.e.,

$$\nu_i = t_i^{-1} \tag{3}$$

where $\nu_i$ is the release rate for the ith level and $t_i$ is the time.

We can make this more useful by realizing that $\nu_i$ is related to the energy $E_i$ of the level, by

$$\nu_i = \nu_0 \, e^{-(E_c - E_i)/kT} \tag{4}$$

where $\nu_0$ is a constant frequency, k is Boltzmann's constant, and $E_c$ is the energy of the mobility edge. Equation (4) is the form always found when a rate is determined by activation over a barrier. In this case, the barrier is the energy needed to raise the electron from its trap at energy $E_i$ to the mobility edge at $E_c$, so the barrier is $E_c - E_i$. Putting Eqs. (3) and (4) together, we see that the release time can be used to measure the energy of the localized level.

It is possible at any time t to divide the levels into two classes, which we call fast states and slow states. The fast states are those with $t_i \ll t$, and the slow states are those with $t_i \gg t$, where t is the time of observation. The fast states have released almost all their trapped electrons. Any electrons in fast states have been trapped and released many times, and are, therefore, in thermal equilibrium with the electrons above the mobility edge. On the other hand, the slow states have not yet released an appreciable fraction of their trapped carriers. In fact, they are accumulating more carriers by trapping as time proceeds. We can use Eqs. (3) and (4) to find the energy which separates fast from slow, or, as we usually call them, shallow from deep levels. Shallow levels have $E_i$ above a demarcation energy $E_d$, and deep levels have $E_i$ below $E_d$. Using Eqs. (3) and (4) and taking the logarithm, we find

$$E_c - E_d = kT \ln \nu_0 t$$

or
$$\tag{5}$$
$$E_d = E_c - kT \ln \nu_0 t$$

which shows that $E_d$ gets deeper logarithmically with time at fixed temperature and linearly in temperature at fixed time.

In an amorphous material, the density of states is not a set of discrete levels, but, rather, is continuous as in Fig. 1. Nonetheless, the demarcation energy of Eq. (5) still separates deep from shallow states. Figure 2 shows what happens to a bunch of electrons introduced at time t = 0 into the states above $E_c$. In this figure, several changes were made so that it describes a wide variety of cases. The density of states is drawn with a peak.

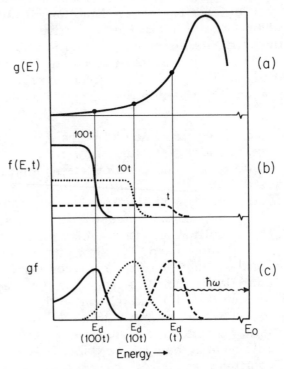

Fig. 2    (a)  Sketch of continuum of bound states in the gap which
          rapidly trap electrons after photoexcitation.  Carriers
          empty out of shallow states and are retrapped uniformly
          so that the mean occupation number evolves as in (b).
          For a density of states which decreases rapidly below
          $E_O$, the carriers are distributed in a packet, shown in
          (c), localized near $E_d(t)$.  The arrow indicates the
          threshold energy for the PA transition which increases
          as the carriers thermalize.

That might be the situation if the amorphous material had discrete localized states from impurities or defects as well as the tails on the bands shown in Fig. 1. In addition, $E_c$ is given the label $E_0$, and is shown at some arbitrary position at the high energy end of the scale. These complications need not be of concern. At first, the electrons are trapped into the shallowest states because the number of these is largest and the probability of getting trapped in them is, therefore, highest. As time proceeds, however, carriers will be released from states above $E_d$ and trapped into states below $E_d$. Because the density of states decreases rapidly with energy into the gap, the probability of trapping into deep states is highest for the states just below $E_d$. As a result, the carriers reside in a narrow energy range close to $E_d$. The packet of carriers move deeper in energy logarithmically in time and linearly in temperature. In Fig. 2(b) is a sketch of how the probability of a state being occupied changes with energy and time. At a fixed time, all states with $E < E_d$ have equal probability of being occupied, and this probability grows with time as more carriers are trapped in these deep states. The probability drops rapidly with increasing energy for $E > E_d$ because carriers are released from the shallow states and are trapped in deep ones. The step in the probability (more accurately, the mean occupation number) moves an equal energy for every equal factor in time, i.e., logarithmically. By multiplying the probability of occupancy (Fig. 2(b)) by the density of states (Fig. 2(a)), we find the actual distribution of electrons in energy (Fig. 2(c)). The electrons form a packet moving deeper in energy as time proceeds, or at fixed time if the temperature is increased.

Figure 2(c) indicates how one could test this model of carrier thermalization. It shows how photons (light) with sufficient energy ($h\nu = \hbar\omega$) could re-excite the trapped electrons. At any time and temperature, a minimum photon energy would be required to raise the electrons to the mobility edge. This minimum energy is predicted to get larger as time proceeds at fixed temperature, or as T is raised at fixed time. The prediction, therefore, is that there will be a new optical absorption when electrons are excited, say, by a pulse from a laser which excites electrons across the band gap. This photoinduced optical absorption (PA) is predicted to have a threshold (edge) which shifts to higher energy linearly in T and logarithmically in t.

For our experiment, we make a sample of $As_2Se_3$ glass. I still believe that the physics of chalcogenide glasses, the materials toward which Stan Ovshinsky first directed attention, is more interesting than that of a-Si:H. Since the latter material has proved more important technologically so far, many more people are studying it and the same experiments have been done, some at Energy Conversion Devices, on a-Si:H as those I will discuss for amorphous (a-)As$_2$Se$_3$. On the surface of our slab of glass, we paint two graphite electrodes between which we apply a voltage for measuring photocurrent. It turns out that this is easier than TOF measurements and gives much of the same information. We place the sample in a cryostat or oven to control its temperature. The part of the sample between the electrodes is excited by a pulse of light from a dye laser. At the same time (or sequentially), we can measure the transmission through the excited part of the sample with an infrared probe beam. This beam comes from a tungsten lamp and monochromator. The probe beam must be infrared or it will not be transmitted through the sample even in the absence of laser excitation. Various photodetectors are used to measure the small change in transmission that results when the laser pulse hits the sample. Using a digital transient recorder, we can measure how the absorption disappears after the laser pulse. Alternatively, we can use a boxcar integrator to measure the spectrum at a fixed delay time after the laser pulse.

The results of such time-resolved PA measurements are shown in Fig. 3. It is clear that there is a threshold for the induced absorption. In Fig. 3(a), the spectra are given for four time delays after the laser pulse: $10^{-5}$s, $10^{-4}$s, $10^{-3}$s, and $10^{-2}$s. The edge shifts to higher energy by roughly equal amounts for each decade of time. This is the logarithmic time dependence we predicted. In Fig. 3(b), the spectra are shown for four temperatures. The prediction of a linear shift of the threshold with temperature is also confirmed. This is a dramatic demonstration that the thermalization process that I have described is actually going on. We are actually seeing the electrons falling into progressively deeper states as time proceeds.

I have assumed that every level has the same probability of trapping an electron regardless of its energy. This is not a necessary assumption, but it makes the discussion much simpler. Recently, D. Monroe and I have published a treatment of the multiple-trapping problem [7] which includes the possibility that this probability depends on the energy of the level.

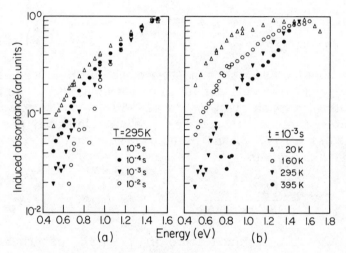

Fig. 3    The PA spectrum, (a) for deveral time delays
          following pulsed excitation, and (b) at $t = 10^{-3}$s
          for various temperatures.  In both cases, the
          shift to higher energies is in qualitative
          agreement with the MT process for which the
          demarcation energy shifts as $kT \ln \nu_0 t$.

Now that we have established how the thermalization process occurs, it is simple to predict how the current will vary with time after the excitation pulse. There is one more fact we need to use to determine the number of electrons above the mobility edge (proportional to the current) as a function of time. If there are two sets of states with different energies in any system, the thermal equilibrium densities of electrons in the two sets is determined not only by the energy difference between the levels but is also proportional to the number of states at each level. The relationship between the number of conducting electrons (those above $E_c$) and the number trapped at $E_d$ is given, approximately, by

$$\frac{n}{N_o} \simeq \frac{g(E_c)}{g(E_d)} \; e^{-(E_c - E_d)/kT} \tag{6}$$

where we have used the fact that the numbers of states at $E_c$ and $E_d$ are proportional to the densities of states at those two energies. The number of trapped electrons is essentially the same as the total number injected $N_o$ because so few reside at any energy far from $E_d$. Using Eqs. (5) and (6) and the fact that the current is proportional to n gives

$$I \propto \frac{(\nu_o t)^{-1}}{g(E_c - kT\ln\nu_o t)} \tag{7}$$

which shows the important feature of TPC or TOF experiments for a multiple-trapping system that the current gives a spectroscopy of the density of states. That is, we can measure the transient current at a variety of times and temperatures and obtain $g(E_c - kT\ln\nu_o t)$.

For many amorphous materials, as I already mentioned, the current decays as a power law $t^{-1+\alpha}$. This requires that $g(E)$ be exponential near $E_c$:

$$g(E) = N_L \; e^{-(E_c - E)/kT_o} \tag{8}$$

where $kT_o$ is the energy width of this band tail of localized states. Substituting this into Eq. (6) gives

$$I \propto (\nu_o t)^{-1+T/T_o} \tag{9}$$

which shows the power law behavior but requires that $\alpha = T/T_o$. To recapitulate: the power law decay implies that $g(E)$ is exponential but that, in turn, leads to the prediction that $\alpha$ is proportional to the temperature.

Figure 4 shows the results of TPC experiments at several temperatures. The power law decay is beautifully demonstrated. It is clear that the power law gets less steep as T is increased. In fact, when we extract $\alpha$ from the slope of the lines at each T and plot $\alpha$ against T in the inset of Fig. 4, we find $\alpha = T/T_0$, as predicted, with $T_0 \simeq 550K$ for a-As$_2$Se$_3$. This is dramatic evidence that the multiple trapping model is correct and that we can use transient photoconductivity to measure the density of states.

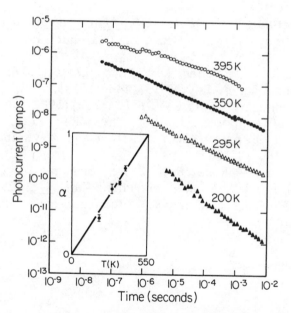

Fig. 4     Time decay of PC after pulsed excitation with band gap light at several temperatures. At each temperature, the decay of the photocurrent accurately follows $t^{-1+\alpha}$. The temperature dependence of $\alpha$ is quite linear as shown by the inset.

We have worked out many more aspects of the transient photocurrent: the steeper power law decay of the current after the transit time, the effects of recombination and we have extracted the microscopic parameters from the data. All of this, as well as a more accurate treatment of the multiple trapping process, is contained in the papers by Orenstein, Kastner and Vaninov (see ref. 5) and that by Monroe and Kastner (see ref. 7). We are seeing the transient photoconductivity used more and more as a spectroscopy of the density of states in amorphous materials. At Energy Conversion Devices, Huang, Guha and Hudgens have done this beautifully for a-Si:H [8]. I believe this approach will continue to be important in the future.

This work was supported by NSF grants Nos. DMR 81-15620 and DMR 81-19295, and the Joint Services Electronics Program through Contract No. DAAG-29-83-K-0003.

REFERENCES

1.    N.F. Mott and E.A. Davis, Electronic Processes in Non-crystalline Solids, 2nd edition, Clarendon Press, Oxford, (1979).
2.    G. Pfister and H. Scher, Adv. Phys. 27, 74(1978).
3.    J. Noolandi, Phys. Rev. B 16, 4466(1977).
4.    F.W. Schmidlin, Phys. Rev. B 16, 2362(1977).
5.    J. Orenstein and M.A. Kastner, Phys. Rev. Lett. 46, 181 (1981); J. Orenstein, M.A. Kastner and V. Vaninov, Phil. Mag. B 46, 23(1982).
6.    T. Tiedje and A. Rose, Solid State Commun. 37, 49(1981).
7.    D. Monroe and M.A. Kastner, Phil. Mag. 47, 605(1983).
8.    C.Y. Huang, S. Guha and S.J. Hudgens, Phys. Rev. B 27, 7460(1983).

# GEMINATE RECOMBINATION AND INJECTION CURRENTS:

## DIAGNOSTICS FOR EXTENDED STATE MOBILITY

M. Silver

Department of Physics and Astronomy
University of North Carolina
Chapel Hill, North Carolina 27514

## I.    INTRODUCTION

In this paper I will raise the question, "Is the extended state mobility in a-Si:H low ($\simeq 10 cm^2/V$-s) or high ($> 200 cm^2/V$-s)?" This will be done by examining two seemingly independent phenomena: (1) geminate recombination, and (2) injection currents. The former refers to a primary photo-excitation process leading to a production of free carriers. This process involves the probability of escape of an excited pair from each other. Among the various factors governing this primary yield is the extended state mobility which determines how far apart the pair will be when they become thermalized. If the mobility is low, the mean free path is small, and they diffuse apart during thermalization. On the other hand, if the mobility is high, the mean free path is large, and they move apart ballistically in the range of the Coulomb interaction. In the former case, the separation is generally small, and the yield would be expected to be small. In the latter case, the separation would be large and the yield approaches unity.

Injection currents may either be electrode limited or bulk limited. The time dependence of the injection current can be interpreted in terms of each process; in the presence of traps, electrode limited currents will rise as a function of time until a steady current is reached, while bulk limited currents will decay. In either case, the time dependence will yield a value for the extended state mobility based on a model for trapping and the density of traps.

Thus, these seemingly diverse phenomena depend upon the band mobility, and therefore must be interpreted in a consistent manner.

## II.    GEMINATE RECOMBINATION

Geminate recombination is the name given to the process whereby two thermalized oppositely-charged particles attempt to escape from each other in the presence of their mutual Coulomb attraction. L. Onsager [1] in 1938 first discussed this question when considering the probability that two oppositely-charged ions will dissociate in a weak electrolyte after a short-ranged charge exchange reaction. The problem involves the escape of a particle over a barrier. Chandrasekhar [2] presents a beautiful and complete description of this and other random walk problems.

To help vizualize the process, Fig. 1 shows the Coulomb potential surrounding a localized positive charge. (For simplicity, this positive charge is assumed to be stationary.) After the charge exchange and thermalization, the positive and negative carriers are separated by a distance $r_0$. Shown in Fig. 1 is the case where this separation is due to light excitation of energy $h\nu$. This will be discussed later. The particles will be considered to have escaped when their separation is $r_c$, the distance at which the binding energy of the pair is $kT$, i.e., their total energy is zero. Another important distance is $r'$ where the charges are separated by a nearest-neighbor distance. Again for simplicity, the charges will be considered to reside at this distance until they literally recombine or until they start to separate again.

One can easily guess at the solution of this problem from the following simple argument. When the two charges are initially separated by a distance $r_0$, the binding energy of the pair is:

$$U(r_0) = -e^2/4\pi\epsilon\epsilon_0 r_0. \tag{1}$$

The escape probability is then the probability that the total energy of the pair is zero. In this case,

$$P_{esc} = \exp\left[\frac{U(r_0)}{kT}\right] = \exp\left[-\frac{e^2}{4\pi\epsilon\epsilon_0 kTr_0}\right]. \tag{2}$$

This expression can be simplified by letting $e^2/4\pi\epsilon\epsilon_0 kT = r_c$, and Eq. (2) becomes:

$$P_{esc} = \exp\left[-r_c/r_0\right]. \tag{3}$$

It is instructive to calculate Eq. (2) from a simple solution of the
continuity of current equation (rather than formally as was done
by Onsager [1] and Chandrasekhar [2]). This was done by Silver
and Jarnagin in 1968 [3]. The advantage of their approach is that
one can vary the boundary conditions in a simple way and take
into account a slow final step in the recombination process. For
spherical symmetry:

$$j = \frac{I}{4\pi r^2} = -D\frac{dn}{dr} + \frac{n\mu e}{4\pi \epsilon \epsilon_0 r_0^2}, \tag{4}$$

where D is the diffusion constant, $\mu$ the mobility, and n is the
density of carriers. The boundary conditions used are: (1)
$n(\infty) = 0$, (2) $n(r_0) = n_0$, and (3) $n(r') \frac{4}{3}\pi r'^3 k_r = I_r$, where $k_r$ is the
final step recombination rate, r' is the nearest-neighbor distance,
and $I_r$ is the recombination current. For simplicity, we make

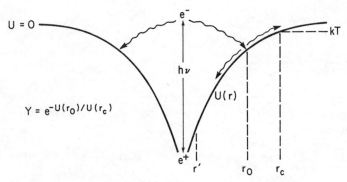

Fig. 1     Schematic diagram for geminate recombination. Shown
is the Coulomb interaction, U(r), between a localized
hole $e^+$ and a localized electron $e^-$. The electron is
excited by a photon of energy $h\nu$. The electron is assumed to be thermalized to the potential at a position $r_0$.
The distance, $r_c$, represents the position where the
Coulomb energy is kT and r' is the nearest-neighbor
distance. In some sense, r' could be thought of as the
lowest energy charge transfer state.

$k_r = \nu/f_w$, where $\nu$ is the jump frequency between nearest-neighbor equivalent sites, which makes:

$$n(r') = \frac{f_w}{\nu}\frac{3}{4\pi}\frac{I_r}{r'^3}.$$

By integrating Eq. (4) from $r_o$ to $\infty$, one obtains the dissociation current $I_d$. By integrating from $r_o$ to $r'$, one obtains the recombination current $I_r$. The yield, $P_{esc}$, is then:

$$P_{esc} = \frac{I_d}{I_r + I_d}. \tag{5}$$

For $f_w = 1$, we obtain $P_{esc} \simeq e^{-r_c/r_o}$ as the simple view predicted. For large $f_w$:

$$I_r = \frac{4\pi D r_c n_o}{1 + \frac{3Dr_c}{r'^3\nu}\frac{-r_c/n_e' r_c/r_o}{f_w e}} \tag{6}$$

and

$$I_d = \frac{4\pi D r_c n_o e^{-r_c/r_o}}{1 - e^{-r_c/r_o}} \tag{7}$$

For $P_{esc} \simeq I$, $I_D \gg I_r$, or:

$$f_w \gg \frac{r'^3 \nu e^{r_c/r'}}{3Dr_c} \tag{8}$$

(Dispersion would have the effect of reducing D thereby requiring an even higher $f_w$.)

If one makes a further simple approximation $D = \frac{\nu r'^2}{6}$, then:

$$f_w \gg \frac{2r'}{r_c}e^{r_c/r'} = \frac{2r'}{r_c}e^{-U(r')/kT}. \tag{9}$$

It is clear from Eq. (9) that at low temperatures $e^{-U(r')/kT}$ is very large [remember $U(r')$ is negative], and so, even with a slow final recombination rate, particles could not escape and the yield will be small. At high temperatures $e^{-U(r')/kT}$ is small, so a slow final step would increase $P_{esc}$. However, the question is, "How can we distinguish a slow final step limitation, as indicated in Eq. (9), from a $P_{esc}$ determined by $\exp[-e^2/4\pi\epsilon_o kTr_o]$?" The

time dependence of the luminescence yield at intermediate temperatures can distinguish the two possibilities. If $f_w$ is very large,
then the luminescence decay will be exponential; on the other
hand, if $f_w$ is small, but $r_c/r_o$ is large, then the luminescence
decay will follow a power law. Ries, Bassler and Silver [4] have
shown the expected time response of the two cases by performing
a Monte Carlo simulation. Their results are shown in Fig. 2A. It
is seen that when $f_w \approx 10^3$, the decay is exponential for at least
three decades. These simulated results can be compared with the
experimental data of Tsang and Street [5]. Their results are shown
in Fig. 2B. The comparison shows that the experimental data resembles the $f_w = 1$ curve, and does not exhibit significant exponential decay. This comparison creates doubt on the notion that a
slow final step is limiting the geminate recombination.

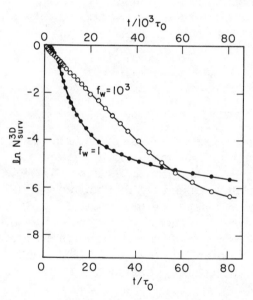

Fig. 2A     A Monte Carlo simulation for luminescence for a slow
and a fast final step. Shown is the log of the number of
survivors vs time for a fast final step $f_w = 1$ and a slow
final step $f_w = 10^3$. The luminescence decay will be
proportional to the decay of number of survivors. Notice
that the time scale for the $f_w = 10^3$ curve is also $10^3$
times larger than for $f_w = 1$. Consequently, the curves
do not cross.

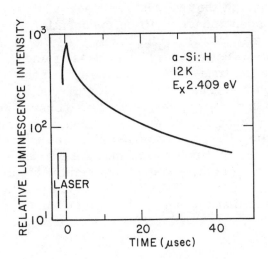

Fig. 2B   Luminescence decay in a–Si:H.  These data were obtained
          by Tsang and Street [5].  Notice that little, if any, ex-
          ponential decay is evident.

Now that we see that a slow final step is not likely to be
a significant factor in geminate recombination in a–Si:H, we explore $r_0$ after a photo-excitation.  As shown in Fig. 1, the most
likely process is the subsequent thermalization just after excitation.  We, therefore, separate the problem into two parts:  (1)
thermalization, and (2) diffusion over the  Coulomb barrier.  To
date, no one seems to have done a thorough analysis of (1).  However, Knights and Davis [6] have presented a simple picture which
we can use as a guide.  They assumed that the time to thermalize
was:

$$t_{th} = \frac{h\nu - E_g}{h\nu_p^{\,2}} ,$$
(10)

where $\nu_p$ is an optical phonon frequency and $E_g$ is the band gap.
Vardeny and Tauc [7], using picosecond photo-induced absorption
techniques, have estimated $t_{th}$ to be approximately $10^{-12}$ s for an
excess energy of 0.1eV.

We use this value now to estimate $r_0$. For an inherently low extended state mobility, the m.f.p. of the excited electron will be small, and it will diffuse as it thermalizes. On the other hand, if the mobility is high, the m.f.p. will be large, and the electron will move ballistically. The separation between these two regions occurs when the m.f.p. $\lambda \simeq r_c$. In amorphous silicon $r_0 \simeq 50\text{Å}$ at room temperature. Consequently, band mobilities below $50\text{cm}^2/\text{V-s}$ will be described by diffusive motion since $\lambda \leqslant 50\text{Å}$.

Using the mobility suggested by Tiedje et al. [8], $\simeq 10\text{cm}^2/\text{V-s}$ gives:

$$r_0 = \sqrt{\frac{\mu kT}{e}}\, t_{th} \simeq 50\text{Å} . \tag{11}$$

Consequently, $P_{esc} = e^{-1}$, and should be observable. On the other hand, if the band mobility was very large as proposed by Silver et al. [9], then the motion is ballistic, and:

$$r_0 \simeq \left(2\frac{\Delta E}{m}\right) t_{th}^{1/2} . \tag{12}$$

In this case, $r_0 \simeq 10^{-5}\text{cm}$ which is very large compared with $r_c$ and $P_{esc} \simeq 1$.

From these simple arguments, geminate recombination should be observed for the field-free case in organic crystals where $\mu \simeq 1.0\text{cm}^2/\text{V-s}$, photoconducting polymers where $\mu \simeq 10^{-4}\text{cm}^2/\text{V-s}$, and in some amorphous chalcogenides, such as Se, where $\mu \simeq 1\text{cm}^2/\text{V-s}$. Before presenting experimental results, we give a brief discussion on the effect of electric fields.

An external field has the effect of lowering the barrier in the direction of the field. This would have the effect of increasing the escape probability. But this is not a one-dimensional problem since escape in the direction perpendicular to the applied field would not be affected. Batt, Braun and Hornig [10] have presented a simple picture for the yield. A schematic diagram of the effect is shown in Fig. 3A. Also shown in Fig. 3B is the angular dependence of $P_{esc}$ for various applied fields [10]. As is seen, the yield is obtained from an averaging over all directions.

Pai and Enck [11] give the results of this averaging and their results are shown in Fig. 4A for small $r_0$. In Fig. 4B, we show simulation results by Silver, Adler, Madan and Czubatyj [12]

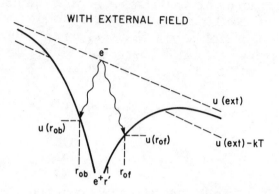

WITH EXTERNAL FIELD

Fig. 3A  Schematic diagram for geminate recombination. Here, the
diagram is similar to Fig. 1 except that effect of an applied external field is included. The subscripts f and b
refer to the forward and backward directions of the applied
external force field.

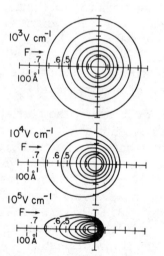

Fig. 3B  Angular distribution for
$P_{esc}$ for various applied
fields. In each curve,
we show the results of
a calculation of $P_{esc}$
for various $r_o$ and various applied fields.
Since the yield in the
backward and trans-
verse directions are not
affected very much by
the applied field, one
must average over all
directions and not use
a one-dimensional ana-
logue.

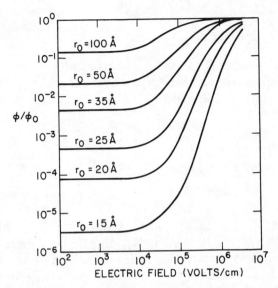

Fig. 4A    Field dependence of $P_{esc}$. Shown are the results of a
calculation where $r_0$ is relatively small so that the zero
field yield is small. Notice the yield approaching unity
(> .9) is not achieved until the applied field is greater
than $10^5$ V/cm.

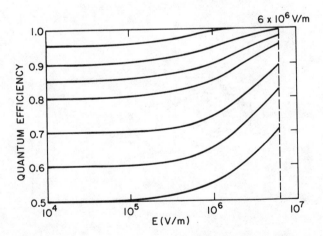

Fig. 4B    Field dependence of $P_{esc}$. Shown here are results of
calculations for $r_0$ relatively large so that the zero
yield is not very small. Notice that very high fields
are needed to make $P_{esc}$ approximately unity.

for larger $r_0$'s. The interesting result is that if $r_0$ is small ($r_0 < r_c$), then the applied field needed to make $P_{esc} > .95$ is at least $10^6 V/cm$, which is a very high electric field.

The data showing that the Onsager model accurately predicts geminate recombination comes from experiments on organic crystals, a-Se, and the photoconducting polymers. Each of these classes of materials show:

1.  $P_{esc}$ ↑ as E ↑

2.  $P_{esc}$ ↑ as T ↑

3.  $P_{esc}$ ↑ as $(h\nu - E_g)$ ↑.

Figure 5A shows the results in anthracene [13], Fig. 5B for a-Se [11], and Fig. 5C for PVK [14]. In all three cases, the predicted field and temperature dependence and the expected photon energy dependence are observed. This is very interesting, considering that the results refer to three widely different materials: an organic crystal (low $\mu$), a-Se (low $\mu$), and a photoconducting polymer (low $\mu$).

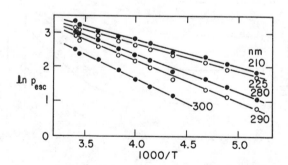

Fig. 5A     Geminate yield in anthracene. Shown is the temperature
            dependence for $P_{esc}$ for various photon energy excita-
            tions. Each curve represents experimental results for
            different excess kinetic energy after excitation.

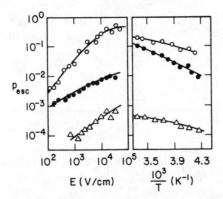

Fig. 5B     Geminate yield in a-Se. Shown are the field and
temperature dependence of $P_{esc}$ for various photon
energy excitations.

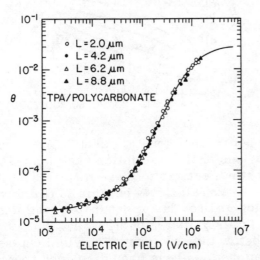

Fig. 5C     Field dependence of $P_{esc}$ in a photoconducting polymer.
The solid line is a theoretical fit to the experimental
data.

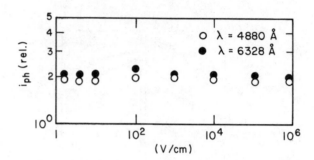

Fig. 6A    Field dependence for $P_{esc}$ in a-Si:H.  Shown are results
           for two different photon excitations.

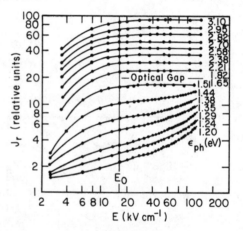

Fig. 6B    Field dependence for $P_{esc}$ in a-Si:H.  Shown are results
           for various photon excitation energies.  There are two
           interesting features to these results:  (1) the yield de-
           creases at small fields but this is due to volume recom-
           bination and not due to geminate recombination, and (2)
           there is a significant value for $P_{esc}$ at excitation ener-
           gies below the band gap.  The yield curves for various
           photon energies have been displaced to more clearly
           show the lack of field dependence.  In all cases where
           there is no field dependence, the absolute value for
           $P_{esc} > 0.95$.

Noticeably absent from the list of materials exhibiting geminate recombination is a-Si. There is, in fact, a wealth of information showing that at room temperature geminate recombination is not observed in a-Si:H. Two examples are given in Figs. 6A and 6B. These results were obtained by Silver, Adler, Madan and Czubatyj [12] and by Curasco and Spear [15]. Notice that there is no evidence for any electric field, temperature or photo-energy dependence when $h\nu > E_g$.

Perhaps it is only a coincidence, but it is certainly interesting that wherever geminate recombination is observed, the extended state mobility is also low. This is true even in anthracene where it is known that the luminescence lifetime is of order $10^{-8}$s. Consequently, it is difficult to understand why geminate recombination is not observed in a-Si:H if the extended state mobility were indeed low as has been proposed. I find it difficult to accept that a slow final step is responsible because the predicted exponential decay of the luminescence at intermediate temperatures is not observed. Consequently, we are left with a dilemma, "Why is a-Si:H an exception, if it indeed is an inherently low mobility material?" To me, the answer is that the extended state mobility is, in fact, not low but high. In the next section, I will discuss reverse recovery and time-dependent injection experiments which also suggest that the band mobility is high.

III.    TRANSPORT STUDIES BY REVERSE RECOVERY AND INJECTION CURRENTS

With the suggestion from geminate recombination arguments that the extended state mobility in a-Si:H may be high, it is useful to look for other experimental results. This had previously been done by time-of-flight techniques [8]. Those experiments give the drift time and, therefore, the drift mobility. However, the interpretation involves a knowledge of the density of localized states. In its most general form, the trap controlled drift mobility is:

$$\mu_D = \frac{\mu_o n_c}{n_t + n_c} ,$$

(13)

where $n_c$ is the density of extended carriers, $n_t$ is the density of trapped carriers, and $\mu_o$ is the band mobility. For an exponential, density of traps $n_c$ and $n_t$ are functions of time, and so is the effective mobility. In an experiment, one measures $t_d$, the time for the carriers to be collected, and a drift mobility can be calculated [8]. If one has a pure exponential distribution of localized states, then the time-dependent mobility $\mu_D$ is given by:

$$\mu_D(t) \sim \mu_o \, \alpha(1-\alpha)(\nu t)^{\alpha-1} , \tag{14}$$

where $\nu$ is the escape frequency from a trap, $\alpha = T/T_o$, and $T_o$ characterizes the energy distribution of traps (i.e., $N_T(\epsilon) = N_c e^{-\epsilon/kT_o}$). In this case, the apparent mobility is:

$$\mu_a = \frac{\mu_o}{(1-\alpha)^{1/\alpha}} \left(\frac{1}{\nu t_o}\right)^{(\frac{1}{\alpha} - 1)} , \tag{15}$$

where $t_o = \dfrac{L}{\mu_o E}$ . This formula works only for a pure exponential density of traps with $\alpha < 0.8$ [8]. For $\alpha = 1$, Eq. (14) takes on a different form [16], and:

$$\mu_D(t) = \frac{\mu_o}{2 + \ell n \nu t} . \tag{16}$$

For a typical transit time, $\simeq 10^{-7}$s, excitation time of $10^{-8}$s, and $\alpha = 1$, the T.O.F. current would decay by approximately 35% for $\nu \simeq 10^{12} \text{sec}^{-1}$.

If the density of localized states was indeed a pure exponential as proposed by Tiedje [8], then it is simple to extrapolate Eq. (16) and $\mu_o \simeq 13 \text{cm}^2/\text{V-s}$ would indeed be derived. However, recently LeComber and Spear [17], and C.Y. Huang, S. Guha, and S.J. Hudgens [18] have proposed a density of states which approaches the mobility edge more slowly at energies less than $\simeq 0.18$eV from the band edge. Their density of states is shown in Fig. 7 [19]. This makes the extrapolation to obtain $\mu_o$ much more doubtful. For example, if the density of states were flat till an energy $\epsilon_c - \epsilon_o$ and exponentially decreasing below, then:

$$\mu_D = \frac{\mu_o e^{-\epsilon_o/kT}}{2 + \ell n \nu t} . \tag{17}$$

For an $\epsilon_o$ of 0.18eV and $\alpha = 1$, the projected extended state mobility $\mu_o$ would be around 1400 times larger than the prediction from the pure exponential density of states. Lowering $\epsilon_o$ to 0.1eV gives a value 55 times larger. On the other hand, it is unrealistic to make the density of states flat. However, as Fig. 7 shows, it decays more slowly, such as a slow gaussian, $g(\epsilon) = \exp[-2(\epsilon/\epsilon_0)^2]$, or as a slow exponential, $g(\epsilon) = \exp[-2\,\epsilon/\epsilon_0]$, for the region between $E_0$ and the mobility edge. Each of these functions reduces the predicted value of $\mu_0$ obtained from a flat density of states, but only by a factor of approximately 5 to 10. Consequently, care must be taken in extrapolating the density of states toward the mobility edge, and any flattening in a range of 0.1eV or greater could give values of $\mu_0$ 10 to 50 times larger than the pure exponential. Based on these considerations, a $\mu_0$ above $100 cm^2/V\text{-}s$ would be compatible with the density of states shown in Fig. 7, even if the interpretation of T.O.F. experiments were correct [19].

With the uncertainties regarding the estimates for $\mu_0$ presented above, other experiments are needed which yield estimates for $\mu_0$. One of the best of these is related to forward bias injection and reverse recovery in a p-i-n device.

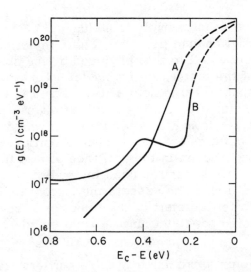

Fig. 7     Proposed density of localized states. Shown are the density of localized states obtained in ref. [17], curve B, and ref. [18], curve A.

There are two general models for the forward bias current in a p-i-n device. The first and best known of these appears in the book by Lampert and Mark (in Section 10.2) [20]. A simple physical picture of the phenomena is that the major portion of the applied voltage appears across the i layer. In this case, recombination limits the current rather than space charge. The amount of charge of either sign of carrier must be approximately the same and is:

$$Q^{\pm} \simeq \frac{CV\bar{\tau}}{\bar{t}_n + \bar{t}_p} \tag{18}$$

where $\bar{t}_n$ and $\bar{t}_p$ are the average transit times, $\bar{\tau}$ is the common average lifetime, and C is the capacitance. This is essentially a gain idea and works only if $\bar{\tau} \gg \bar{t}_p$ and $\bar{t}_n$. In this case, the forward current is:

$$j \simeq 8\epsilon\epsilon_0\bar{\tau}\mu_n\mu_p\frac{V^3}{L^5}. \tag{19}$$

However, if the band mobility is low, then $\bar{\tau}$ is diffusion limited and depends upon $Q^+$ or $Q^-$. It is known that $\bar{\tau}$ is then of order $\bar{t}_n$ and/or $\bar{t}_p$, and:

$$j \simeq 2\epsilon\epsilon_0(\mu_n\mu_p)^{1/2}V^2/L^3. \tag{20}$$

The charge in the sample is $CV(\mu_p/\mu_n)^{1/2}$, where $\mu_p$ is assumed to be smaller than $\mu_n$. The double injection current is thus approximately equal to that produced by the space-charge limited current of the faster moving carrier. Equation (20) is a result of strict space-charge neutrality. If this is relaxed for $Q^+ < CV$, then $j \propto (\mu_n + \mu_p)$. This result is exactly what one would have expected from a simple picture in which both the hole and the electron currents are space-charge limited. This argument appears to be correct even in the presence of fast trapping except that the drift mobilities must be used instead of band mobility. Under these conditions, one expects the current to decay as a function of time after the start of the injection because more and more traps are being populated until a final steady state current is reached.

Because only approximately CV/e carriers are in the sample, the transit time of the majority carrier is obtained from $CV/i_f$, where $i_f$ is the forward bias current.

The second model is due to Benda, Hoffman and Spenke [21] and was formulated in 1965. They assume the voltage drop across the device is primarily across the p-i and i-n junctions with only approximately $2kT/e$ across the i layer. The important part of their theory is that the lifetime of the carriers is approximately equal to the transit time under the small voltage $2kT/e$.

One, therefore, expects two results: (1) the current will increase with time as the voltage across the i layer decreases due to the build up of the injected plasma in the i layer. This increases the voltage across the junction and, consequently, increases the injection current. A final steady current is reached when most of the applied voltage appears across the junction. A complication can arise when there is significant trap filling. This will give an additional increase in the current because the transit time will decrease. This will be discussed later. (2) the charge in the i layer will be $i_f \times t_o$, where $t_o$ is the transit time of the carriers under a voltage of $2kT/e$. It follows that a measurement of the charge in the i layer and the forward current yields the transit time.

The next question relates to the transient response when the injection current is removed. This may be done by reversing the applied voltage. This process is called reverse recovery. If the Lampert and Mark [20] model were applicable, the reverse current would decay as the conductance of the i layer decays. In other words, the device would act as a resistor and the voltage drop across it would be simply $V_{ReV} - i_{ReV}R_c$, where ReV refers to the reverse voltage and current, and $R_c$ is the circuit resistance. The conductance in the bulk can decay no faster than the transit time of the carriers under the reverse field. Therefore, one would expect a linearly decaying reverse current, with perhaps a long tail due to trap emptying. This argument applies when the forward bias resistance of the i layer is larger than $R_c$ and includes the effect of multiple trapping.

In the Benda, Hoffman and Spenke model, there is almost no voltage across the i layer, so all the voltage is across the junction. Hence, when the applied voltage is reversed, the voltage across the junction and, consequently, the voltage across the device, does not change. The initial reverse current in the circuit would be:

$$i_{ReV} \simeq (V_{ReV} - V_{jf})/R_c, \tag{21}$$

where $V_{jf}$ is the voltage across the device under forward bias. Notice that in this case, the $i_{ReV}$ depends on the circuit resistance and $V_{ReV}$. Since $V_{jf}$ does not change initially, the device acts as

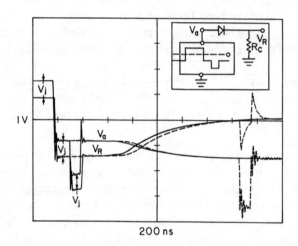

200 ns

Fig. 8    Reverse recovery in x-Si. Shown in the insert is the simple circuit to make measurements of the reverse recovery. The voltage sequence is also shown. Plotted is $V_R = (R = 50\Omega)$ and $V_a$ vs time for two different voltage sequence. $V_a - V_R$ represents the voltage across the device. Upon reversing $V_a$, $V_j$ does not change. Also notice that when $V_a$ is made more negative before the reverse recovery is complete, the recovery is faster and there is little or no discernible displacement current due to the switching of the voltage. However, when the voltage is rapidly changed after recovery is complete, the standard RC displacement current is observed. Thus, $V_j$ not changing until after reverse recovery is complete clearly shows that the Benda, Hoffman, Spenke model applies to x-Si p-i-n devices.

a battery and not as a resistor. It then follows that the $i_{ReV}$ will persist until all the charge in the i layer has decayed. This time, therefore, is only a function of $i_{ReV}$ and is not related to the lifetime of the carriers. (It is also worth noting that since the voltage across the device does not change on switching from forward to reverse bias, there is essentially no displacement current.)

With the markedly different predictions of the two models, the applicable one will be easy to discern from the experimental results. Figure 8 shows the results in x-Si. It should be apparent that the voltage across the device, V-iR, does not change upon switching. We intentionally put a large capacitance across the device to mimic an amorphous film. Nevertheless, there is no displacement current on switching until after the device has been discharged. This is seen when we apply an additional reverse voltage pulse well after the reverse current has decayed. When we apply the additional reverse voltage before the reverse current has decaysed, we see only an additional current due to the fact that $V_{ReV}$ has changed.

These results show that the Benda, Hoffman, Spenke [21] model applies to x-Si. Further, the value obtained for $t_0$ from $Q/i_f$ is the transit time for a voltage of 2kT/e across the i layer, and agrees with published data for $\mu_0$. In addition to the forward bias voltage and current dependence, the time and temperature dependences also agree with the predictions.

"Does the Benda, Hoffman, and Spenke model apply to a-Si?" This is difficult to say, but there are some indications that it does. However, further experiments are needed. Nevertheless, forward bias and reverse recovery experiments suggest that the extended state mobility is high. Perhaps the simplest indication comes from the magnitude of the forward bias current that can be obtained in a p-i-n device. We have observed current densities as large as $16A/cm^2$ for a voltage of 5.0V across a $3.5\mu m$ device. Using Eq. (20), the effective drift mobility is at least $28cm^2/V$-s. Of course, using the Benda, Hoffman, Spenke model, the derived mobility would be considerably larger since the voltage across the i layer is much less than the applied voltage. Note, however, that this is not the best method to estimate the mobility because we do not know how the applied voltage is distributed between the junctions and the i layer. We can make a better estimate by examining the time dependence of the forward bias current and the reverse recovery. The typical forward bias

current in a 3.5$\mu$ device with approximately 7.5V across the
p-i-n is shown in Fig. 9A, where we see a slow rise of the injection current. It is also apparent that when switching off the forward bias to a small, ~ -1.0V reverse bias, except for some
ringing in the circuit, the current decays very fast. Electronically, we can subtract the extraneous currents due to switching and
ringing, and what remains should be the conduction current in the
sample. These results are shown in Fig. 9B, where the reverse
current decays in approximately 15ns. Another indication that
the current decays quickly can be seen when one looks at even
higher forward bias currents. We have made the pulse width very
narrow and the repetition rate very high. Again, shown in Fig. 9C,
the current decays very quickly. We have not subtracted the
switching current in Fig. 9C.

We suggest that the slow rise in the forward current shown
in Fig. 9A is due to a trap-filling process, and the rapid decay of

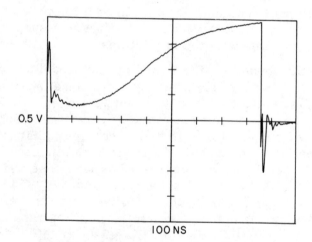

Fig. 9A    Time dependence for a high forward current level in
a-Si:H. Shown is the current vs time. We have made
the pulse width short in order to show the reverse recovery on the same time scale. The vertical scale is
0.5V/div across a 50$\Omega$ resistor.

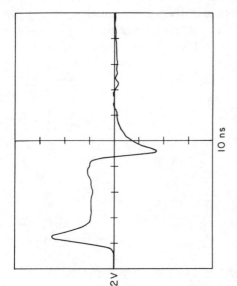

2V                        10 ns

Fig. 9C    Forward bias and reverse recovery
in a-Si:H. Here, the forward bias
was made very large so that a
steady forward current could be
obtained in a short time. The displacement current has not been
subtracted. Notice that the reverse recovery has decayed in a
very short time. The reverse bias
was -1.0V.

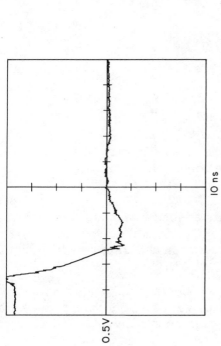

0.5V                       10 ns

Fig. 9B    Reverse recovery in a-Si:H. Shown is
the reverse recovery of Fig. 9A. The
scale has been expanded to 10ns/div
and the ringing and displacement currents have been subtracted. Notice
that the major part of the reverse recovery has decayed in less than 20ns.
If one estimates $t_0$ from $Q/i_f$ and
$\mu_O = eL^2 i_f/2kTQ$, one obtains
$\mu_O \simeq 900 \text{ cm}^2/\text{V-s}$.

the reverse recovery current is due to a high extended state mobility. If this is the case, we can then estimate the transit time, and therefore the extended state mobility from the following simple argument. The injection current, $i_j$, branches between trapping and transit across the i layer, or:

$$i_j = \frac{dQ_T}{dt} + i,$$
(22)

where $\frac{dQ_T}{dt}$ is the rate of trapping and i is the measured current. It also follows that:

$$\frac{dQ_T}{dt} = \frac{neL}{b(N_O - N_T)},$$
(23)

where $N_O$ is the density of trap available in the material, $N_T$ is the density that has been filled, b is the trapping rate constant, n is the density of free carriers, and L is the thickness of the i layer. It is assumed that there is no trap emptying so that this is not a multiple trapping model. These traps are, therefore, very deep. Data on trap emptying vs temperature suggest that the traps are $\sim 0.5\text{eV}$ from the mobility edge. Further,

$$i = \frac{neL}{t_o},$$
(24)

where $t_o$ is the transit time across the i layer. From Eqs. (22), (23), and (24),

$$i(t) = i_j[1 + t_o b(N_O - N_T)]^{-1}.$$
(25)

When all the traps are filled, the measured current is $i_j$ because $N_T = N_O$. A theoretical calculation on this process gives a time dependence similar to that shown in Fig. 9A. Also, we assume that $i_j$ does not change with time. If the Benda, Hoffman, Spenke model were applicable, then this assumption would be questionable. Since the ratio of $i(o)/i_j \approx 0.1$ from Fig. 9A, $t_o b(N_O) \approx 9$, or $t_o$ is approximately 9 times longer than the trapping time for the forward bias voltage shown in Fig. 9A.

We can use four possible models for $bN_O$ and calculate $t_o$ from these. The caveat is that $t_o$ must be less than a few ns since the reverse recovery current decays so fast even at low reverse bias. The models are: (1) diffusion limited trapping to neutral traps (low $\mu_o$), (2) diffusion limited trapping to charged traps (low $\mu_o$), (3) ballistic trapping to neutral traps (high $\mu_o$), and (4) ballistic trapping to charged traps (high $\mu_o$). The estimates of $t_o$ from these models are given in Table 1.

From Table 1, (4) is most compatible with the experimental results. Case (2) is borderline, but measurements at lower voltage show this also to be inappropriate.

If the transit time is $3 \times 10^{-11}$s, the derived value for $\mu$ is approximately 550cm$^2$/V-s. This estimate is only as good as the estimate of $N_o$, the voltage across the i layer, and the cross-section of the charged centers. These might be in error by factors of 2 or so, but not likely more. In other samples, we have estimated values of $\mu$ lying between 200 and 800cm$^2$/V-s. These values are almost 2 decades larger than the previous estimates based on a pure exponential density of localized states. Therefore, reinterpretation of T.O.F. experiments and closer examina-

Table 1. Estimates for $t_o$.

| Trapping Kinetics | Relationship for $bN_o$ | $bN_o$ | $t_o$ |
|---|---|---|---|
| (1) Diffusion limited neutral traps | $4\pi DN_o R_T$ | $1.6 \times 10^8$ | 60 ns |
| (2) Diffusion limited charged traps | $\mu e N_o/\epsilon$ | $1.6 \times 10^9$ | 6 ns* |
| (3) Ballistic neutral traps | $N_o \sigma v$ | $3 \times 10^9$ | 3 ns** |
| (4) Ballistic charged traps | $N_o 4\pi (\frac{e^2}{4\pi\epsilon\epsilon_o kT})^2 v$ | $3 \times 10^{11}$ | $3 \times 10^{-11}$ |

\* But $t_o = 16$ ns from $L^2/\mu V$.
\*\* But derived $\mu = 5.5$cm$^2$/V-s which is too small to be consistent with a ballistic model.

$N_T \approx 10^{16}$cm$^{-3}$     $V = 7.5$V     $\mu = 1$cm$^2$/V-s for diffusion limited cases

$R_T = 5 \times 10^{-8}$cm     $i_j/i(o) = 10$     $L = 3.5 \times 10^{-4}$cm

tion of reverse recovery and forward bias injection experiments suggest that $\mu_0$ could be very large, and the failure to observe geminate recombination at room temperature is a result of a high band mobility.

Because the reverse recovery is so fast, we have not been able to determine if the Benda, Hoffman, Spenke model is completely applicable to a-Si:H. We suspect that the field distribution in the i layer lies between that predicted by Lampert and Mark and by Benda, Hoffman and Spenke. Nevertheless, either model suggests a high band mobility.

In conclusion, the data and arguments presented suggest that the extended state mobility in a-Si:H alloys may be large. Should this idea be correct, then not only are geminate recombination results understandable but also the response time of a-Si films could be much shorter than previously expected. It would then allow a-Si devices to work at very high frequencies provided the deep traps can be filled.

REFERENCES

1.      L. Onsager, Phys. Rev. 54, 554(1938).
2.      S. Chandrasekhar, Rev. Mod. Phys. 15, 1(1943).
3.      M. Silver and R.C. Jarnagin, J. Mol. Cryst. 3, 461(1968).
4.      B. Ries, H. Bassler and M. Silver, to be published.
5.      C. Tsang and R.A. Street, Phys. B 19, 3027(1979).
6.      E.A. Davis and J.E. Knights, J. Phys. Chem. Solids 35, 543(1974).
7.      Z. Vardeny and J. Tauc, Phys. Rev. Lett. 46, 1223(1981).
8.      T. Tiedje, T.M. Cebulka, D.L. Morel and B. Abeles, Phys. Rev. Lett. 46, 1425(1981); also, AIP Conference Proceedings #73, 197(1981).
9.      M. Silver, N.C. Giles, E. Snow, M.P. Shaw, V. Cannella and D. Adler, Appl. Phys. Lett. 41, 935(1982).
10.     R.H. Batt, C.L. Braun and J.F. Hornig, J. Appl. Opt. Supp. 3, 20(1969).
11.     D.M. Pai and R.C. Enck, Phys. Rev. B 11, 5163(1975).
12.     A. Madan, W. Czubatyj, D. Adler and M. Silver, Phil. Mag. B 42, 257(1980).
13.     R.R. Chance and C.L. Braun, J. Chem. Phys. 64, 3573 (1976).

14.    P.M. Borsenberger and A.I. Ateya, J. Appl. Phys. 49,
         4035(1978).
15.    F. Curasco and W.E. Spear, Phil. Mag. B 47, 495(1983).
16.    M. Silver, L. Cohen and D. Adler, Appl. Phys. Lett. 40,
         261(1982).
17.    P.G. LeComber and W.E. Spear, Amorphous Semiconduc-
         tors, Topics in Applied Physics, edited by M.H. Brodsky,
         Springer-Verlag, Berlin, Vol. 36, 1979, p. 251.
18.    C.Y. Huang, S. Guha and S.J. Hudgens, Phys. Rev. B 27,
         7460(1983).
19.    W.E. Spear and H. Steemers, Phil. Mag. B 47, L107(1983).
20.    M. Lampert and P. Mark, Current Injection in Solids,
         Academic Press, New York, 1970.
21.    H. Benda, A. Hoffman, and E. Spenke, Sol. Stat. Elect 8,
         887(1965).

# LIGHT-INDUCED EFFECTS IN HYDROGENATED AMORPHOUS SILICON ALLOYS

S. Guha*

Energy Conversion Devices, Inc.

1675 West Maple Road, Troy, Michigan 48084

## I.    INTRODUCTION

Staebler and Wronski [1] observed that prolonged light exposure causes significant changes in the dark and photoconductivity of hydrogenated amorphous silicon alloys (a-Si:H). The effect was found to be reversible; annealing at temperatures above 150°C was found to restore the original values. Later work showed [2,3] light-induced reversible changes in many other properties of the material. Changes were observed in (1) photoluminescence, (2) density and energy distribution of gap states, (3) electron spin resonance, (4) sub-band gap absorption, (5) IR spectrum, (6) diffusion length, and (7) solar cell performance. Several models have also been put forward [4-6] to explain this phenomenon.

In this paper, we shall review the experimental observations of light-induced effects in a-Si:H, and we shall discuss the various models in the light of the experimental observations. It has been shown that in some cases [7,8] surface effects may affect or dominate the changes due to light exposure. By suitable annealing of samples in vacuum and carrying out measurements in situ, one may minimize or eliminate surface effects, and in

*On leave from Tata Institute of Fundamental Research, Bombay 400005, India

423

the discussions to follow, we shall consider the experimental
results which reflect changes in the bulk properties only.

## II.    EXPERIMENTAL OBSERVATIONS

### A.    Dark Conductivity

As we have mentioned before, Staebler and Wronski observed
a large change in the dark conductivity after light exposure.  The
experiment was repeated by many other groups on samples prepared under a variety of conditions [9-13], and a wide dispersion
in the magnitude of the light-induced changes was observed.  In
some cases, the dark conductivity was found to be unchanged
after light exposure, and this was interpreted as the signature for
a stable material which does not show light-induced changes.
Systematic studies, however, have shown [14] that changes in
the dark conductivity depend on the initial Fermi level position in
the material.  In slightly n-type material, the Fermi level is
pushed down toward the center of the gap; in p-type material, the
Fermi level is pushed up.  In compensated material, the dark conductivity may go up after light soaking.

All the above results indicate that light soaking creates a
distribution of gap states (both donor- and acceptor-like) below
and above the Fermi level.  Since the dark conductivity, or the
dark Fermi level, depends on the distribution of the donor- and
acceptor-like states, it is not surprising that, depending on the
distribution of the new states, the dark conductivity may increase,
decrease, or remain unaltered after light soaking.  In other words,
measurement of changes in dark conductivity alone is not at all a
sensitive tool to study light-induced effects.

### B.    Photoconductivity

Light soaking has been found to decrease the photoconductivity of undoped and lightly n-doped material [1,15].  This indicates that the $\mu\tau$ product for the majority carriers goes down
after light exposure.  Since $\tau$ is inversely proportional to the number of recombination centers, the results indicate that light exposure gives rise to new states in the gap which act as recombination centers.  Light-induced changes in photoconductivity are
small in materials of poorer quality, or in p-type materials where
the initial value of photoconductivity is low.  Since these materials already have a large number of recombination centers to
start with, addition of new light-induced states is not expected

to change the photoconductivity, and the results are understandable.

In order to have a more rigorous understanding of photoconductivity, one should consider the capture rates at the recombination centers also. To a first approximation, in an undoped material in which the number of photogenerated electrons exceeds that of the holes, photoconductivity is given by $\sigma_{ph} = K/b_r N(E_{tn})$, where K is a constant, $b_r$ = capture cross-section of the hole-occupied valence band states, and $N(E_{tn})$ is the number of states at the trap-Fermi level for the electrons. A change in photoconductivity after light soaking will take place only if either $b_r$ or $N(E_{tn})$ changes. In principle, light soaking may cause new states in the gap without altering the gap state density at the trap-Fermi level for the electrons. In that case, even though metastable effects have taken place in the material, the photoconductivity will not show any change. On the other hand, even if $N(E_{tn})$ increases, if $b_r$ for the new states created is lower, photoconductivity will remain unaltered. Therefore, mere observation of absence of a change in photoconductivity after light soaking is not adequate to demonstrate that metastable defects have not been produced in the material.

C.    Photoluminescence

A light-induced decrease in the luminescence peak corresponding to the main luminescence transition at 1.2eV has been observed [16]. Simultaneously, an enhancement in luminescence occurs [17] at 0.8eV. These effects are metastable and indicate creation of light-induced radiative centers in the gap.

D.    Direct Determination of Gap State Distribution

Since the measurement of dark conductivity, photoconductivity, and photoluminescence give information about the gap state distribution only in an indirect way, a direct measurement of gap state density as a function of energy will be more useful for the study of light-induced effects. There are a number of methods [18] by which the gap state density and distribution can be determined directly. These measurements produced evidence that there is a wide distribution of gap states created by light exposure. There has been some controversy as to the location of the new states in the mobility gap. DLTS experiments [19] on phosphorus-doped materials show that new states are created only in the lower half of the mobility gap, while the distribution of the

gap states in the upper half remains unchanged. Capacitance
voltage [20,21] and transient current measurements [22] on undoped and phosphorus-doped materials, however, show that new
states are created in the upper half of the mobility gap. Note
that these measurements do not probe the lower half. In a recent
study, Guha et al. [23] have probed the upper half of the mobility
gap of light-soaked undoped samples using transient photoconductivity, frequency dependence of capacitance of Schottky diodes
and space charge limited conduction techniques [24]. The results
show (Fig. 1) unambiguously that new states are created in the
upper half of the mobility gap.

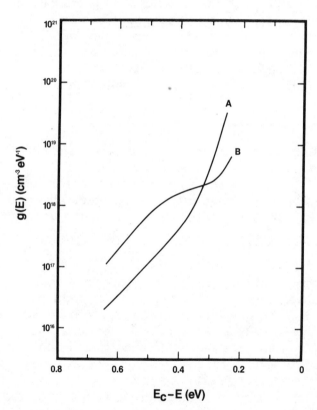

Fig. 1    Gap state distribution in the upper half of the mobility
          gap in hydrogenated amorphous silicon alloys in the
          A) heat-dried, and B) light-soaked state.

There is a great deal of evidence that changes take place in the lower half of the mobility gap as well. Field effect measurements [25] show that new states are created in both the upper and lower half of the gap. Light soaking has been observed to cause a reduction in the diffusion length of the minority carriers [26]. Since diffusion length depends on the number of gap states at the trap Fermi level for holes, this shows that new states are created at the bottom half of the mobility gap. Direct evidence for creation of states in the lower half of the gap also comes from measurement of sub-band gap absorption [27]. Light soaking results in an increase of absorption at photon energy between 0.8 to 1.4 eV, which suggests creation of new states in that energy range below the conduction band edge.

## III.    ORIGIN OF THE LIGHT-INDUCED STATES

We have now seen that the new states are created over a wide range of energy. This suggests that there must be different types of light-induced centers present giving rise to states at different energy regions. Direct evidence that there are different types of light-induced centers has come from measurements of photoconductivity at above-band gap and sub-band gap light [28]. The above-band gap photoconductivity is a measure of the density of states above the Fermi level, whereas the sub-band gap photoconductivity gives information about the states below. It has been observed that the two photoconductivities show different annealing behaviors indicating that the defects responsible for the gap states above the Fermi level are of a different kind than those giving rise to states below the Fermi level.

Based on electron spin resonance measurements, one kind of defect caused by light has been identified [29]. Light soaking has been found to increase the ESR signal with the spin density increasing from $9.10^{15} cm^{-3}$ to $1.8.10^{16} cm^{-3}$. This shows that light soaking causes dangling bonds in the material. Since dangling bonds give rise to states in the gap, the observation is consistent with the other light-induced effects discussed earlier. Dersch et al. [29[ explained the creation of new dangling bonds by assuming that weak Si-Si bonds are broken after light exposure. Since the line width of the ESR signal is large, they conclude that the light-induced spins must be at least 10Å apart. They postulate that hydrogen diffusion from neighboring Si-H bonds takes place after the weak bond is broken, thus moving the dangling bonds further apart.

Recent experiments [30] have shown that the annealing behavior of defects that are responsible for the degradation of photoconductivity and increase in the gap state density at the Fermi level depends on the temperature at which light soaking is carried out. The results are shown in Table I. It is observed that defects created at higher temperatures are more difficult to anneal out. For example, most of the defects created by light exposure at 100K anneal out at room temperature itself. Defects created at 373K, however, can be annealed out only by heating at 493K for one hour. The results can be readily explained by the model suggested by Dersch et al. [29]. For light soaking at low temperature, the defects cannot diffuse apart, and hence can be readily annealed out. At higher temperatures, however, the defects move further apart, and hence they cannot be annealed out easily. We should, however, mention that the mechanism involving hydrogen diffusion is not the only way by which the experimental results can be explained. In general, any model that requires a thermally-activated transition between configuration states representing the annealed and light-soaked conditions is consistent with the experimental findings.

Table I.    Annealing behavior of room temperature photoconductivity, $\sigma_{ph}$, at an intensity of 10 mW/cm$^2$ of white light and density of states at Fermi level, $g(E_F)$. The suffices represent the values (a) after annealing at 493K for one hour, (b) after light soaking at 100 mW/cm$^2$ of white light for one hour, and (c) after light soaking and subsequent annealing at 423K for one hour.

| Light Soaking | $\sigma_{phB}/\sigma_{phA}$ | $\sigma_{phC}/\sigma_{phA}$ | $g(E_F)_B/g(E_F)_A$ | $g(E_F)_C/g(E_F)_A$ |
|---|---|---|---|---|
| 373K | 0.092 | 0.36 | 2.25 | 1.8 |
| 300K | 0.058 | 0.63 | 2.5 | 1.4 |
| 100K | 0.39 | 0.94 | 1.26 | 1 |

We, thus, find that the model involving breaking of weak Si-Si bonds and subsequent hydrogen diffusion does provide a satisfactory explanation for many of the experimental observations. However, there are several problems with the model. It is not easily understood how a photon of energy 2eV can break a Si-Si bond. The energy available, in fact, is even lower since it has been shown [31,32] that degradation is caused by recombination of excess carriers. The recombination event takes place between a free carrier and a trapped carrier of the opposite sign and can release an energy of 1.6eV only or even lower. The complete reversibility of the light-induced effect upon annealing is also not easily understood [6]. If hydrogen diffusion is responsible for moving the spins apart, why do they not go further away after annealing at high temperatures? A possible situation in which hydrogen will like to come back near the original weak bond site is the case when all the changes take place in a grain or a tissue of relatively smaller size. In this case, even at the high temperatures, there is always a possibility that hydrogen will come back near the weak bond site. A convincing test for the hydrogen diffusion model may be obtained from the following experiment. If the linewidth for the spin signal at low temperature is measured as a function of light soaking at different temperatures, in the framework of the above model one expects a narrower linewidth for light soaking at lower temperatures.

There are several other models which have been put forward to explain the light-induced effects. Crandall [31] has identified an electron trap with nearly the same energy for electron capture or release as the defect responsible for the light-induced effects. In this model, the electron trap captures electrons and transfers them to metastable states outside the energy gap, thereby changing the Fermi level. This model, however, does not explain how new states are created both above and below the Fermi level. There is another problem with this model. It has been shown [32] that, unlike the case of pin diodes, forward bias dark current soaking of Schottky diodes on n and p-type material does not result in degradation of the photovoltaic properties. This shows that single carrier trapping alone cannot cause degradation; recombination of electron-hole pairs plays the key role.

We should also mention that any model which ascribes the light-induced effects to a movement of Fermi level alone cannot explain many of the observed light-induced changes. As we have mentioned earlier, light-induced changes do not always cause the

Fermi level to move. Guha et al. [23] have reported substantial
changes in the gap state distribution in samples in which the
Fermi level remained unaltered after light soaking. The observed
changes in sub-band gap absorption after light soaking also cannot be explained in terms of a shift in the Fermi level alone [27].

Adler [6] has proposed a model where pairs of oppositely-
charged dangling bonds ($T_3^+$ and $T_3^-$) get converted into neutral
dangling bonds ($T_3^0$) by trapping electrons and holes. The model
assumes a negative correlation energy for the $T_3$ defect, a hypothesis which for a-Si:H is not yet proven. We have mentioned
earlier that single carrier trapping alone does not cause light-
induced effects. The implication of this observation on Adler's
model is also not properly understood.

In general, a convenient way of describing light-induced
phenomena is through a configurational-coordinate model (Fig. 2).
The two configurational states A to B are separated by an energy
barrier. Transitions from state A and B take place through bond
breaking or bond switching. Bond switching may involve hydrogen
or other impurities in different configurations [2]. Transitions
from state B to A takes place by annealing. Since neither A → B
or B → A transitions are governed by a single activation energy,
there must be several configurational states with different energy
barriers.

IV.    ROLE OF IMPURITIES

The experimental observation that the magnitude of light-
induced effect varies from sample to sample and also is different
in samples grown in different laboratories leads one to suspect
that the effect may be associated with impurities in the material.
There is, indeed, a large amount of impurities present in the films.
Apart from hydrogen, the content of which varies from 5 to 20%,
other impurities like oxygen (0.1 to 1%), nitrogen (0.1 to 0.5%),
carbon (0.05 to 0.1%), chlorine (a few ppm), and sulfur (a few ppm)
are present. There is evidence [2] from very precision measurement
of infrared absorption that prolonged illumination increases the
integrated intensity of Si-H stretch modes at $2000 cm^{-1}$ by a few
percent. This indicates that some changes involving hydrogen are
taking place. Note, however, that these experiments were carried
out on samples with relatively high oxygen content ($\simeq 3\%$), and
one does not know whether similar effects take place in purer
material or not. There have been suggestions that oxygen plays a

crucial role in determining the magnitude of the light-induced effects. Crandall [33] has observed a systematic dependence of electron trap density on the oxygen content; the same traps were attributed by him to be responsible for the light-induced effects. It has been suggested that an $SiO_4$ complex could give rise to such an electron trap. Carlson [34] has shown that the degradation in solar cells is larger when the intrinsic layer has oxygen contamination. Degradation, however, has been observed in cells with very low levels of oxygen content ($< 10^{18} cm^{-3}$) in the intrinsic layer, and the role of oxygen in light-induced effects is still not properly assessed.

Even less is known about the role of other impurities in affecting the light-induced changes. It has been shown that some impurities like Cl or N can give rise to states in the gap [35,36], but whether these states can act as the defect site for light-induced effects is not yet known.

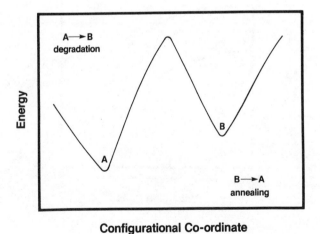

**Configurational Co-ordinate**

Fig. 2        Configurational coordinate diagram for light-induced effect.

## V.    ABSENCE OF RECIPROCITY

To a first approximation, one would expect that the light-induced changes should follow reciprocity, i.e., the effect of a given exposure will depend only on the product of the light intensity and exposure time. Preliminary experiments [37] at RCA, indeed, showed such a behavior. Later experiments [38], however, showed convincingly that reciprocity is not obeyed. Degradation is larger at high intensity light exposure for a shorter time than at low intensity light exposure for a longer time even though the product of the exposure time and flux is kept a constant. In Table II, we show the change in photoconductivity after light exposure at different intensity keeping the total integrated flux the same. We notice that degradation is larger at higher intensity light exposure. A similar trend is observed in degradation of solar cell efficiency. From Fig. 3 we notice that exposure to AM1 light for 16 hours causes much larger degradation than AM1/10 for 160 hours.

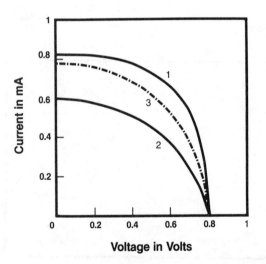

Fig. 3      Lighted current-voltage characteristic for a-Si:H solar cell in 1) heat-dried state, 2) after light soaking at 100 mW/cm$^2$ of AMI light for 16 hours, and 3) after light soaking at 10 mW/cm$^2$ of AMI light for 160 hours.

Table II. Values of photoconductivity measured under AMI illumination before and after light exposure for different time and at different intensities.

| Illumination Level | Time | $\sigma_{phB}/\sigma_{phA}$ |
|---|---|---|
| AMI | 0.6 h | 0.22 |
| AMI/10 | 6 h | 0.36 |
| AMI/100 | 60 h | 0.60 |

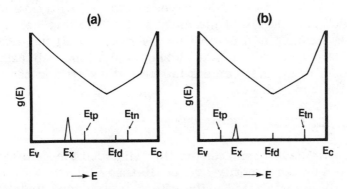

Fig. 4 Density of states diagram showing the positions of the trap-Fermi levels $E_{tn}$ and $E_{tp}$ at a) low, and b) high intensity. $E_{fd}$ is the position of dark Fermi level, and $E_x$ is the energy location for the special center.

One possible way of explaining the absence of reciprocity is through the concept of special recombination centers which cause the light-induced effects [38]. We have shown [32] before that recombination of electron-hole pairs is necessary to induce the light-induced changes. If we assume that the degradation is related to some impurity or defect present in the materials, these defects/impurities would give rise to states at some specific energy location. In that case, only if a recombination event takes place through that localized state, degradation will take place. Now recombination takes place only in the energy region between the trap-Fermi levels for electrons and holes (Fig. 4). If the special center is located far away from the dark Fermi level, at low intensity, there will be a negligible recombination traffic through that center. Degradation will, therefore, be very small. At higher intensities, as the trap Fermi level sweeps through $E_x$, the energy location of the special center, there will be greater degradation. Since for any light intensity and temperature, $E_{tn}$ and $E_{tp}$ can be precisely determined, study of degradation as a function of light intensity and temperature can, in principle, be used as a spectroscopic tool to determine the energy location of the special center. At very high light intensity, when the trap-Fermi levels are far removed from the special center, there may even be a reversal of reciprocity since the special centers would have to compete with many other recombination centers, which would not give rise to degradation. In that case, for a given integrated flux density, one would expect lesser degradation at higher intensity. We have carried out experiments [38] at temperatures down to 200K and high light intensity so that the trap-Fermi levels are about 0.3 to 0.4 eV from the band edges. In all the cases we have seen larger degradation at higher intensities, indicating that if the concept of special centers is valid, they are located within 0.4 eV of the band edges.

## VI.   DEGRADATION OF SOLAR CELLS

Since we have seen that the light-induced states affect both the majority and the minority carrier lifetime [39], the efficiency of solar cells is also affected. The effect of prolonged light exposure on the performance of solar cells has been studied in several laboratories [40]. Degradation has been found to be less in cells kept under reverse bias [31], where the high field is expected to reduce recombination. For the same reason, thinner cells also show less degradation [41]. Incorporation of small amounts of boron in the intrinsic layer has also been found to reduce degradation [41-43].

This could result from the fact that in highly-compensated material, photo-induced changes are small [44]. Moreover, B-doping alters the field profile suitably so that carrier recombination is minimized. Light degradation is also smaller in tandem cells. Since in such cells the top cell, which receives the maximum amount of light, is thin, even after degradation the drift length is larger than the thickness of the intrinsic layer of the top cell and degradation is minimal. It has also been reported [41] that degradation in SSnip ITO cell is less than SSpin ITO structures. This could result from different field profiles in the two structures; further theoretical and experimental work will be necessary to understand this problem.

We, thus, find that although significant changes in material properties take place due to light exposure, by clever engineering designs, the effect can be minimized in solar cell performance. Stability is still a problem when one deals with cells of higher efficiency (8-10%), and further work, both on materials and device design, will be necessary to obtain 10% cells stable against prolonged light exposure.

## VII.   CONCLUDING REMARKS

Over the period of the last five years, a great deal of work has been carried out on light-induced effects. We have learned several important things.

i.      We know that new states are created in the gap over a wide energy range affecting both the majority and minority carrier lifetime.

ii.     All these states are not the same kind.

iii.    Dangling bonds are created.

iv.     Recombination of excess carriers causes the light-induced changes [45].

We still do not know exactly how recombination can break a bond. We do not know whether the effect is intrinsic to the material, or it is impurity-related. And we do not know how to obtain a material with the desired photovoltaic quality which will not show this effect. A satisfactory solution of this problem will have far-reaching technological implications.

ACKNOWLEDGEMENTS

     I gratefully acknowledge helpful discussions with H. Fritzsch
C.-Y. Huang, S.J. Hudgens, M.A. Kastner and R. Tsu, and constant encouragement from S.R. Ovshinsky.

REFERENCES

1.     D.L. Staebler and C.R. Wronski, Appl. Phys. Lett. $\underline{31}$,
       292(1977).
2.     D.E. Carlson, Solar Energy Mats. $\underline{8}$, 129(1982).
3.     J.I. Pankove, Solar Energy Mats. $\underline{8}$, 141(1982).
4.     S.R. Ovshinsky, J. Noncryst. Solids $\underline{32}$, 17(1979).
5.     S.R. Elliot, Phil. Mag. $\underline{39}$, 349(1979).
6.     D. Adler, J. de Phys. $\underline{42}$, C4-3(1981); Solar Cells $\underline{9}$, 133
       (1983).
7.     I. Solomon, T. Dietl and D. Kaplan, J. de Phys. $\underline{39}$, 1241
       (1978).
8.     M.H. Tanielian, H. Fritzsche, C.C. Tsai and E. Symbalisty,
       Appl. Phys. Lett. $\underline{33}$, 353(1978).
9.     S. Guha, K.L. Narasimhan and S.M. Pietruszko, J. Appl.
       Phys. $\underline{52}$, 859(1981).
10.    B.A. Scott, J.A. Reimer, R.M. Plecenik, E.E. Simonyi and
       W. Reuter, Appl. Phys. Lett. $\underline{40}$, 973(1982).
11.    D.I. Jones, R.A. Gibson, P.G. LeComber and W. E. Spear,
       Solar Energy Mats. $\underline{2}$, 93(1979).
12.    Y. Mishima, Y. Ashida and M. Hirose, J. Noncryst. Solids
       $\underline{59-60}$, 707(1983).
13.    B.Y. Tong, P.K. John, S.K. Wong and K.P. Chik, Appl.
       Phys. Lett. $\underline{38}$, 789(1981).
14.    H. Fritzsche, Solar Energy Mats. $\underline{3}$, 447(1980).
15.    C.R. Wronski and R.E. Daniel, Phys. Rev. B $\underline{23}$, 794(1981).
16.    K. Morigaki, I. Hirabayashi, N. Nakayama, S. Nitta and
       K. Shimakawa, Solid State Commun. $\underline{33}$, 851(1980).
17.    J.I. Pankove and J.E. Berkeyheiser, Appl. Phys. Lett. $\underline{37}$,
       705(1980).
18.    See, for example, S. Guha, Solar Energy Mats. $\underline{8}$, 269
       (1982).
19.    D.V. Lang, J.D. Cohen, J.P. Harbison and A.M. Sargent,
       Appl. Phys. Lett. $\underline{40}$, 474(1982).
20.    D. Jousse, P. Viktorovitch, L. Vieux-Rochaz and A. Chenevas-
       Paule, J. Noncryst. Solids $\underline{35-36}$, 767(1980).

21. D. Jousse, R. Basset and S. Delionibus, Appl. Phys. Lett. 37, 208(1980).
22. J. Beichler and H. Mell, EC Photovoltaic Solar Energy Conference, Stressa, Italy, May 1982.
23. S. Guha, C.-Y. Huang and S.J. Hudgens, Phys. Rev. B 29, 5995(1984).
24. C.-Y. Huang, S. Guha and S.J. Hudgens, Phys. Rev. B 27, 7460(1983).
25. M.H. Tanielian, N.B. Goodman and H. Fritzsche, J. de Phys. 42, Suppl. 10, C4-375(1981).
26. J. Desner, B. Goldstein and D. Szostak, Appl. Phys. Lett. 38, 998(1980).
27. A. Skumanich, N.M. Amer and W.B. Jackson, Bull. A.P.S. 27, 146(1982).
28. D. Han and H. Fritzsche, J. Noncryst. Solids 59-60, 397(1983).
29. H. Dersch, J. Stuke and J. Beichler, Appl. Phys. Lett. 38, 456(1981).
30. S. Guha, C.-Y. Huang and S.J. Hudgens, Appl. Phys. Lett. 45, 50(1984).
31. R.S. Crandall and D.L. Staebler, Solar Cells 9, 63(1983).
32. S. Guha, J. Yang, W. Czubatyj, S.J. Hudgens and M. Hack, Appl. Phys. Lett. 42, 588(1983).
33. R.S. Crandall, Phys. Rev. B 24, 7457(1981).
34. D.E. Carlson, J. Vac. Soc. Tech. 20, 290(1982).
35. S.M. Pietruszko, K.L. Narasimhan and S. Guha, Phil. Mag. B 43, 357(1981).
36. A.E. Delahoy and R.W. Griffith, J. Appl. Phys. 52, 6337 (1981).
37. D.L. Staebler, R.S. Crandall and R. Williams, Appl. Phys. Lett. 39, 733(1981).
38. S. Guha, Appl. Phys. Lett. (in press).
39. R.A. Street, Appl. Phys. Lett. 42, 507(1983).
40. See, for example, Proc. of SERI Workshop on Light-Induced Effects in a-Si:H and its Effect on Solar Cell Stability, San Diego, 1982; Solar Cells 9, (1983).
41. Y. Uchida, M. Nishiura, H. Sakai and H. Haruki, Solar Cells 9, 3(1983).
42. S. Tsuda, N. Nakamura, K. Watanabe, T. Takahama, H. Nishiwaki, M. Ohnishi and Y. Kuwano, Solar Cells 9, 25(1983).

43.  W. Kruhler, M. Moller, H. Pfleiderer, R. Plattner and
     B. Rauscher, EC Photovoltaic Solar Energy Conference,
     Stressa, Italy, (1982).
44.  N.M. Amer, A. Skumanich and W.B. Jackson, J. Noncryst.
     Solids 59-60, 409(1983).
45.  Some recent experimental results [Nakamura et al., J. Non-
     cryst. Solids 59-60, 1139(1983)] indicate that hole trap-
     ping also causes metastable changes in the material.  The
     effect, however, is small.  It is possible that trapping of
     the holes weakens the bonds which then break up more
     easily in the presence of recombination.